The Geology of S

A Field Guide

Gareth George

Frontispiece A. Imposing exposure of late Namurian–early Westphalian strata at the Henrhyd Falls, North Crop of the South Wales Coalfield. This locality holds the key to the sequence stratigraphy interpretation of the succession (Chapter 5, locality *3e*).

Frontispiece B. Impressive chevron anticline with fracture cleavage developed in turbidites and mudstones belonging to the Cwm-yr-Eglwys Mudstone Formation (Caradoc) at Newport Sands (Chapter 7, locality *8a*).

The Geology of South Wales:
A Field Guide

Gareth T. George

Published by gareth@geoserv.co.uk

The Geology of South Wales: A field guide

Copyright © Gareth T. George 2008

All Rights Reserved: no part of this book may be reproduced in any form, by photocopying or by any other photographic or mechanical means, including information storage or retrieval systems, without permission in writing from both the copyright owner and publisher of this book.

First published in 2008 by gareth@geoserv.co.uk

Copies of the book are available from:

G. T. George
17 Aviemore Gardens
Bearsted
Kent ME14 4BA

Telephone: 01622 739135
E-mail: gareth@geoserv.co.uk

ISBN: 978-0-9559371-0-1

Printed and bound by printondemand-wordwide.com Peterborough

Contents

Preface

Acknowledgements

Chapter 1 **Introduction** 1
1.1 Transport 1
1.2 Accommodation and food 2
1.3 Landscape and scenery 3
1.4 Prehistory and historic buildings 6
1.5 Cultural attractions 6
1.6 Exploring on foot 8
1.7 Wildlife 9
1.8 Sports and leisure activities 9
1.9 Concluding remarks 10.

Chapter 2 **Overview of the geology of South Wales** 10
2.1 Introduction 10
2.2 Structural and stratigraphical framework 14
  2.2.1 Precambrian 14
  2.2.2 Early Palaeozoic 15
  2.2.3 Caledonian(Acadian) Orogeny 21
  2.2.4 Old Red Sandstone 22
  2.2.5 Carboniferous (Dinantian and Silesian) 22
  2.2.6 Variscan Orogeny 26
  2.2.7 Mesozoic and Palaeogene–Neogene (Tertiary) 27
  2.2.8 Quaternary (Pleistocene and Holocene) 28
2.3 Field itineraries and safety guidelines 32.

Chapter 3 **Vale of Glamorgan** 34
3.1 Introduction 34
3.2 Geological history 34
3.3 Itineraries 36
  3.3.1 Ogmore-by-Sea 36
  3.3.2 Seamouth and Dunraven Bay 45
  3.3.3 Cwm Marcross and Nash Point 50
  3.3.4 Whitmore Bay (Barry Island) 53
  3.3.5 Sully Island and Sully Bay 57
  3.3.6 Lavernock and St. Mary's Well Bay 60

Chapter 4 **Gower** 66
4.1 Introduction 66
4.2 Geological history 67
4.3 Itineraries 71
  4.3.1 Rhossili 72
  4.3.2 Three Cliffs Bay 77
  4.3.3 Caswell Bay 82
  4.4.4 The Mumbles and Swansea Bay 86

Chapter 5 **Headwaters of the River Neath and River Tawe**   90
    5.1   Introduction 90
    5.2   Geological history 90
    5.3   Some aspects of sequence stratigraphy 92
          5.3.1   Eustatic sea-level changes 93
          5.3.2   Sequence stratigraphy terminology and concepts 94
    5.4   Itineraries 96
          5.4.1   Upper Neath Valley 97
          5.4.2   Craig y Dinas 105
          5.4.3   Penwyllt and Ynyswen 109

Chapter 6 **South Pembrokeshire and Carmarthen Bay**   117
    6.1   Introduction 117
    6.2   Geological history 118
    6.3   Itineraries 120
          6.3.1   Marloes Sands 120
          6.3.2   Freshwater West 126
          6.3.3   Manorbier 132
          6.3.4   Laugharne 136
          6.3.5   Skrinkle Haven and Lydstep 137
          6.3.6   Tenby (South Beach) 141
          6.3.7   Stackpole Quay 144
          6.3.8   Stack Rocks to St Govan's Head 148
          6.3.9   West Angle Bay 154
          6.3.10   Tenby (North Beach) to Waterwynch Bay 156
          6.3.11   Ragwen Point to Telpyn Point and Amroth 161
          6.3.12   Amroth to Wiseman's Bridge 171
          6.3.13   Saundersfoot to Monkstone Point 175
          6.3.14   Broad Haven 179
          6.3.15   Little Haven 181
          6.3.16   Nolton Haven 184

Chapter 7 **North Pembrokeshire and Cardigan Bay**   187
    7.1   Introduction 187
    7.2   Geological history 188
    7.3   Itineraries 189
          7.3.1   St Non's, Caerfai and Caerbwdy bays 190
          7.3.2   Porth Clais 194
          7.3.3   Solva 195
          7.3.4   Whitesands Bay 197
          7.3.5   St David's Head 201
          7.3.6   Abereiddy and Porthgain 204
          7.3.7   Fishguard and Strumble Head 208
          7.3.8   Newport Bay and Ceibwr Bay 212
          7.3.9   New Quay and Aberaeron 215

Appendices 1–8   224

Glossary   232

Bibliography   239

Preface

My first introduction to the captivating geology of South Wales was on field trips led by the late Professor T R Owen when I was an undergraduate student at the University College of Swansea in the 1960s. I was so impressed that I stayed at Swansea to complete a Ph.D thesis under the supervision of Professor Gilbert Kelling, who fostered my interests in clastic sedimentology. Since this time, I have myself led many field trips to South Wales for undergraduates, Open University groups, oil company personnel and tourists. In addition I have spent many holidays in Pembrokeshire with my parents, wife and children who came to know many of the geological localities more intimately than they may have wished.

South Wales is an ideal place to study geology and geomorphology because the rocks are well exposed along impressive sea-cliffs, mountain crags and river sections, which are located in areas of exceptional natural beauty, including the Brecon Beacons and Pembrokeshire Coast National Parks, Gower (Area of Outstanding Natural Beauty) and the Vale of Glamorgan Heritage Coast. The geology is also extremely varied, with classic sections in rocks of Precambrian, Palaeozoic, Triassic, Early Jurassic and Quaternary age. Many of the localities have been designated, or are being considered for notification, as 'Sites of Special Scientific Interest' (SSSI) and/or Regionally Important Geological and Geomorphological Sites (RIGS). A large proportion of the rocks are of sedimentary origin and in some areas they have been intensely folded and faulted as a result of tectonic movements within the earth's crust (Caledonian, Armorican and Alpine orogenies). There are also many fine examples of intrusive and volcanic igneous rocks present in the Precambrian and Lower Palaeozoic strata of Pembrokeshire and the Old Radnor and Builth Inliers.

Logistically it was necessary to limit the number of localities described in this guide. The difficult decision of which sites to include was governed not only by the quality of their geology, but also by access and safety considerations. Thus, those sites with good public access were favoured over those that were isolated, located on private land or were in working or recently abandoned quarries. I have revisited all of the localities described in the guide within the past three years to assess their suitability for inclusion. Site descriptions have been augmented with many illustrations including geology and locality maps, cross sections, graphic logs, depositional models and a selection of the photographs I have taken over the last 40 years, which I hope have made the guide user-friendly.

Gareth George
Llanmadoc, March 2008

Acknowledgements

I would like to thank my colleagues and friends who have accompanied me on field trips and research visits to South Wales over the last 40 years. I would particularly like to mention my geological colleagues at Swansea University (1960–70) who now form members of the, still active, Half Moon Club. Many of these 'old boys' contributed chapters for two previous geological guides of South Wales: *Geological excursions in South Wales and the Forest of Dean* (1971) edited by Douglas and Michael Bassett and *Geological excursions in Dyfed, South-West Wales* (1982) edited by Michael Bassett. These are now out of print; they were known affectionately as the Black Book and Blue Book, respectively, and were invaluable sources of field data, which I gratefully acknowledge. Members of group (Brian Williams, Gilbert Kelling, David James and John Stead) have provided updates in unpublished field guides for reunion field trips to South Wales.

I would like to acknowledge the encouragement and valuable suggestions given by Roger Jones during the early stages of the preparation of the book. My wife, Jean, edited the text and provided many useful comments.

The author is grateful to Her Majesty's Stationary Office for permission to reproduce some of the fossils illustrated in Figures 2.5b, 2.7b and 7.9 (taken from George 1970) (Licence Ref. V2008000326). Every possible attempt has been made to contact the copyright holder of the remaining fossil sketches in the above figures (taken from the British Museum Natural History 1969). English Nature granted permission (Ref. AR 26967) to publish Figure 2.2 in its original form (taken from Atkinson 1959). All of the other figures, based on published works, have been modified and/or redrawn and the relevant authors have been cited in the captions and are listed in the Bibliography. All of the photographs are the copyright of the author.

Finally, I would like to dedicate the book to my late father, my mother, wife and children for their patience, encouragement and company during many field visits to South Wales.

Chapter 1

Introduction

South Wales is a popular tourist destination, due largely to its stunning scenery, engrossing history, rich culture and plentiful leisure and holiday amenities. These aspects of the region are briefly reviewed here for the benefit of the families and travelling companions of Earth scientists, and for the committed geologist who also needs time to relax and enjoy the wider aspects of a visit to South Wales.

The geographical definition of South Wales presents a challenge. To the west and south it is enclosed by Cardigan Bay and the Bristol Channel respectively; to the east it extends to the English border counties of Gloucestershire and Herefordshire. However, its northern limit is defined only loosely, but in this book it is taken as a line from Aberaeron on the west coast to Llandrindod Wells in the east. Thus, the region includes the southern halves of the counties of Ceredigion and Powys, and the whole of Pembrokeshire, Carmarthenshire, Monmouthshire and Glamorganshire. Three-quarters of the population of Wales, of just under three million, live in and around the industrial cities and towns of the south. Only about 20 per cent of the total population speak Welsh, with the highest concentration being in North Wales, Ceredigion and Carmarthenshire.

Although it is often thought that the green grass of Wales is attributable to its high rainfall, many parts of the lowlands of South Wales have average or below-average rainfall, mild winters and they even boast some of the sunniest spots in Britain. South Pembrokeshire has a particularly mild climate and it produces the earliest new potatoes grown in mainland Britain.

1.1 Transport

South Wales is accessible from outside the United Kingdom by direct flights to Cardiff International Airport or on flights via Amsterdam or Paris. It is also possible to fly into Heathrow, Gatwick or Bristol airports and continue to South Wales by train, coach or in a hire car. There are internal flights to Cardiff from Glasgow, Edinburgh, Aberdeen, Belfast, Dublin and Cork. Visitors from the Republic of Ireland can sail from Rosslare to Fishguard (2 hour crossing) or Pembroke Dock (4 hour crossing) and from Cork to Swansea (10 hour crossing). Visitors, who enter the UK through the Channel Tunnel, or on the cross channel ferries, can continue the journey to South Wales by road or rail. Intercity trains from London (Paddington) take just over 2 hours to reach Cardiff and continue farther west to Bridgend, Neath, Swansea, Carmarthen, Haverfordwest and Fishguard.

The M4 motorway provides a fast road link from London via the Severn Bridge to Cardiff, Bridgend, Neath and Swansea. A section of motorway and dual carriageway continues farther west to Carmarthen (A48), Haverfordwest and Fishguard (A40) and to Pembroke Dock (A477). From Birmingham the M5 links up with the M4 for the Severn Bridge crossing or you can take the M50 to Ross-on-Wye and enter South Wales through Monmouth and Abergavenny. The A465 (Heads of the Valleys road), which runs south of the Brecon Beacons and north of the coalfield, connects Abergavenny with Ebbw Vale, Merthyr Tydfil and Neath. From Merthyr Tydfil the A470 cuts north through the picturesque scenery of the Brecon Beacons to Brecon, and south through the Taff Valley and Pontypridd to link with the M4.

Although the most convenient way to travel to and around South Wales is by car, public transport can also be recommended for selected day trips and tours using the local or special bus services. From Swansea, the Gower Explorer buses provide a comprehensive service to the main beaches and villages throughout the year. Likewise, in Pembrokeshire, the coastal bus services (Puffin Shuttle, Strumble Shuttle, Poppit Rocket, Celtic Coaster and Coastal Cruiser) are very convenient for accessing the towns and beaches, and for planning walks. Timetables for these services are available from the bus companies and tourist offices, and those for Pembrokeshire are also published in the free bimonthly tourist newspaper *Coast to Coast*. Visitors can purchase a Freedom of Wales Flexi Pass, which provides unlimited access to all the mainline train services and almost every bus service in the Principality, and also offers discounts to many tourist attractions.

1.2 Accommodation and food

There is no shortage of accommodation to suit all needs and pockets, from hotels and motels in the towns and cities, country hotels, inns, guesthouses, farms, study centres, youth hostels and bed & breakfast accommodation, as well as an extremely good selection of self-catering cottages. There are also many caravan and camping sites particularly around the main tourist resorts. The Wales Tourist Board grades the accommodation on a 1–5 five star rating and comprehensive lists are available from the tourist offices and websites.

Good quality fresh food produce is available from butchers, greengrocers, delicatessens, farm shops and markets throughout South Wales. A favourite Welsh delicacy is lava bread, which is seaweed (*Porphyra*) usually eaten with bacon or cockles for breakfast or as a snack. Swansea market has a group of stalls devoted to selling lava bread and Penclawdd cockles. Locally caught crab, lobster, bass, sewin (sea trout) and wild salmon are other specialities on restaurant and bistro menus. Wales is also well known for its delicious mountain and saltmarsh lamb, Welsh Black beef, Brecon venison and Pembrokeshire turkey. Cawl is a robust stew made from poorer cuts of lamb, leeks and potatoes and it is delicious with crusty bread. The local dairy products include, a variety of hard cheeses (Caerfai, Caerphilly, Llanboidy and Llangloffan), goats' cheeses, butter and icecream. Pastries such as bara brith (fruit bread) and Welsh cakes are good for packed lunches. Welsh wines from vineyards at Tintern, Monmouth, Abergavenny, Vale of Glamorgan and south Pembrokeshire, and Welsh whisky from Penderyn, are being successfully produced and marketed.

1.3 Landscape and scenery

The spectacular and beautiful landscape of South Wales has evolved over a long period of time as a result of erosion associated with major changes in sea level, ice advances, fluvial activity and weathering. These surface processes have been greatly influenced by the bedrock geology seen today, and past covers of Mesozoic strata. The landscapes vary from plateaux, moorlands, escarpments, and rolling hills incised by deep valleys, to coastal lowlands and dunes, wetlands, estuaries, tidal flats–saltmarshes, and some of the most beautiful beaches and islands in the British Isles. Much of this landscape has been recognized as unique by the designation of National Parks and Areas of Outstanding Natural Beauty. South Wales has two national parks, the Pembrokeshire Coast and the Beacon Beacons. Recently (2005), the western part of the Brecon Beacons National Park has become a UNESCO Geopark, recognizing the area as one of the most geologically interesting landscapes in southern Britain. The Fforest Fawr (Great Forest) Geopark incorporates a wealth of attractions, including the Welsh National Showcaves (Dan yr Ogof), the Black Mountain, Pen y Fan, Carreg Cennen Castle and the Craig-y-nos Country Park. The Gower peninsula was the first Area of Outstanding Natural Beauty (AONB) to be designated in the United Kingdom and in 2006 it celebrated its 50th anniversary. Ridge (1999) described Gower as "a happy accident of geological forces, climatic conditions and human enterprise; a harmonious blend of landscape, nature and history". Nowhere is this statement more pertinent than in the area around Rhossili Bay and Worms Head (Fig. 1.1).

Large parts of the coastlines of the Vale of Glamorgan, Gower, Pembrokeshire and Cardigan Bay have also been designated Heritage Coasts. These special (non-statutory) coastal strips, are managed so that their natural beauty is conserved and, where appropriate, accessibility is improved. Substantial parts of the above areas are owned and administered by the National Trust, which, together with the Countryside Council for Wales, the South Wales Wildlife Trust and many other organizations, support and encourage conservation and tourism. South Wales also has many Sites of Special Scientific Interest (SSSI) designated for their importance in both the earth and biological sciences.

The South Wales coastline displays some of the best coastal scenery in Europe. Cliffs with prominent headlands and peninsulas protect a large part of the coastline, interrupting glorious sandy bays, secluded coves, dunefields, river mouths and estuaries. Many of the tidal inlets were formed as a result of increased erosion when melt-waters from the last (Late Devensian) glaciation drained into the existing rivers. Subsequently, during the post-glacial (Flandrian) rise in sea level, many of the sinuous valleys were drowned to form rias, Milford Haven being a spectacular example. The limestone coasts of the Vale of Glamorgan, Gower and south Pembrokeshire display a wide variety of landforms, including extensive wavecut platforms, caves and blowholes, natural arches, sea stacks, dolines (swallow holes) and dry valleys. A very prominent feature of the coastal zone is the development of a relatively flat platform (60 m platform), which was cut by marine erosion during the late Tertiary (Pliocene) sea level high, about 5 million years ago (Fig. 1.1).

To the north of Milford Haven the coastline is more rugged, being composed of a wide variety of sedimentary and igneous rocks, their contrasting colours, weathering and attitudes adding further interest to the landscape, which includes the Marloes and Dale peninsulas and the islands of Skomer and Skokholm. Farther to the north,

St Brides Bay marks the western termination of the softer rocks of the Pembrokeshire Coalfield; farther north again, hard igneous rocks form much of the bare crags of the St David's Peninsula, Strumble Head and Mynydd Preseli, which rise above the 60 m coastal platform. The highly indented coastline of north Pembrokeshire contrasts with the smooth sweep of Cardigan Bay. Much of this coastal landscape is best appreciated from the sea and there are many sightseeing boat trips along the coast and around the islands, which also provide opportunities to see the varied wildlife for which South Wales is famous.

The inland scenery is also diverse and breathtaking. The Cambrian Mountains form the backbone of mid-Wales and occupy the area between Snowdonia in the north and the Brecon Beacons to the south. This unspoilt and sparsely populated landscape is an upland plateau dissected by river valleys and gorges, and is dotted with small lakes. It is the most important watershed in Wales, supplying the sources of many large rivers, including the Severn, Wye, Tywi and Teifi. The topography of the Beacon Beacons, culminating at Pen y Fan (886 m OD), consists of north-facing scarps with dramatic erosional scallops (cwms) formed as a result of glacial erosion. On the southern margin of the National Park, the headwater regions of the River Neath and River Tawe are noted for their large cave systems (Dan-yr-Ogof, Porth-yr-Ogof and Ogof Ffynnon Ddu), karstic landforms and beautiful wooded valleys with impressive waterfalls (Frontispiece A). The waterfalls have attracted and inspired many famous artists, including J. M. W. Turner (1775–1851) and J. B. Smith (1848–1884), and the area is often referred to as "waterfall country". The area was also praised by Alfred Russel Wallace (1823-1913), the naturalist famed for his pioneering work on evolution and natural selection, who wrote in his autobiography, "I cannot call to mind a single valley that in the same extent of countryside comprises so much beautiful and picturesque scenery and so many interesting special features as the Vale of Neath".

To the east along the northern rim of the coalfield are the Valleys (Rhondda, Taff, Rhymney etc.), which epitomize the past industrial landscape of South Wales. The steep-sided river valleys that extend southeastwards across the coalfield determine the linear settlements of terraced houses, many of which were originally occupied by coalminers and ironworkers. North of Merthyr Tydfil, just off the A470, is Cyfarthfa Castle, one of the most impressive monuments to the Industrial Revolution in South Wales. William Crawshay (1788–1867) financed the building of this grand castellated mansion in 1824, from the great wealth acquired by his family from the Cyfarthfa ironworks. George Borrow, who passed through Merthyr Tydfil in 1854 on his epic journey across Wales, described the ironworks in his travelogue *Wild Wales* (first published in 1862) as "remarkable edifices, though of a gloomy horrid Satanic character" and he went on to describe the line of blast furnaces as "a house of reddish brick with a slate roof–four horrid black towers behind, two of them belching forth smoke and flame from their top–holes like pigeon holes here and there–two immense white chimneys standing by themselves". Today the castle is a museum and art gallery surrounded by 65 hectares of parkland. Some of the remains of Cyfarthfa ironworks are at present being restored while the remains of four furnaces and the engine house of Ynysfach ironworks can be seen at the southern end of the town. In the centre of town, at 4 Chapel Row, a small museum commemorates the birthplace of Joseph Parry (1841-1903) the composer of the hymn-tune "Aberystwyth" and the choral favourite "Myfanwy". During his childhood (9–13 years of age) he worked in a coalmine and at Cyfarthfa ironworks,

Fig. 1.1 Bough timbers of the *Helvetia* exposed on Rhossili sands with Worms Head in the background. Note the flat top of the Inner Head and the 60 m platform extending inland.

before he and his family emigrated to Pennsylvania. In 1874 he was appointed as the first professor of music at the University College of Wales, Aberystwyth.

On the eastern rim of the coalfield is the Blaenavon World Heritage Site, where the Industrial Revolution landscapes are preserved around the 18th century ironworks. Nearby is Big Pit, the National Mining Museum of Wales, where you can go on an enthralling underground tour. Alexander Cordell (1914–1997) made the unique industrial landscapes of the Valleys famous through his best-selling novels (*Rape of the fair country, Hosts of Rebecca, Song of the Earth, The fire people, This sweet and bitter Earth,* etc). Details of the towns and landscapes referred to in these novels, comprising what is often referred to as Cordell Country, are described in a series of four interesting tour guides, which are available at tourist offices.

The Brecon and Monmouth Canal (built 1797–1812), runs for 50 km between Brecon and Pontypool, through the idyllic scenery of the National Park and the Usk Valley. The canal was originally used to transport limestone, processed lime and other products from Brecon to Newport up until 1930. Recently the canal has been restored and reopened to the public (1970), and it is now used for informal recreation, including canal-boat holidays, canoeing and fishing, and there are walks along the full length of the towpath. Along the restored canal there are six locks and several public houses, and it passes close to the large abandoned limestone quarries at Trefil and Llangattock, the old ironworks in the Clydach Gorge and the ruins of the disused limekilns at Talybont-on-Usk. A unique trip can also be taken on the narrow-gauge Brecon Mountain Railway, which makes a short journey (65 minutes) through the National Park along the full length of the Taf Fechan reservoir.

The Wye Valley, which straddles the Wales and England border, is recognized as an Area of Outstanding Natural Beauty for its spectacular limestone gorge and wooded ravines, which follow the meandering River Wye from Hereford to Chepstow. This anglicized corner of South Wales includes the historic towns of Chepstow and Monmouth, many picturesque villages, and the early medieval Abbey at Tintern, located by one of the most spectacular stretches of the River Wye.

1.4 Prehistory and historic buildings

South Wales, particularly north Pembrokeshire, is rich in prehistoric sites. The bare and atmospheric Preseli Hills are littered with hill forts, burial chambers, stone circles and standing stones, and are famed for being the source of the bluestones used for the construction of the inner circle and horseshoe of Stonehenge in late Neolithic times (about 2500 BC). Pentre Ifan, probably the finest burial chamber in Wales, dates back to about 4000 BC (Fig. 1. 2). Nearby at Castell Henllys, an Iron Age hill fort has been excavated, and thatched roundhouses have been reconstructed

Fig. 1.2 Pentre Ifan burial chamber (cromlech), Mynydd Preceli; the capstone is 5 m long.

Fig. 1.3 Laugharne Castle and the path leading to the Dylan Thomas Boat House.

6

on the original foundations. There are other more elaborate tombs at Tinkinswood (near Dyffryn Gardens) and Parc le Breos (near Parkmill on Gower). Along the coastline many Iron Age forts are located on precipitous promontories, and bone caves are also common, including the famous Paviland Cave, Minchin Hole and Bacon Hole on the Gower, and Hoyle's Mouth at Tenby. Information regarding the excavations of these sites and collections of their bones and artefacts can be seen in the small museums at Swansea and Tenby.

South Wales has a wealth of ancient monuments and historic buildings, many of which are cared for by Cadw (Welsh: to keep), the equivalent of English Heritage. Some of the most popular castles are: Chepstow, Raglan, Cardiff, Caerphilly, Carreg Cennen, Kidwelly, Laugharne (Fig. 1.3), Manorbier, Carew, Pembroke and Cillgerran. Other historic sites on the tourist route include the Roman fortress, amphitheatre and museum at Caerleon, and the ruins of Tintern and Strata Florida Abbeys. In the tiny city of St David's, the magnificent cathedral has been a site of pilgimage and worship since the 6th century; it hosts an annual nine-day festival of classical music in late May and many other choral and orchestral concerts. Adjacent to the cathedral is the impressive Bishop's Palace, which is being extensively restored by Cadw.

1.5 *Cultural attractions*

Cardiff, the historic and modern capital city of Wales, is home to many of the principality's major tourist attractions. Cardiff Castle, situated in the heart of the city, dates from Roman times. It has a magnificent Norman keep and rooms lavishly decorated in the Gothic revival style by the third marquis of Bute and the architect William Burges. Burges also designed and built the nearby fairytale-style red castle (Castell Coch) for the marquis in the 1870s. The National Museum has one of Europe's finest art collections and excellent geology, natural history and archaeology sections. It runs a busy programme of exhibitions, lunchtime talks and family activities throughout the year. There are two international concert venues; St David's Hall and the recently opened Wales Millennium Centre (Canolfan Mileniwm Cymru). The latter, located on the Cardiff Bay waterfront, provides a wide range of concerts and cultural entertainments and is the home of the Welsh National Opera Company. While in this area you can tour the many attractions of the newly redeveloped Cardiff Bay area, on foot or via the waterbus service. The Millennium Stadium, the first sports stadium in the UK with a retractable roof, was built on the site of the old Arms Park stadium and is the new home of the Welsh International Rugby Union. In March 2008, a capacity crowd watched Wales win the grand slam in the Six Nations Championship.

Just 7 km west of Cardiff is the Museum of Welsh Life, an open-air museum sited in 40 hectares of parkland surrounding St Fagans Castle. This very popular tourist attraction has 45 period buildings on display, which have been transported from various locations around Wales and painstakingly reconstructed on the site. The indoor museum has galleries illustrating Welsh costumes, musical intruments, farming implements and other artefacts, and the manor house is also open to the public. About 30 minutes drive away, at the northern end of the Rhymney Valley in the village of Nelson, is Llancaiach Fawr medieval manor. Here you can take a guided tour of the fortified manor accompanied by servants dressed in period costume who relate local gossip, ghost stories and events relating to the Civil War.

Dyffryn Gardens, which is currently being restored, is one of the premier gardens of Wales and has been awarded a grade 1 status in the Cadw register of lansdscapes. Farther afield is the National Botanic Garden of Wales, a short distance from the M4 motorway at Llanarthe in Carmarthenshire. The 230 hectare site, set in the former 18th century regency park of Middleton Hall, was opened in May 2000 and has developed into a fascinating garden with exotic plants housed in a spectacular oval glasshouse, designed by Sir Norman Foster.

The Land of Song is also famous for its male voice choirs, such as the Treorchy, Pontarddulais, Morriston Orpheus and, not so famous, but from my hometown, Côr Meibion Maesteg. Many of these local choirs hold concerts in chapels, pubs and rugby clubs, as well as in larger venues all over South Wales. At any of these concerts you are certain to hear renditions of choral favourites such as Myfanwy, Hen Wlad fy Nhadau (Land of My Fathers) and We'll Keep a Welcome. Festivals of music and poetry, known in Wales as eisteddfodau, are also an important aspect of Welsh culture. The International Eisteddfod is held annually at Llangollen, whereas the National Eisteddfod of Wales and the Urdd Eisteddfod (youth festival) are held at different venues across Wales during the summer.

There are many galleries displaying the work of contemporary artists and photographers. Graham Sutherland (1930–1980), one of the most significant British artists of the mid-20th century, derived much of his inspiration from the rugged landscapes and seascapes of Pembrokeshire. A large collection of his work can be seen at the National Museum Cardiff and there is talk of opening a Sutherland centre in St David's to replace the foundation gallery originally sited in Picton Castle (closed in 1989). Augustus John (1878–1961) was born in Tenby and painted many portraits, including those of his fellow countrymen Dylan Thomas and Richard Burton.

1.6 *Exploring on foot*

By far the best way to explore and appreciate the landscape and scenery of South Wales is by walking the coastal and inland public footpaths and trails. Details of many of the best walks can be found in pamphlets available at the tourist information centres and the National Trust. The splendid Pembrokeshire Coast Path opened in 1970, extends for 300 km from Amroth in the south to St Dogmaels in the north, along cliff-tops, across beaches and through seaside towns and quaint villages. Most of the trail, suitable for walkers of all abilities, provides a unique opportunity to access the largely unspoilt landscape of the Pembrokeshire Coast National Park, which is internationally important for its nature reserves and geology. Most people walk selected parts of the trail on a series of daytrips, but it is also possible to complete the whole walk in a week, if you are fit. The official guide to the trail (John 2001) contains a wealth of information and it is well illustrated with Ordnance Survey maps and colour photographs.

One of the best walks in Ceredigion is along the Heritage Coast from Llangranog to Cwmtydu and New Quay. When in New Quay pick up the pamphlet *Dylan Thomas' New Quay,* which is a town walk covering the houses the poet lived in and his favourite pubs. Inland there are interesting walks along the Teifi Valley to view the Cenarth Falls and Cilgerran Castle. The official trail of the Brecon Beacons National Park is the Beacons Way, an east–west trek of 163 km, taking about eight days. Details of shorter walks and other events are available at the information centres at Libanus and Craig-y-nos. The Libanus mountain centre, located 8 km to the southwest of Brecon, has stunning views of Pen y Fan (the highest mountain in South Wales), and the adjacent moorland of Mynydd Illtyd offers some fine walks. The Craig-y-nos Country Park, near Abercrave, is sited in the landscaped gardens of the former stately home of the famous opera diva Adelina Patti. Pontneddfechan is an ideal centre to explore the spectacular waterfalls located along the upper tributaries of the River Neath; the information centre, opposite the Angel Inn, provides leaflets giving details of the various walks. Many of the waterfalls, including Sgwd Gwladys, Sgwd yr Eira and Sgwd Clun Gwyn, and the Aberdulais, Melincourt and Henrhyd Falls, are important geological localities. The Wye Valley Walk is a long (218 km) but relatively low-gradient route from Chepstow to Rhyader, but becomes steeper as it leads to the source of the River Wye at Pumlumon in the Cambrian Mountains.

1.7   Wildlife

Many of the islands around the coast of South Wales are important nature reserves for seabirds and seals. Skomer Island is a National Nature Reserve noted for its seabird colonies, with up to 500 000 birds, including puffins, Manx shearwaters, guillemots and razorbills, present during the spring and summer months. There are regular sailings (weather permitting) from Martin's Haven to Skomer and also less frequent guided trips to Grassholm and Skokholm. Daily sailings depart from St Justinian's lifeboat station to Ramsey Island, which has one of the largest Atlantic grey seal colonies in the UK. Seals and their pups can be seen from August to December on the beaches of the Marloes Peninsula and Skomer Island. In spring and summer the cliffs are alive with guillemots, razorbills and kittiwakes, while shoals of porpoise are a common sight in Ramsey Sound. Choughs, ravens and peregrine falcons can be observed along the cliffs of the islands and the mainland, which in spring and early summer are covered in wild flowers. Red kites, almost extinct in Wales at the turn of the 20th century, are now quite common, with an estimated 250 breeding pairs. To be sure of seeing these superb birds of prey you can visit the feeding stations at Llanddeusant in the Black Mountain or Gigrin Farm near Rhyader.

1.8   Sports and leisure activities

South Wales offers an extensive range of sporting and leisure activities. Angling is particularly popular and it enjoys a large local following as well as supporting a healthy tourist trade. Sea fishing is excellent from the sandy beaches, estuaries, rocky promontories, piers, and from boats. In summer months, pollack, mackerel, tope and flatfish are common, with large bass coming inshore to feed during the autumn. During the winter months, cod and whiting are plentiful, while bass and turbot are particularly common around Cardigan Bay. Sea trout, known as sewin in

Wales, feed in the Irish Sea and run up the estuaries to their native rivers to spawn from early spring to late summer. Good catches of brown trout and salmon have been recorded from the Wye, Usk, Towy and Teifi rivers. There are also many lakes and reservoirs stocked with fish that provide a range of coarse fishing.

The varied coast offers endless opportunities for all types of water-sports including, sailing, canoeing, kayaking, diving and snorkelling. The best surfing beaches are Rest Bay (Porthcawl), Rhossili, Newgale and Whitesands Bay; windsurfing is popular on the long sandy beaches at Rhossili, Pembrey and Pendine. Coasteering, a relatively new activity that involves traversing cliff sections and leaping into pools, is popular along the rocky cliffs of north Pembrokeshire. This group activity requires a qualified guide and the use of wetsuits. Paragliding, from the summit of Rhossili Down, is also an established activity. Rock climbing is popular, particularly on the limestone cliffs of the south Pembrokeshire coast between Stack Rocks and St Govan's Head, and on Stackpole Head, where there are 17 named climbs. Information boards and a pamphlet provide details of the agreed seasonal climbing restrictions to protect the nesting and feeding sites of important colonies of seabirds, choughs, peregrines and ravens.

There are around 80 golf courses to choose from and some companies run special golfing holidays. The premier golf course, at the Celtic Manor Resort, just off the M4 near Newport, hosts the Wales Open golf competition and will be the venue for the 38th Ryder Cup in 2010.

There are a variety of attractions for children besides the glorious sandy beaches. These include funfares at Barry Island and Porthcawl, the Oakwood theme park near Haverfordwest, outdoor activity centres, coastal and inland pony trekking and riding schools, and many of the castles hold historical exhibitions and re-enactments. Cycling is another popular tourist activity and there are plenty of off-road trails suitable for mountain bikes. A fasinating day can be spent at the Dolaucothi goldmines near Pumsaint (five saints), Carmarthenshire, where you can take underground tours of the mines, try your hand at gold panning or explore the varied landscape of the Cothi Valley. Gold was first extracted by the Romans around 75 AD using mainly opencast methods. Much later during the late 18th and early 19th centuries ore was extracted from adits and deeper shafts before the mine closed in 1938, due to difficulties in extracting the gold from the ore. In recent times most Welsh gold, including that used for the wedding rings of the royal family, has come from the Clogau mine near Barmouth (closed in 1995) and also from the Gwynfynydd mine at Dolgellau (closed in 2007).

1.9 Concluding remarks

In this chapter I have summarized relevant information for those visiting South Wales, and highlighted some of the region's outstanding features. This is only a small sample of what South Wales has to offer, and visitors should consult the many guidebooks and brochures available at bookshops, information centres and websites for fuller accounts. The remaining chapters of the book concentrate on the varied and interesting geology of South Wales, which can be enjoyed by Earth science students, professional and amateur geologists, as well as their families and friends.

Chapter 2

Overview of the Geology of

South Wales

2.1 Introduction

The important role that geology has played in influencing the landscape, natural resources and culture of South Wales was briefly mentioned in the previous chapter. Landscape and scenery, for example, are directly related to variations in the composition and structure of the underlying strata and the prolonged effects of weathering, and erosion by ice, rivers and the sea. The varied geology of South Wales has also provided a wealth of natural resources, which have been exploited over a long period of time. Exploitation began during Prehistoric times with the collection of flints to make tools, and the use of local rocks for building settlements and erecting stone circles, monoliths and tombs. In this respect, Mynydd Preseli appears to have been particularly important as is shown by its wealth of megalithic remains, and for being the source area of the bluestones of Stonehenge (Fig. 2.1). The collection and transportation of the bluestones, each weighing up to 4.5 tonnes, from Mynydd Preseli to Stonehenge (a distance of about 390 km) can be considered to be the earliest and most remarkable series of geological field trips to South Wales (Fig. 2.2). The seventy-nine bluestones, identified as spotted dolerite (69), rhyolite (5) and volcanic ash (5), and a single 17 tonne-block of micaceous sandstone (Altar Stone), probably derived from the Senni Formation (Lower ORS), obviously had a special significance for the Beaker people of Salisbury Plain. Much later, field surveys by Roman geologists led to the rediscovery and extraction of the gold at Dolaucothi, the iron ore at Mumbles and many other mineral deposits developed throughout the Principality.

In more recent times the rich deposits of coal, black-band iron ore and limestone, present in both the South Wales and Pembrokeshire coalfields, fuelled the Industrial Revolution. Although much of these non-renewable resources (excluding the limestone) are now largely exhausted, their past exploitation has determined the pattern of human settlement and much of the existing culture of the "Valleys". Today, the beautiful scenery, the varied and interesting geology and the rich culture and biodiversity of South Wales are of paramount importance in attracting tourists and Earth scientists.

Fig. 2.1 Spotted dolerite exposed on the crags of Carn Meini the probable site from which the bluestones were extracted by the Beaker geologists. The burial chamber emphasizes the importance of completing risk assessments!

Fig. 2.2 A party of Beaker people transporting a bluestone block by raft along the Pembrokeshire coast—the first geological field trip to South Wales! (Reconstruction by Alan Sorrell in Atkinson 1959).

In this chapter the geological history of South Wales is summarized and where necessary put into its wider stratigraphical framework. Only brief references are made to global scale processes, such as plate tectonics, continental drift, eustatic sea-level changes and biological evolution, and further information regarding these topics can be sourced in modern texts on historical geology (e.g. Woodcock & Strachan 2000; Brenchley & Rawson 2006). A summary of the main stratigraphical units, discussed in the itineraries, and their numerical ages (based on Gradstein *et al.* 2004), is presented in Table 2.1.

Era	System / Series / Stage			Age Ma	Groups / Formations etc		
Quaternary	Holocene		Flandrian	11.8 ka	Post-glacial: peat, blown sand, beach sand, marsh, alluvium etc.		
	Upper Pleistocene		Devensian Ipwichian	126 ka	Heterogeneous glacial deposits formed during the Last Glacial Maximum & proglacial deposits and formed during ice retreat; solifluction (head). Raised beach deposits, cave deposits (Hoxnian - Late Devensian). Older Glacial deposits (poorly dated)		
	Middle		Wolstonian Hoxnian Anglian				
Mesozoic	Jurassic	Lower	Sinemurian	189.6 +/-1.5			
				196.5 +/-1.0	Lias Group	Porthkerry Member Lavernock Shale Member St Mary's Well Bay Member	
			Hettangian				
	Triassic	Upper	Rhaetian	199.6 +-0.6 ~203.6	Penarth Group	Lilstock Formation Westbury Formation	
			Norian	~216.5	Mercia Mudstone Group	Blue Anchor Formation Branscombe Mudstone Fm	
				299 +/-0.8	Permian (299 - 251 ma) to early Late Triassic (c. 251 - 216 Ma) absent		
Upper Palaeozoic	Carboniferous	Silesian			Pennant Sandstone Fm		
					S. Wales Coal Measures Gp		
				~318	Marros Group	See Appendix 1	
		Dinantian			Pembroke Limestone Group		
				359.2 +/-2.5	Avon Group		
	Devonian	Upper			Skrinkle Sandstone Group	Quartz Conglomerate Subgroup	
				385.3 +/-2.6			
		Middle		397.5 +/-2.7			
		Lower	Emsian Pragian Lochkovian	411 +/-2.8	?	Ridgeway Conglomerate Freshwaterwest Formation	Brownstones Red Marls / Senni Fm St Maughans Formation
				416 +/-2.8			
	Silurian	Pridoli		418.7 +/-2.7	Sandy Haven Formation Albion Sands Formation	Moors Cliff Formation Freshwater East Formaton	
		Ludlow		422.9 +/-2.5	Red Cliff Formation		
		Wenlock		428.2 +/-2.3	Gray Sandstone Group		
		Llandovery	Telychian Aeronian Rhuddanian		Coralliferous Group Skomer Volcanic Group	Aberystwyth Grits Group	Trefechan Fm Mynydd Bach Fm Borth Fm
				443.7 +/-1.5		Caban Conglomerate / Cwmere fms etc	
Lower Palaeozoic	Ordovician	Upper	Ashgill		Llandeilo - Caradoc Shales	Yr Allt Formation Nantmel Mudstone Fm Cwm-yr-Eglwys / Dinas Island fms Penyraber Mudstone Fm	
			Caradoc	460.9 +/-1.6			
		Middle	Llanvirn		Caerhys Shale / Castell Limestone Llanrhian Volcanic Formation Penmaen Dewi / Abermawr Fms Ogof Hen Formation	Fishguard Volcanic Group	
			Arenig				
		Lower	Tremadoc	478.6+/- 1.7			
				488.3 +/-1.7			
	Cambrian	Upper	Merioneth	~501	Lingula Flags Menevian Beds		
		Middle	St David's		Solva Beds		
		Lower	Branch Placentia	~510	Caerbwdy Sandstone St Non's Sandstone		
				542 +/-1.0			
Neoproterozoic	Ediacaran				Pebidian Super Group (~600 - 580 Ma)	Coomb Volcanic Formation	
					St David's Granophyre (587 +25 -14 Ma)		
				~630 Ma			
	Cryogenian				Johnston Complex (643 +5 -28 Ma)	Stanner Hanter Complex (702 +/-8)	
				850 Ma			

Table 2.1 Chart and numerical ages of the main onshore stratigraphical units referred to in the itineraries. Shaded areas represent unconformities; numbers in bold are Global Standard Section and Point (GSSP) values.

2.2 Structural and stratigraphical framework

2.2.1 Precambrian (Neoproterozoic)

The most extensive outcrops of late Precambrian rocks in South Wales occur in Pembrokeshire, where they are represented by the Pebidian Supergroup (*circa* 600 – 580 Ma), and the Johnston Complex (643 +5 / –28 Ma) (Table 2.1, Fig. 2.3a). Thus both of these groups of basement strata fall within the upper part of the Neoproterozoic Era (1000–542 Ma). The Pebidian Supergroup is composed of a thick (> 2 km) pile of volcanics (basic lavas and tuffs) interbedded with acidic pyroclastics and volcaniclastics, which are intruded by acid plutons and minor basic sheets. They crop out in the cores of two ENE-trending Caledonian anticlines, namely, the St David's and Hayscastle anticlines (Fig. 2.3b). In the former anticline the Pebidian is intruded by the St David's Granophyre, which has provided a radiometric date of 587 +25 / –14 Ma. Farther to the south, diorite–granodiorite–granite gneisses and volcanics, belonging to the Johnston Complex, occur within an allochthonous Variscan thrust (Johnston–Benton Fault Zone).

On the eastern extremity of central Powys, small fault-bounded slivers of late Precambrian igneous rocks (Stanner–Hanter Inlier) and sediments (Old Radnor Inlier) occur within the Church Stretton Fault Zone (for site descriptions see Woodcock 2000; Jones 2000). Here the intrusive igneous rocks have been dated at 702 ± 8 Ma, which makes them the oldest Precambrian basement rocks present in England and Wales. However, the sediments are much younger and have been correlated with the Longmyndian Supergroup of Shropshire.

Another small, but important, outcrop of latest Precambrian (Ediacaran Period 630–542 Ma) is located to the southwest of Carmarthen in the Llangynog Inlier, where the succession consists of about 1000 m of subaqueously deposited volcanic and volcaniclastic rocks belonging to the Coomb Volcanic Formation (see Cope 1982, 2000; Cope & Bevins 1993; Bevins 2000). Some of the bedded tuffs contain fossil medusoids (jellyfish), referred to as an Ediacaran fauna, the name being derived from the famous type locality in Australia (Ediacara Hills) where these soft-bodied fossils were first described in the 1940s.

During the late Neoproterozoic, the temporary accretion of Avalonia with the northern margin of Gondwana was accompanied by complex tectonics (Cadomian Orogeny). Detailed tectono-stratigraphical mapping and geochemical studies have shown that the Neoproterozoic rocks of southern Britain can be assigned to a number of fault-bounded terranes, each of which record episodes of subduction and arc magmatism, related to the plate tectonic evolution of eastern Avalonia (for reviews see Phoraoh & Carney 2000; McIlroy & Horák 2006). In this context the Precambrian rocks located within and to the east of the Welsh Borderland Fault System, including those of the Llangynog Inlier and the Johnston Complex, are assigned to the Wrekin Terrane. Conversely, the Pebidian rocks located to the west of this fault and to the south of the Menai Strait Fault System, are assigned to the Cymru Terrane. This latter terrane, which forms the basement rocks of the Early Palaeozoic Welsh Basin, was folded, faulted, metamorphosed, uplifted and deeply eroded as a result of the aforementioned late Precambrian orogeny.

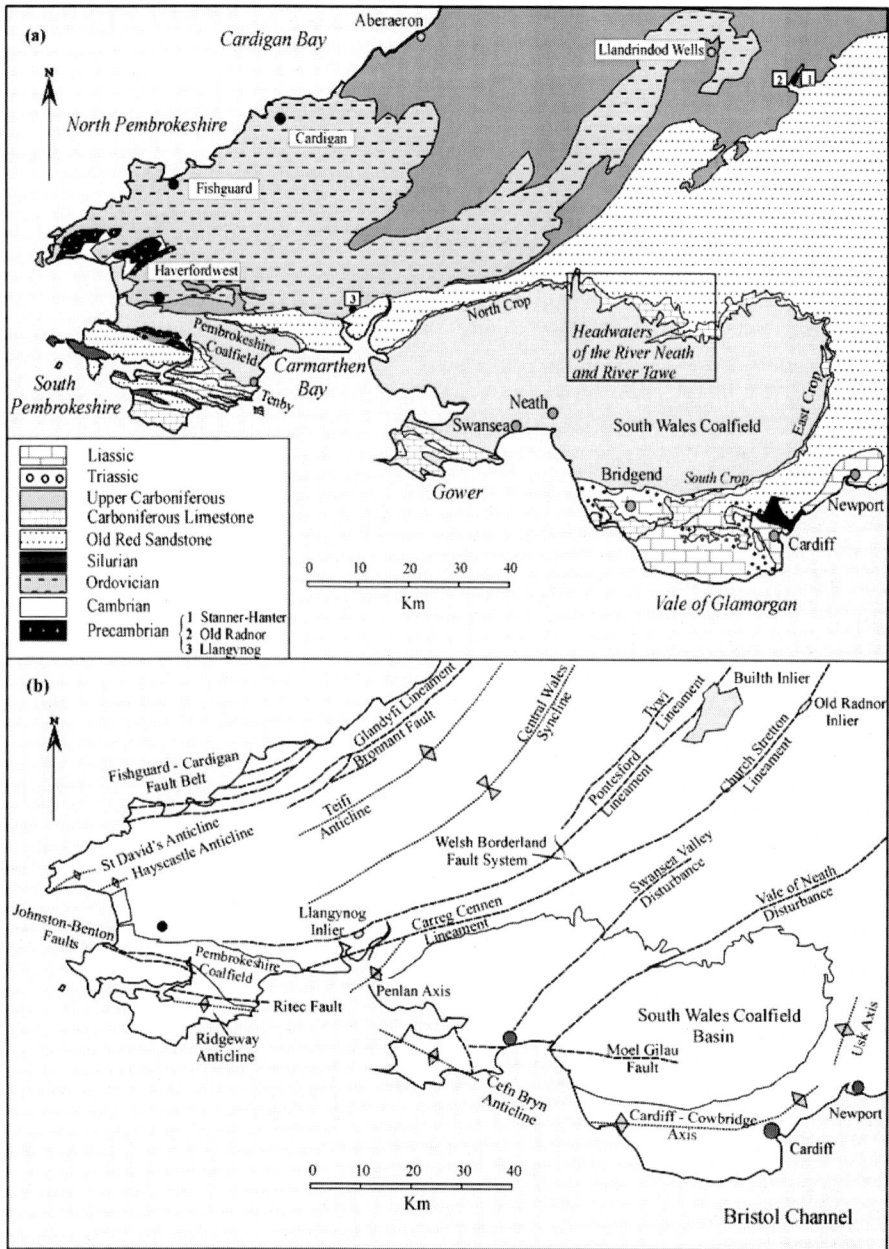

Fig. 2.3 (a) Simplified geological map of South Wales showing the areas covered in the field itineraries (see Howells 2007 for a colour map). (b) Map showing the main structural elements in the region.

2.2.2 Early Palaeozoic (542–416 Ma)

The Lower Palaeozoic strata of Pembrokeshire consists of an extremely thick (~12 000 m) succession of marine sediments deposited on the southern margin of the

Iapetus Ocean, which was created as a result of the break-up and drifting apart of Laurentia and a small fragment of northern Gondwana known as Avalonia (Fig. 2.4a). During this time the Welsh Basin went through a period of rapid evolution as a result of extensional tectonics and eustatic sea-level changes. The Welsh Borderland Fault System was a very important palaeogeographic feature that defined the hinge-zone, separating the tectonically subsiding basin from a more stable shelf area known as the Midland Platform (a northeastwards-extending salient of the southern landmass of Pretannia) (Fig. 2.5a). Within the basin, further deep-seated fractures, including the Bala Fault, Fishguard–Cardigan Fault Belt, Bronnant Fault–Glandyfi Lineament and the Central Wales Lineament, also controlled sedimentation at various times. The interplay of syn-sedimentary tectonics, eustatic sea level changes and periodic volcanic activity has resulted in very complex vertical and lateral variations in rock types; reflected in the regionally diverse stratigraphic nomenclatures. In this section only some of the more important aspects of the evolution of the Welsh Lower Palaeozoic Basin are documented; more detailed accounts can be consulted in Brenchley *et al.* (2006) and Cherns *et al.* (2006).

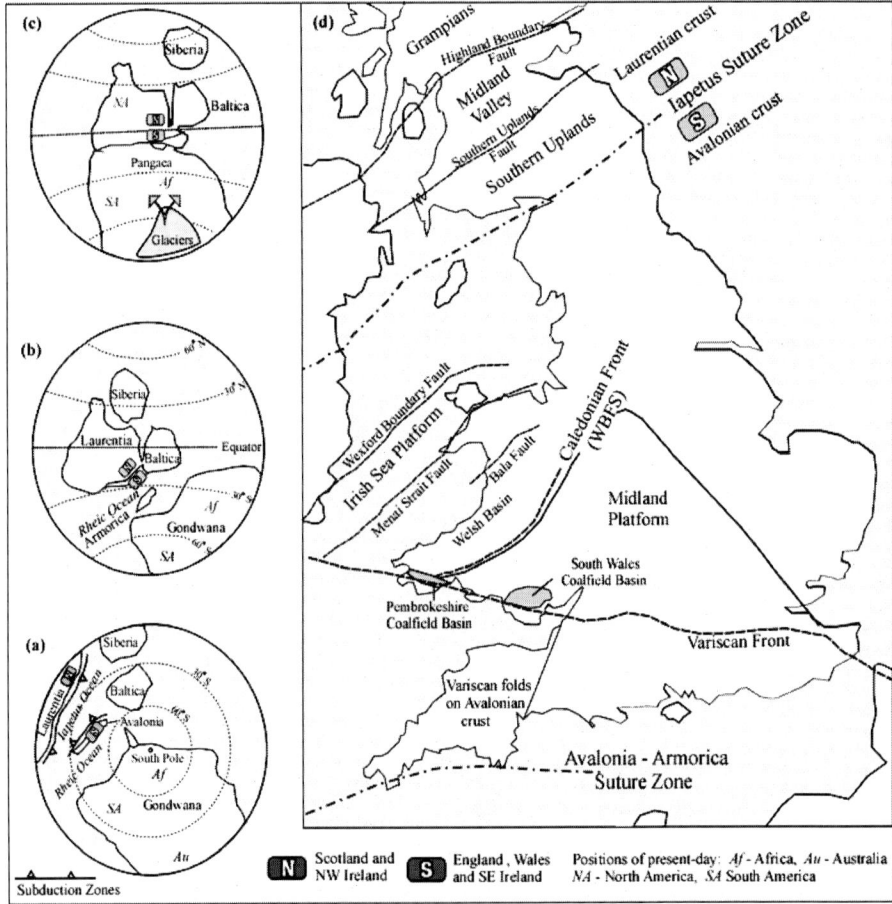

Fig. 2.4 Global palaeo-continental reconstructions for selected intervals during the Palaeozoic Era: **(a)** Mid-Ordovician ~470 Ma. **(b)** Mid-Silurian ~425 Ma. **(c)** Late Carboniferous ~300 Ma. **(d)** Simplified terrane map of Britain (modified from Holdsworth *et al.* 2000).

In southwest Wales Cambrian sedimentation was initiated by the deposition of a basal conglomerate, which onlaps southeastwards onto folded and weathered Precambrian basement rocks (Fig. 2.5a). The remainder of the Cambrian succession is composed of about 1600 m of nearshore and offshore marine clastic sediments, containing well-established faunas of trilobites, brachiopods, molluscs and sponges (Fig. 2.5b). Many of these Cambrian facies display sets of hummocky cross stratification (HCS) indicating that they were deposited on a storm-influenced (southwards-inclined) shelf environment. Farther north, deep-water turbidite facies and manganese-rich shales were deposited in a more rapidly subsiding graben, defined by the Bala and Menai Strait Faults.

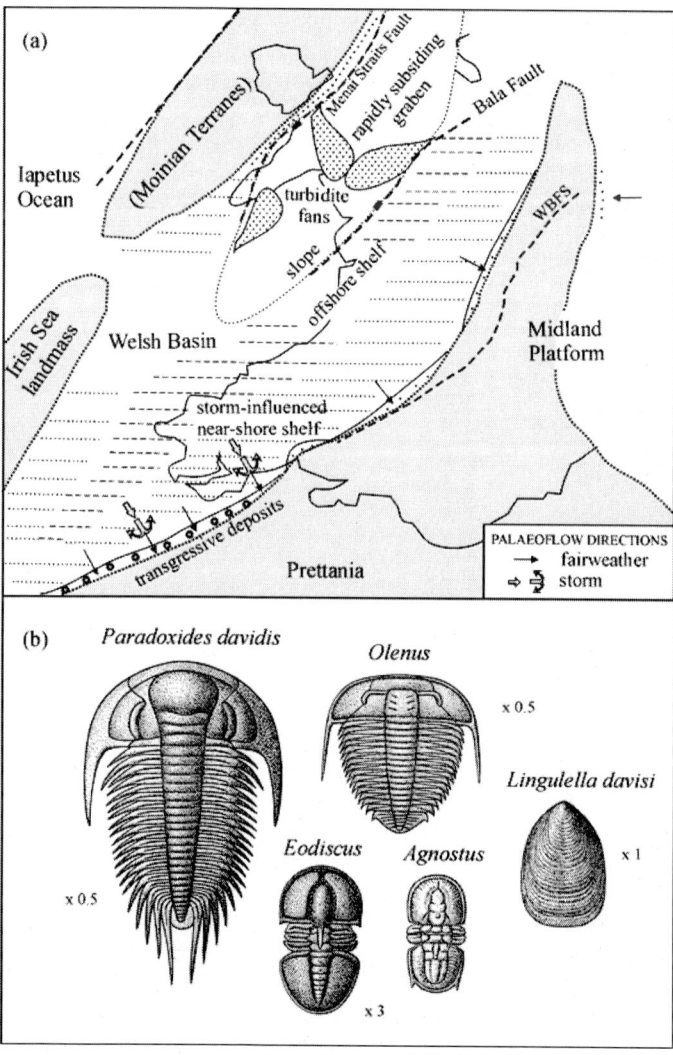

Fig. 2.5 (a) Early Cambrian palaeogeographical map of Wales (based on Brasier *et al.* 1992). (b) Typical Cambrian fossil trilobites and *Lingula* (reproduced from British Museum (Natural History) 1969; George 1970).

In southwest Wales, the top of the Cambrian and early Ordovician (Tremadoc) strata are absent due to a major unconformity, which was formed by a combination of thermal uplift of the basin margin related to arc volcanism and glacio-eustatic sea-level fall. Thick shallow marine sequences that accumulated on other parts of the platform during the early Ordovician, were later uplifted and gently folded. A major transgression, which occurred during the early Caradoc at the base of the *Nemagraptus gracilis* Biozone (GSSP 460.9 ±1.6 Ma; Table 2.1), resulted in an expansion of deeper-water marine sedimentation on the platform. Towards the end of the Ordovician (late Ashgill) a short-lived Gondwana glaciation caused a rapid, high-magnitude fall (~100 m) in sea level. As a consequence of these changes in relative sea level, the marginal shelf sequences are characterized by unconformities (often associated with coastal onlap) and minor depositional hiatuses. Within the deeper-water environments of the Welsh Basin, graptolitic shale and turbidite sequences were deposited throughout most of the Ordovician (Llanvirn–Ashgill). Recent sedimentology studies of the Ordovician succession, in the Cardigan and Dinas Island area, have shown that sand-rich submarine fans and coarse debrites were deposited from mass flows sourced from the uplifted shoulders of a graben, the margins of which were defined by syn-sedimentary faults (Davies *et al.* 1997) (Fig. 2.6). In this setting the late Ordovician glacio-eustatic sea-level fall caused an expansion of slope facies, including the slump and slide deposits seen along the coast at Llangranog (see Davies *et al.* 2006, Plates 1, 2). In basinal areas isolated from coarse clastic supplies, eustatic changes in sea level resulted in the deposition of grey bioturbated (oxic) shales during lowstands, as opposed to black graptolitic shales (anoxic) during highstands; a trend also seen in similar Silurian facies.

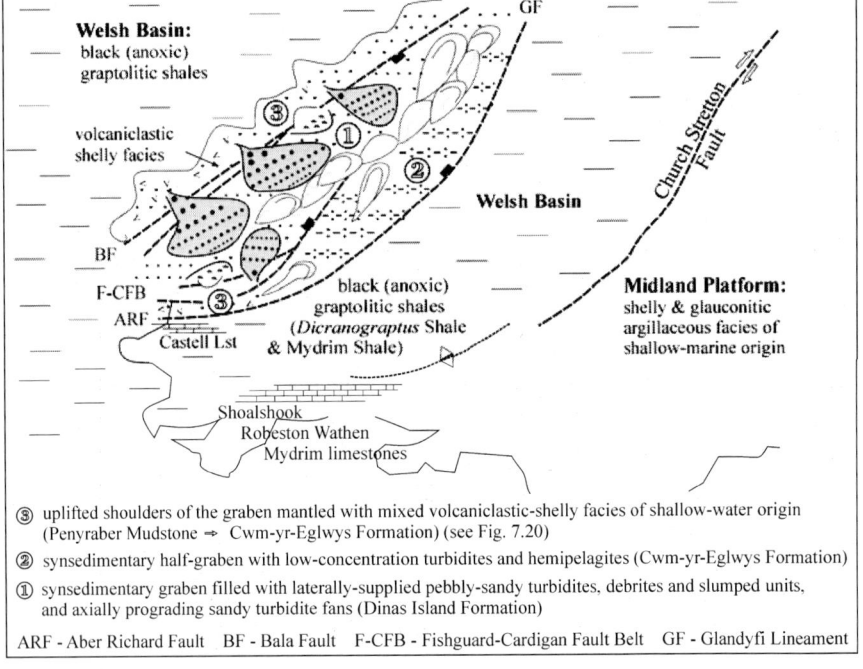

③ uplifted shoulders of the graben mantled with mixed volcaniclastic-shelly facies of shallow-water origin (Penyraber Mudstone → Cwm-yr-Eglwys Formation) (see Fig. 7.20)
② synsedimentary half-graben with low-concentration turbidites and hemipelagites (Cwm-yr-Eglwys Formation)
① synsedimentary graben filled with laterally-supplied pebbly-sandy turbidites, debrites and slumped units, and axially prograding sandy turbidite fans (Dinas Island Formation)

ARF - Aber Richard Fault   BF - Bala Fault   F-CFB - Fishguard-Cardigan Fault Belt   GF - Glandyfi Lineament

Fig. 2.6 Late Ordovician (Caradoc) palaeogeography showing the main syn-sedimentary faults defining the Fishguard - Cardigan graben (generalized from Davies *et al.* 2002).

During the following early Silurian transgression, shelf facies were deposited much farther east, beyond the Welsh Borderland Fault System; here the bathymetry of the platform has been estimated by reference to the distribution of depth-related brachiopod faunas (Fig. 2.7). In southwest Wales tectonic uplift resulted in the exposure of the shelf, which was incised and bypassed by a fluvial system that transported coarse detritus northwards into the basin. Marine shelf sedimentation continued through Wenlock and Ludlow times with the deposition of mixed clastic facies, calcareous shales and impure limestones, which often contain rich shelly faunas. These marginal sequences are punctuated by local unconformities, for example in the Old Radnor Inlier the richly fossiliferous early Wenlock Dolyhir Limestone (30 m thick) rests unconformably on Precambrian sediments. In the deeper basin further extensional tectonics, resulted in the development of a northeast-trending trough, the eastern margin of which was controlled by syn-sedimentary faults located along the Glandyfi and Central Wales lineaments. This depositional setting is typified by the classic turbidite facies of the Aberystwyth Grits Group, which were deposited in sandy–muddy turbidite lobes that prograded northeastwards through the axis of the basin (Fig. 2.8). Marine sedimentation came to an end at various times between the Ludlow and Přídolí, due to a combination of eustatic sea-level fall, increasing sedimentation rates, and a decrease in basin accommodation, which accompanied the closure of the Iapetus Ocean and the collision of Avalonia and Laurentia (Fig. 2.4b).

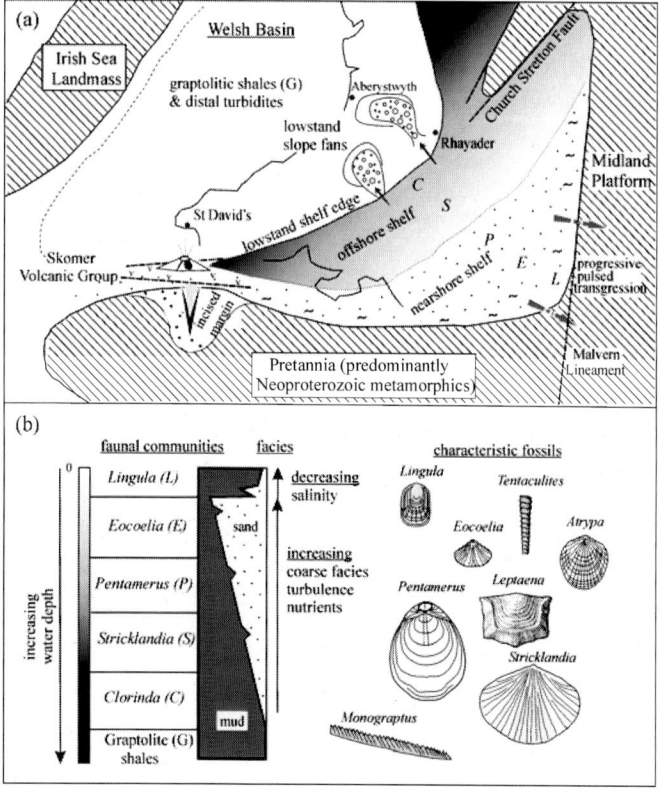

Fig. 2.7 (a) Early Silurian palaeogeography (after Bassett *et al.* 1992). (b) Silurian fossils, faunal communities and palaeo-ecological controls.

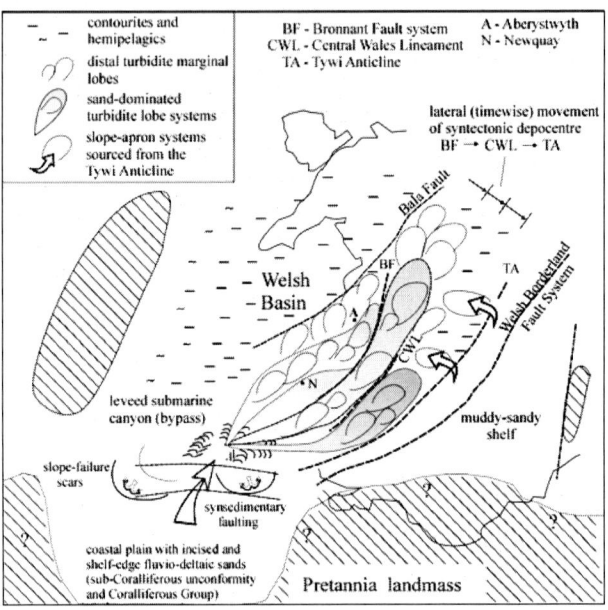

Fig. 2.8 Palaeogeographic map for Wales during the deposition of the Aberystwyth Grits Group and related Llandovery (Telychian)–early Wenlock facies (generalized from Bassett *et al.* 1992; Davies *et al.* 1997).

The Ordovician was also characterized by intense igneous activity, which was confined to specific centres. The earliest, Trefgarn Volcanic Group (Tremadoc–Arenig), is composed of basic, intermediate and acid rocks with a calc-alkaline chemistry, which were extruded in an island arc setting during the early stages of the southward subduction of the Iapetus crust below Avalonia (Fig. 2.9). Later (Llanvirn), igneous activity is recorded by basaltic pillow lavas, rhyolite lavas and acidic tuffs, exposed in Mynydd Preseli (Sealyham Volcanic Group), Strumble Head–Fishguard (Fishguard Volcanic Group) and on Ramsey Island. These mixed (bimodal) basic and acid volcanics have a more tholeiitic geochemistry and are believed to have formed in a marginal basin (Fig. 2.9). The small quantities of intermediate and acid volcanics are believed to have formed as a result of low-pressure fractional crystallization of tholeiitic basalts. Further early Ordovician volcanics of acid to intermediate composition occur in the Builth–Llandrindod inlier (Builth Volcanic Group), which is located within the Welsh Borderland Fault System. Ordovician volcanic activity was associated with the intrusion of countless sills and dykes mainly of dolerite–gabbro composition. On Skomer Island and the Marloes Peninsula, basalts flows and pillow lavas, flow-banded rhyolites, tuffs and interbedded sediments belonging to the early Silurian (Llandovery) Skomer Volcanic Group, were formed in half-grabens delineated by a series of east-west trending, inverted normal faults (the Wenall, Ritec and Benton faults; Fig. 2.7a).

On a global scale, Early Palaeozoic igneous activity can be related to the plate tectonic processes of sea-floor spreading, subduction, volcanic arc and back-arc formation, and continental rifting. The extrusion of great volumes of magma resulted in the global atmosphere having an abnormally high level of carbon dioxide and a very low oxygen level, which accompanied by a glacio-eustatic sea-level fall, caused a significant faunal extinction at the end of Ordovician times.

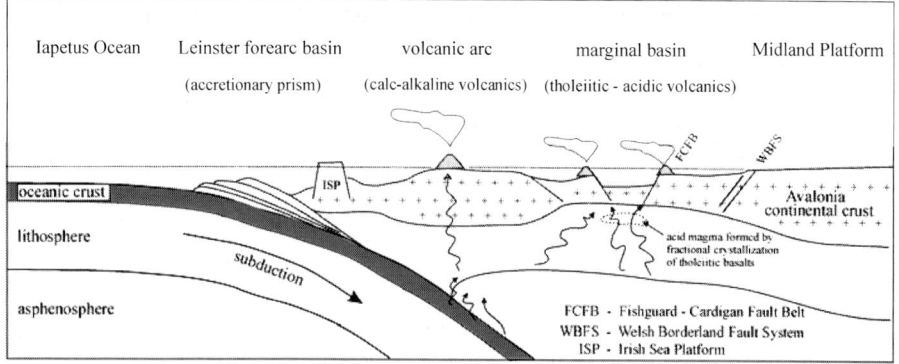

Fig. 2.9 Simplified cross section showing the Early Palaeozoic plate tectonic setting and modes of magma generation on the northern flank of Avalonia and the Iapetus Ocean.

2.2.3 Caledonian (Acadian) Orogeny

All of the Early Palaeozoic rocks of the region were deformed during an orogeny (or orogenies), which began during the late Silurian and extended into the mid Devonian. The Caledonian Orogeny was the result of the closure of the Iapetus Ocean, which led to the docking of Avalonia with Laurentia during mid–late Silurian times (Fig. 2.4b). The much later Acadian Orogeny, believed to be related to the closure of the Rheic Ocean, produced the main folding and cleavage in the Cambrian–early Devonian rocks of England and Wales. This Acadian deformation was most extreme to the west of the Welsh Borderland Fault System, where the rocks are faulted, highly cleaved and deformed into tight northeast–southwest-orientated folds. This Acadian tectonic trend is well illustrated by the Fishguard–Cardigan Fault Belt and the complex folding along the Teifi Anticline and the Central Wales Syncline (Fig. 2.3b). On the Midland Platform deformation was less severe and is characterized by gentle folds and localized faulting. These rocks also record evidence of low-grade regional metamorphism, which occurred at temperatures of 200–300° C, well above those reached during burial diagenesis. In general the degree of metamorphism increases westwards away from the Welsh Borderland Fault System, with the highest temperature zone occurring in north Pembrokeshire. This lateral variation can be related to the thicker sedimentary successions, greater deformation and the presence of igneous centres, which caused higher thermal gradients in the western parts of the Welsh Basin.

Besides these structural features, the orogeny is also recorded in some areas by an unconformity at the base of the Lower Old Red Sandstone (ORS). This relationship is well seen to the south of the Ritec Fault in Pembrokeshire and farther eastwards into Carmarthenshire, the Black Mountains and the Brecon Beacons. To the north of the Ritec Fault and in the Welsh Borderlands there is a transitional upward passage from marine Silurian into continental ORS facies. In these transitional successions (Ludlow–Přídolí series) the Ludlow Bone Bed was previously used to mark the Silurian–Devonian boundary. More recently this datum has been raised to a position at the top of a distinctive pedogenic carbonate (calcrete), previously referred to as the *Psammosteus* Limestone; now known as the Chapel Point Calcrete in southwest Wales and the Bishop's Frome Limestone in the Welsh Borderlands (see Hillier & Williams 2007, Fig.2). This horizon marks a prolonged period (around 10 000 years) of non-deposition and carbonate soil formation. In South Wales and the Welsh Borderlands the Townsend Tuff Bed has also been used to mark the informal position of the Silurian–Devonian boundary.

2.2.4 Old Red Sandstone

The late Palaeozoic began with the deposition of thick redbed facies belonging to the Old Red Sandstone (ORS), which occur to the east and southeast of the Welsh Borderland Fault System. The Lower ORS consists of clastic sediments derived from the newly uplifted Caledonian mountains and subsequently deposited in a variety of arid to semi-arid continental environments, including estuaries, broad alluvial plains, ephemeral braided–meandering rivers and alluvial fans. This succession has an average thickness of about 1800 m, but increases to over 4500 m in the hanging-wall of the Benton Fault, and decreases to 440–1440 m to the south of the Ritec Fault. These redbeds are practically devoid of fossils and are consequently extremely difficult to date and correlate, although the Chapel Point Calcrete and three named Plinian airfall tuffs provide useful marker bands. The presence of hummocky bedforms in the tuffs suggests that the violent Plinian eruptions were possibly accompanied by tsunamis, which inundated the continental plain. During middle Lower ORS times (late Lochkovian–Pragian) landmasses in the Bristol Channel, composed of Neoproterozoic/Lower Palaeozoic rocks, provided the source areas for the two poorly dated alluvial fan facies – the Llanishen Conglomerate Formation developed in the Cardiff area, and the Ridgeway Conglomerate Formation of Pembrokeshire, which reaches a maximum thickness of over 350 m in the hangingwall of the Ritec Fault (Fig. 2.10a). Throughout the Anglo-Welsh Basin the Middle ORS is absent due to one or more unconformities that record episodes of uplift and erosion related to the Acadian Orogeny.

Within the basin the Upper ORS is represented by a relatively thin (100–300 m) succession composed largely of siltstones, sandstones and conglomerates of braid plain origin. In South Pembrokeshire the Skrinkle Sandstone, developed to the south of the Ritec Fault, displays an upward passage from interbedded continental redbed and green–grey estuarine/marginal marine facies into the fully marine facies that characterize the Lower Carboniferous (Avon Group) (Fig. 2.10b).

2.2.5 Carboniferous (Dinantian and Silesian)

At the beginning of the Dinantian, a major rise in sea level resulted in the flooding of the southern margin of the ORS continent (Wales–Brabant High) and the development of a southward dipping carbonate ramp environment (Fig. 2.11). On this ramp, which later evolved into a shelf environment, a variable succession of carbonate facies containing shelly faunas of corals, brachiopods and molluscs, were deposited in a warm tropical sea over a period of about 35 Ma (Table 2.1, Appendix 1). These limestones are up to 1225 m thick in South Pembrokeshire and Gower but thin onto the northern and eastern margins of the basin. Throughout this period eustatic sea-level changes resulted in the deposition of many transgressive and regressive sequences. A large proportion of these carbonate facies were derived

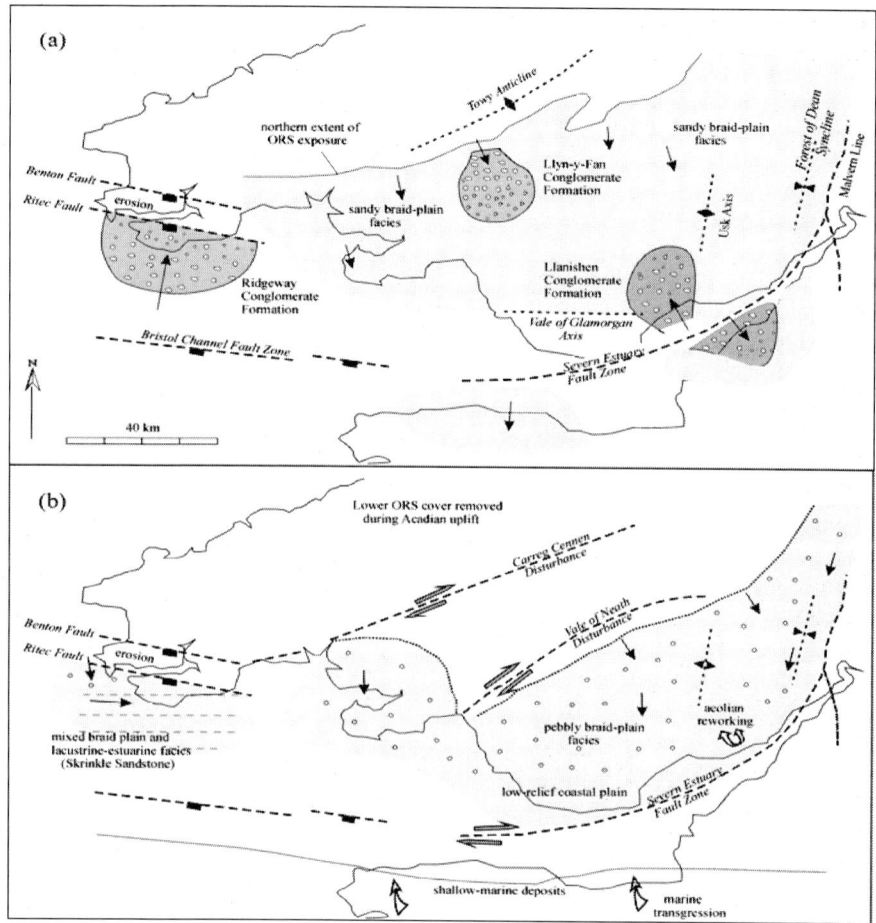

Fig. 2.10 ORS palaeogeography and active tectonic features in South Wales. (a) Lower–Middle ORS extraformational conglomerates. (b) Upper ORS (both modified from Hillier & Williams 2006).

from biogenic products formed within the photic zone of the contemporary marine environment. Consequently, sediment production rates were highest during the transgressions and highstands of sea level, lower during falling stages, and virtually zero during lowstands. During highstands of sea level, oolitic grainstone shoals and barrier bars prograded southwards from inner to mid ramp positions. The build-up of these regressive sequences was sometimes followed by eustatic sea-level falls, which resulted in emergence and subaerial weathering of the early-cemented carbonates to form palaeokarst surfaces and thin palaeosols. Subsequent transgressions resulted in the palaeokarsts being overlain by back-barrier lagoonal facies, and higher-energy bioclastic facies being deposited above shoreface erosional planes (ravinement surfaces) (Fig. 2.11).

A significant rise in sea level occurred during the late Dinantian (Oystermouth Formation) and continued into the early Namurian when thick (~350 m) basinal shales with many goniatite-bearing marine bands were deposited in Gower (Bishopston Mudstone Formation) (Appendix 1). Along the northern and eastern

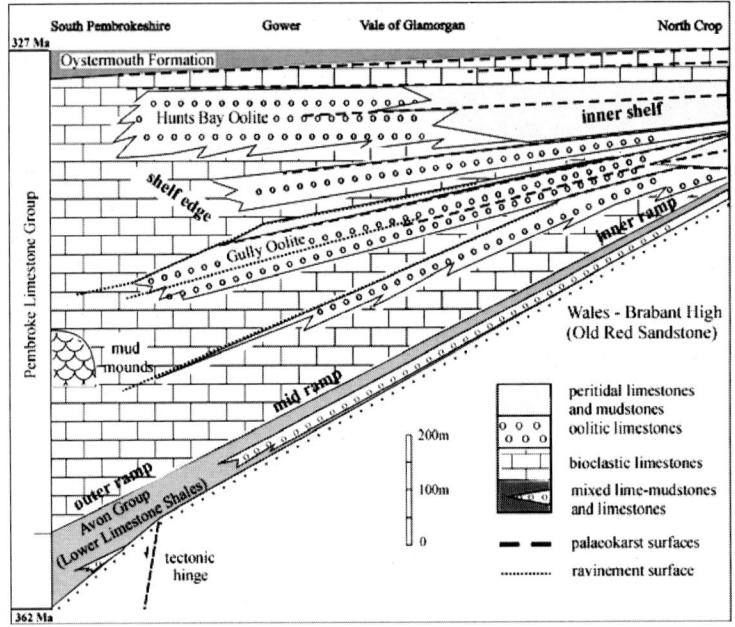

Fig. 2.11 Schematic north–south section of the Lower Carboniferous (Dinantian) ramp environment showing the distribution of the carbonate facies (modified from Tucker & Wright 1990).

margins of the basin, quartz arenite facies (Twrch Sandstone Formation) were deposited in braid-deltas, which prograded onto a slowly subsiding storm-dominated ramp (Fig. 2.12). At this time the clastic ramp was receiving a high sediment influx and was also being subjected to frequent eustatic sea-level changes. Later, increased basin subsidence resulted in the deposition of well-developed coarsening-upwards cycles, capped by transgressive shales containing goniatites, brachiopods, bivalves and trace-fossil assemblages. Two major sea level falls occurred during the late Namurian and early Westphalian, which resulted in the deposition of thick incised valley-fills (Telpyn Point Sandstone and Farewell Rock respectively). These coarse grain fluvio-deltaic sandbodies have regionally developed erosional bases (sequence

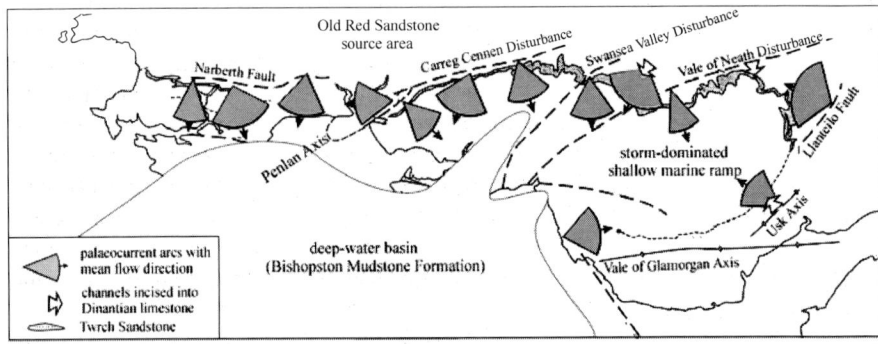

Fig. 2.12 Palaeocurrent map showing the dispersal routes for Namurian (Twrch Sandstone) braid deltas (modified from George 2000).

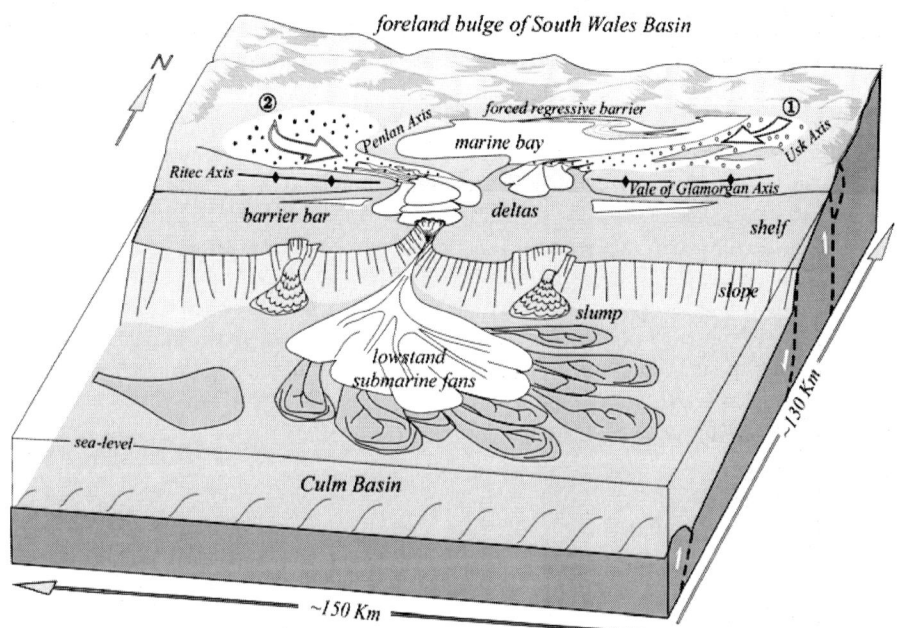

Fig. 2.13 Speculative palaeogeographic model illustrating the possible relationship between late Namurian (Yeadonian) incised fluvio-deltaic systems of the South Wales Basin and the lowstand fan deposits of the Culm Basin (modified from George 2001). System 1 supplied quartz arenites from an eastern source; system 2 supplied lithic arenites (Telpyn Point Sandstone) from a western source.

boundaries) and multi-storey/multi-lateral channel geometries, and they are capped by palaeosols. The much lower base levels at these times also resulted in channels eroding through emergent shelves to reach the shelf-edge, from where they supplied coarse sediment directly to deepwater submarine fans (lowstand fans). This type of incision occurred during the late Namurian lowstand when fluvio-deltaic systems, sourced from the northern margin (forebulge) of the South Wales Basin, supplied large volumes of clastic sediment to lowstand submarine fans in the Culm Basin of Devon and Cornwall (Fig. 2.13).

The Westphalian strata of the South Wales Coalfield consist of up to 4000 m of mixed sandstone, siltstone and argillaceous facies containing important coal seams. Most of the coal seams worked in the past, and those worked in opencast sites occur in the lower part of the succession (Westphalian A and B). Coal prices have risen sharply recently and although the reserves at Tower colliery (closed January 2008) are now exhausted, two nearby collieries at Aberpergwm and Treforgan may be reopened, thus continuing the tradition of mining in the Welsh Valleys. Typically the coal seams are developed above coarsening-upwards fluvio-deltaic and fluvio-lacustrine cycles and they are overlain by mudstones containing non-marine bivalves and occasional marine faunas. Higher in the succession the Pennant Measures are characterized by thick multi-storey fluvial sandbodies deposited by braided–sinuous rivers, which flowed predominantly northwards from the rising Variscan mountains to the south.

Although Britain was located in the tropics during the Carboniferous, its climate was being strongly influenced by Southern Hemisphere (Gondwana) glaciations, which caused the cyclical eustatic sea-level changes so evident in the Namurian and

Westphalian. In these successions major rises in sea level are recorded by marine bands, usually containing goniatites, which are important for correlation and dating. Falls in sea level resulted in the rapid expansion of tropical coastal swamps, which were drained by fluvio-deltaic systems. These swamps provided the ideal environmental conditions necessary for the rapid growth and evolution of club mosses, tree ferns and horsetails and the rapid accumulation of thick peat horizons. Due to the very large amounts of carbon used up in the formation of Dinantian limestones and Silesian coals, as well as an increase in the consumption of carbon dioxide during photosynthesis, the Carboniferous atmosphere contained an exceptionally high level of oxygen, but was depleted in carbon dioxide. The high oxygen level (about 14 percent greater than today), is believed to have stimulated the growth of plants particularly giant club mosses (up to 40 m high), tree ferns and horsetails, and also extremely large insects, including scorpions, millipedes over 1 m long and dragonflies with wingspans of 50–75 cm.

2.2.6 Variscan Orogeny

Throughout the Upper Carboniferous (Silesian) the progressive convergence of the southern continent of Gondwana (including Armorica) and the northern continent of Laurussia (Laurentia, Avalonia, Baltica and Siberia) resulted in the Variscan Orogeny and the consolidation of the supercontinent of Pangaea (Fig. 2.4c). The northward propagation of the Variscan deformation front initiated a foreland basin in SW England, with South Wales forming its northern peripheral zone (Fig. 2.14). A number of features, concerning the evolution and fill of the South Wales Basin, support a Variscan foreland basin origin:

- The progressive northwards migration of the depocentre, increasing subsidence rates, and the gradual change in the direction of clastic supply routes from a northerly (foreland bulge) source to southerly (orogenic) source (Kelling 1988; Burgess & Gayer 2000).
- The stratigraphic evolution of the basin fill from Dinantian carbonates→ Namurian marine clastics→ Westphalian coal-bearing sequences→ Pennant fluvial sandstones (Leveridge & Hartley 2006).
- The east–west orientation of the basin, parallel to the Variscan thrust front (Fig. 2.14).

In southern parts of the basin, particularly those to the south of the Variscan front, the Upper Carboniferous strata are highly faulted and folded, with north–south crustal shortening values of around 30 per cent rising to over 50 per cent in the west. In the Pembrokeshire Coalfield (between the Benton and Ritec faults) WSW-trending folds are asymmetrical with steep to over-turned northern limbs, which are often replaced by thrusts. Many Caledonian structures, such as the Ritec Fault, the Carreg Cennen, Swansea Valley and Vale of Neath disturbances, were reactivated at this time and often had a profound influence on sedimentation. Further examples of these Armorican structures are discussed in the itineraries and more detailed accounts have been published by Hancock *et al.* (1982) and Leveridge & Hartley (2006).

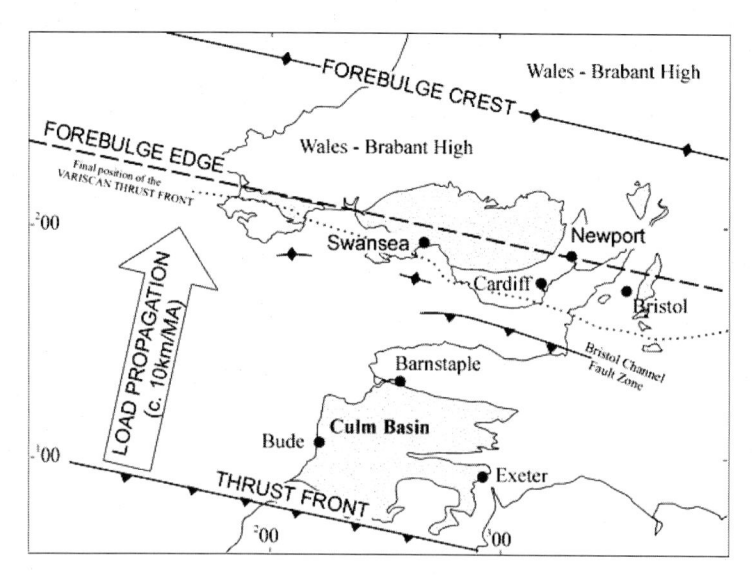

Fig. 2.14 Map showing the relative positions of the main tectonic elements during Armorican deformation in the foreland basin of southwest Britain (modified from Burgess & Gayer 2000).

2.2.7 Mesozoic and Palaeogene–Neogene (Tertiary)

The Variscan unconformity records a long period of time (>55 Ma) when Carboniferous and older strata were uplifted, folded, faulted and eroded. The maximum amount of erosional stripping occurred along the tectonically active zones such as the Vale of Glamorgan Axis, the Usk Axis, Cefn Bryn Anticline, Penlan Axis and Ridgeway Anticline (Fig. 2.3b). During this time (Permian to early Triassic) South Wales was situated around 15–25° north of the equator (a latitude range similar to the present-day Sahara) and was thus experiencing an arid to semi-arid climate. Although there is no record of Permian sediments in onshore South Wales, it is known that a global mass extinction occurred during the latter part of this period (about 250 Ma ago), which resulted in the demise of about 90 per cent of marine life and 75 per cent of land life. It is now thought the extinction was triggered by volcanic eruptions on the supercontinent of Pangaea, which caused a rapid increase in greenhouse gasses, lower oxygen levels and extreme global warming. These climatic conditions are reflected in the overlying red Triassic breccio-conglomerates, seen in the Vale of Glamorgan, which accumulated on scree-slopes and in wadis radiating southwards from Carboniferous Limestone hills. Down palaeo-slope the alluvial fans prograded into a large saline lake where thick clays and occasional carbonates and evaporites accumulated (Mercia Mudstone Group).

From late Triassic to early Jurassic times the arid landscape was progressively flooded by the Tethys Ocean, which resulted in the deposition of the Rhaetic (Penarth Group) facies and the open-marine Blue Lias facies. During this transgression the remaining up-standing limestone hills were onlapped by a variety of high-energy marginal facies. No records of Mesozoic strata younger than Liassic remain on the land surface of South Wales. However, just offshore in the Cardigan

Bay Basin, a thick Triassic–Jurassic succession has been proved in exploration wells (e.g. the Mochras Borehole; BGS, Cardigan Bay, Sheet 52N 06W), and there are also Triassic–Jurassic–Cretaceous successions in the Bristol Channel Basin (Tappin *et al.* 1994, see also the summary in Howells 2007). Published palaeogeographic maps, indicate that the coastal areas of South Wales had a fairly complete covering of Jurassic strata (see Cope 2006), only small amounts of Lower Cretaceous and quite an extensive cover of Upper Cretaceous Chalk (see Rawson 2006). These Mesozoic deposits were removed during a long period of erosion (*circa* 60 million years) that occurred during the Paleogene and Neogene (Tertiary). The Flimston pipe-clay deposits, which have been tentatively correlated with the Bovey Formation of south Devon (Eocene–Oligocene), are remnants of this cover. Note that thick sequences of non-marine Palaeogene sediments have been recorded in boreholes in the St George's Channel Basin (Tappin *et al.*, 1994, Jackson *et al.*, 1995, Howells 2007).

2.2.8 *Quaternary (Pleistocene and Holocene)*

The final episode of sediment accumulation is recorded by a variety of Pleistocene glacial–periglacial–interglacial deposits and recent (Holocene) peat/submerged forest, marsh, dune, beach and alluvial deposits. South Wales was subjected to a number of glaciations in the Middle Pleistocene (Anglian and Wolstonian), and a final one in the Upper Pleistocene (Late Devensian). Middle Pleistocene deposits are very poorly preserved in onshore South Wales, but mixed glacial, incised outwash and fluvial facies (up to 300 m thick) have been recorded in boreholes in Cardigan Bay and the Irish Sea (Tappin *et al.*, 1994, Jackson *et al.*, 1995). During the Ipswichian Stage, which marks the base of the Upper Pleistocene, the warmer climate resulted in a significant rise in sea level and the formation of the *Patella* raised beach (Hunts Bay Beach/Member). During the Late Devensian glaciation (often referred to as the Last Glacial Maximum) ice advanced from Central Wales and the Irish Sea and destroyed or modified the record of earlier glaciation. Consequently, most of the glacial landforms and deposits seen today can be attributed to the Late Devensian glaciation (Dimlington Stadial), which reached its maximum extent 20 000–18 000 years ago. Erosional landforms such as U-shaped valleys and cwms (cirques) are common in mountainous areas of South Wales. Some of the best-developed cwms, which occur on the north-facing scarps of the Brecon Beacons (ORS) and the coalfield (Pennant Sandstone), are occupied by moraine-dammed lakes (e.g. Llyn-y-Fan, Llyn Fawr). Along the Teifi Valley the deep gorges at Cilgerran, Cenarth and Newcastle Emlyn were formed as a result of erosion by ice and subglacial streams. This area is also well known for its Late Devensian deposits formed during the advance and retreat of Irish Sea ice. These deposits include tills and extremely thick (>75 m) glacio-lacustrine / glacio-fluvial sands and gravels (see Howells 2007, Plate 53) and also as finely laminated silty clays (varves). The varves record seasonal freeze-thaw suspension deposition in a large proglacial lake known as Llyn Teifi (Davies *et al.* 2003). To the south of Newport and Fishguard, the beautiful Gwaun Valley was originally interpreted as an overflow channel from Llyn Teifi, but it is now considered to be largely subglacial in origin. Much farther to the southwest the spectacular narrow gorge at Trefgarn, flanked by the rhyolite tors of Maidens Castle and Poll Carn, is considered to have formed as a result of erosion by glacial melt-waters.

In southwest Wales there are also a number of eskers and kames, many of which have been exploited for sand and gravel deposits. One of the best examples of a kame delta sequence in Britain was previously exposed in a pit at Mullock Bridge, on the western side of the Gann Valley, near Marloes. At this site the kame delta was composed of horizontal topsets, inclined and slumped foresets and horizontal varved bottomsets (Fig. 2.15). The sandy facies of topset and foreset origin were characterized by exquisitely preserved sets of climbing ripples and cross strata, which displayed very pronounced vertical and lateral facies changes (Fig 2.16a, b, c). Such features are typical of periglacial sequences deposited as a result of seasonal freeze–thaw cycles. Unfortunately, the pit was closed and landscaped in the early 1980s and is not included in the itineraries.

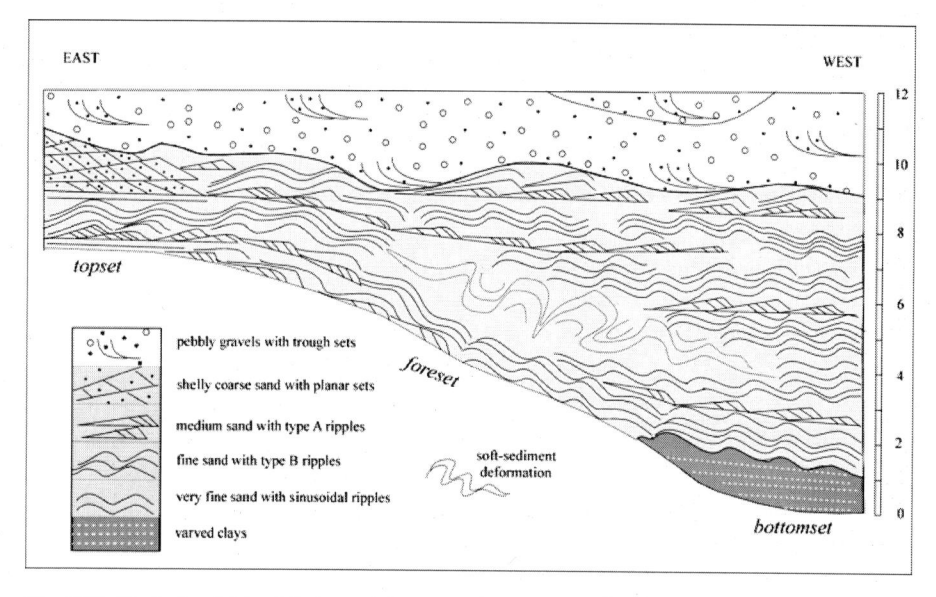

Fig. 2.15 Vertical and lateral facies variations seen in a cross section through a Late Devensian kame delta sequence at Mullock Bridge, Marloes in 1972–1975 (modified from George 1982). Unfortunately the pit was closed and landscaped in the early 1980s.

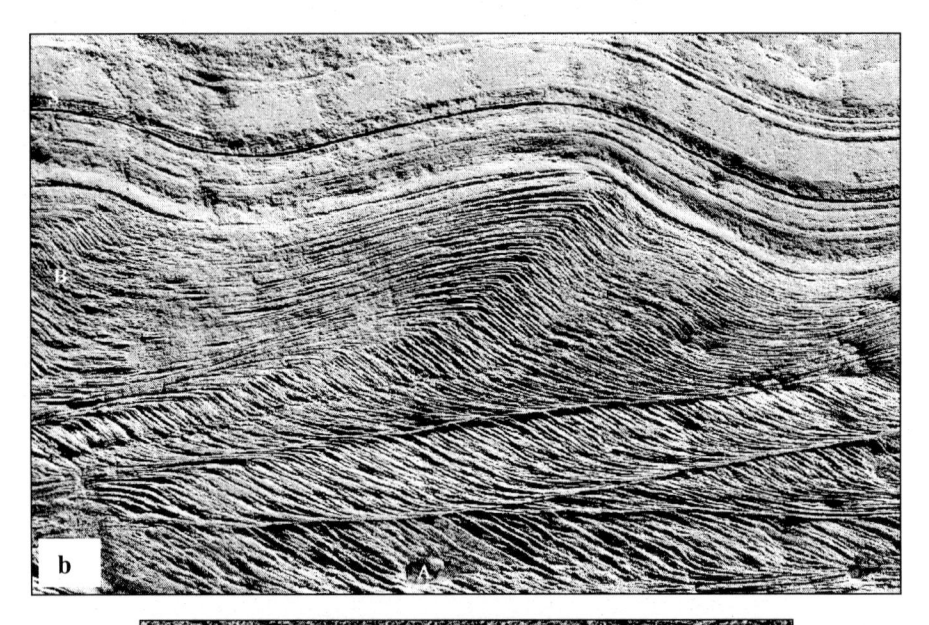

Fig. 2.16 Mullock Bridge kame delta (Late Devensian): **(a)** Sinusoidal ripples out-of-phase enclosing type A sets. **(b)** Ripple-laminated (types A – B – S (sinusoidal) sets deposited as a result of decreasing flow velocity during a seasonal thaw–freeze cycle (resin peel 35 cm wide). Note the regressive and progressive micro-ripples present on the lee-side and stoss-sides, respectively, of the type B set in the centre of the peel. **(c)** Planar cross-sets with graded foresets and some preservation of the topsets (resin peel 40 x 33 cm).

Periglacial conditions, concurrent with the advance and retreat of the late Devensian ice sheets, resulted in the deposition of angular and poorly sorted (frost-shattered) spreads of local bedrock debris, referred to as head. These deposits are preserved in depressions, and at the bases of slopes where they often form well-defined terraces (e.g. Rhossili and Heatherslade on the Gower Peninsula). On Gower, the head deposits often occur directly above the *Patella* raised beach and below Late Devensian till, whereas in other areas they occur above the till and may therefore have formed during the recession of the ice sheet or during a later re-advance (? Lock Lomond Stadial).

A rise in temperature at the beginning of the Holocene (11 800 years ago) marked the beginning of the Flandrian transgression, which resulted in the rapid flooding of the Late Devensian coastline (Fig. 2.17). Coastal plains were inundated to form submerged forests; river valleys, previously deepened by glacial processes, were flooded to form rias (e.g. Milford Haven, Solva). Strong winds reworked glacial sand and transported it inland to initiate the formation of dunes fields, which are scattered around the coastline. Inland the most common Holocene deposits are of alluvium and peat, which often accumulated in bogs located on the sites of former proglacial lakes (e.g. Tregaron Bog located on Llyn Teifi). Offshore tidal sand waves and ridges preserve thick accumulations of sand and gravel, which have been dredged at a number of locations in the Bristol Channel (e.g. Helwick Bank south of Oxwich and Nash Bank south of Porthcawl). Locally there is some concern that the exploitation of these economically important deposits may be causing the depletion of sand from the beaches of south Gower (e.g. Port Eynon) and the Glamorgan Heritage Coast.

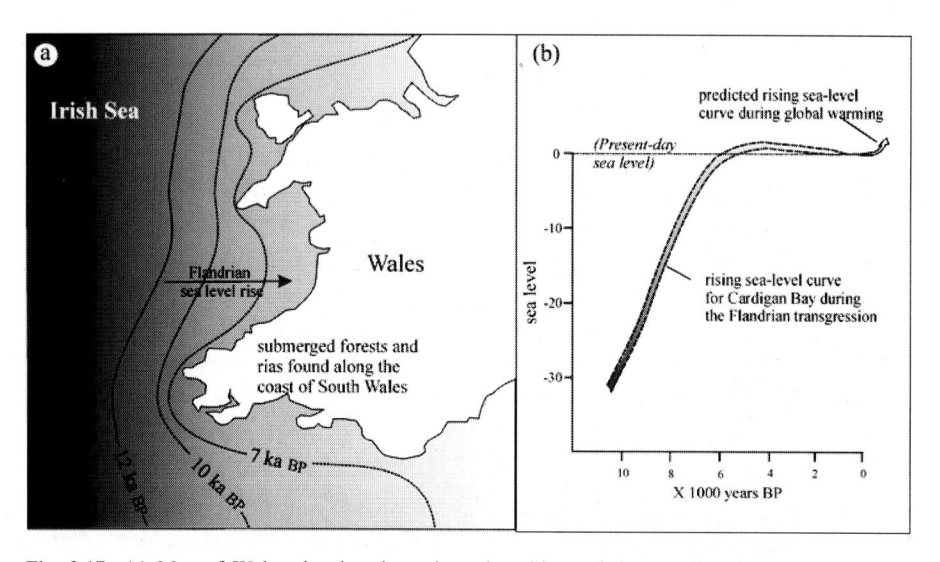

Fig. 2.17 (a) Map of Wales showing the estimated positions of the coastline during the Flandrian transgression between 12 000 years BP and the present day. (b) Sea-level curve for Cardigan Bay during the Flandrian transgression relative to the present day (both modified from Woodcock 2000).

2.3 Field itineraries and safety guidelines

South Wales is an ideal place to study geology and geomorphology because the rocks are well exposed along impressive sea-cliffs and river sections, which are located in areas of outstanding natural beauty. The geology is also extremely varied, with many classic sections through strata of late Precambrian, Palaeozoic, Triassic, Early Jurassic and Quaternary age. Logistically it was necessary to limit the number of localities described in this guide. The difficult decision of which sites to include was governed not only by the quality of their geology, but also by their accessibility. Consequently, those sites with good public access were favoured over those that were isolated, located on private land, or were in working or recently abandoned quarries.

Five regions have been selected to highlight the field geology of South Wales (Fig. 2.3a):

- *Vale of Glamorgan* – a predominantly low-lying region with stratigraphically important coastal exposures of Carboniferous Limestone, Triassic redbeds and Blue Lias facies, which display many classic examples of unconformities.

- *Gower* – noted for its beautiful landscape including rounded hills of Old Red Sandstone and a coastal zone providing impressive outcrops of Carboniferous Limestone, and interesting developments of superficial sediments.

- *Headwaters of the River Neath and River Tawe* – an area that provides some of the best localities in the UK for investigating the relationships between Upper Carboniferous sedimentology and sea-level changes by reference to the relatively new discipline of sequence stratigraphy.

- *South Pembrokeshire and Carmarthen Bay* – an area characterized by stunning coastal geomorphology, captivating Silurian, Old Red Sandstone and Carboniferous stratigraphy/sedimentology and classic Armorican tectonic features.

- *North Pembrokeshire and Cardigan Bay* – a classic region for the study of late Precambrian rocks and Lower Palaeozoic sedimentary successions, including their associated extrusive and intrusive igneous rocks, and Caledonian tectonic structures.

All of the itineraries presented in Chapters 3–7 are illustrated with geology and locality maps, cross sections, stratigraphical columns, graphic logs, depositional models, field sketches and photographs to make the field guide user-friendly. Measured sections of facies sequences have been recorded as graphic logs, which summarize vertical changes in lithology, grain size, sedimentary structures, fossils and trace fossils etc (see Appendix 2). In the field a visual comparison chart is extremely useful for estimating the grain size and textural features of sandstones (see Appendix 3). Collinson *et al.* (2006) provide a detailed account of the description, identification and modes of formation of sedimentary structures, and their use in analysing depositional processes and palaeoflow directions.

To derive the maximum information from the itineraries you should be well prepared for fieldwork. Some of the essential equipment includes: a field notebook, BGS and OS maps, pens and pencils, x10 hand lens, grain size–texture comparator, compass-clinometer, tape measure, camera and binoculars (see further safety items below). A geological hammer is useful, but always keep collecting to a minimum and use your camera to make a permanent record of fossils and specimens whenever possible. No hammering or collecting is allowed at Sites of Special Scientific Interest (SSSI). A GPS (global positioning system) unit is very useful for finding the localities, and is relatively cheap to purchase.

During geological fieldwork proposed in this book, safety precautions need to be followed. Always wear a helmet when working near cliffs and quarry facies and avoid these sites during or just after heavy rain when cliff falls and mudslides are most likely to occur. Consult tide tables before visiting coastal localities, plan your route and make sure that you know the safe exit points from cliff sections. After heavy rainfall do not visit river sections and never enter old mineshafts and adits. Wear appropriate clothing for the prevailing weather conditions, but bear in mind that these can change quickly. Good boots are essential for walking on rough paths and slippery foreshore locations. Keep hammering to a minimum and wear safety goggles when collecting. It is recommended that you carry a backpack with a small first-aid kit, maps of the area, a compass, water and a packed lunch. It is also advisable to carry a mobile phone and to let someone know the localities that you intend to visit and the approximate time of your return. Follow the safety instructions posted on warning signs and always seek permission before entering private land. Check the dates and times of firing at Ministry of Defence ranges and never enter these when the red flags are flying. Party leaders are required to complete risk assessments for all sites visited and this is also recommended for individuals and family groups. Although no working quarries or landfill sites are described in this field guide, it should be noted that risk assessments and written permission are mandatory for visits to these sites.

Further details of specific safety precautions are given in the itineraries. Also, the Geologists' Association has published a *Geological Fieldwork Code*, which provides guidelines on good practice when undertaking general fieldwork, specimen collecting, visiting quarries and field-based research. This leaflet is obtainable from the Geologists' Association, Burlington House, Piccadilly, London, W1J OBG.

Chapter 3

Vale of Glamorgan

3.1 Introduction

The Vale of Glamorgan is generally low-lying country located, informally, to the south of the M4 motorway between Cardiff and Bridgend. The Vale has a magnificent coastline, which extends over 50 km from Sker Point (Porthcawl) in the west to Penarth in the east, and supports many Sites of Special Scientific Interest (SSSI). A 20 km-long strip of this coastline from Newton Point (Porthcawl) to Nash Point and on to Breaksea Point (Aberthaw) has been designated the Glamorgan Heritage Coast. The objectives of this heritage scheme are to conserve the spectacular scenery and unique character of the coastline, and to foster interests in geoconservation, geotourism and related leisure activities. To further these aims, the organization has published information pamphlets and booklet guides, including those covering the geology (Perkins *et al.* 1979), and coastal processes and landforms (Williams *et al.* 1997). There are many delightful walks along the Heritage Coast footpaths, which provide access to the beaches and impressive cliffs with wave-cut platforms and caves. Much of the relatively flat-lying coastal plain represents a remnant of the 60 m Pliocene marine platform. There are two large areas of blown sand to the north and east of Porthcawl (Kenfig Burrows and Merthyr-mawr Warren, respectively).

3.2 Geological history

The Vale lies to the south of the South Wales Coalfield and is composed of Triassic redbeds and Lower Jurassic carbonates, which rest unconformably on inliers of gently folded Silurian, Devonian and Carboniferous strata. Many of the inliers occur along the line of the Vale of Glamorgan Axis (Cardiff–Cowbridge Anticline) and produce the higher ground between 75 and 135 m OD. The oldest rocks discussed in this field guide belong to the Carboniferous Limestone (Dinantian). These predominantly bioclastic and oolitic limestones were deposited on a southward-inclined carbonate ramp located along the southern margin of the Wales–Brabant Massif, sometimes referred to as St George's Land (see Fig. 2.11). During this period of marine carbonate sedimentation many significant falls in sea level occurred, resulting in the emergence of large areas of the ramp and the formation of karstic surfaces and thin soils. Subsequent transgressive events led to marine flooding and renewed shelf sedimentation.

Upper Triassic and Lower Jurassic strata rest unconformably on the Carboniferous Limestone and older strata. This unconformity represents at least 90 million years of Earth history, during which time the reminder of the Carboniferous succession was deposited, uplifted, folded, faulted and eroded during, and after, the Armorican Orogeny. Between 3000–5000 m of Carboniferous strata were removed with the maximum amount of stripping occurring along the tectonically active Vale of Glamorgan Axis. During the latter part of this time period (Permian to Triassic), South Wales was situated 15–25° north of the equator (a latitude range similar to the present-day Sahara) and was thus experiencing arid to semi-arid climate conditions. A major global mass extinction occurred during the late Permian (about 250 million years ago), which resulted in the demise of about 90 per cent of marine life and 75 per cent of land life. Further extinctions occurred during the Late Triassic and at the Triassic–Jurassic boundary (about 200 million years ago). It is now thought that these extinctions were triggered by enormous extrusions of flood-basalts on the super-continent of Pangaea, which caused a rapid increase in greenhouse gasses, lower oxygen levels and extreme global warming. These climatic conditions are reflected in the succeeding Upper Triassic red breccio-conglomerates, which were derived from fault-bounded limestone hills and accumulated on scree-slopes and in wadi channels. Down palaeo-slope the fans prograded into a large saline lake where thick clays and occasional carbonates and evaporites accumulated (Mercia Mudstone Group). From late Triassic to early Jurassic times the landscape, which was characterized by many palaeo-highs, was gradually flooded, resulting in the deposition of lagoonal to restricted marine facies of the Rhaetic (Penarth Group), followed by the open marine facies of the Blue Lias Formation (St Mary's Well Bay Member, Lavernock Shale Member and Porthkerry Member). During this transgression, which was punctuated by minor regressions and still-stands, the archipelago was onlapped by a variety of high-energy marginal facies (e.g. the Sutton Stone and Southerndown Members of the Ogmore-by-Sea area). The stratigraphy and vertical–lateral facies relationships of these successions are illustrated in Figure 3.1.

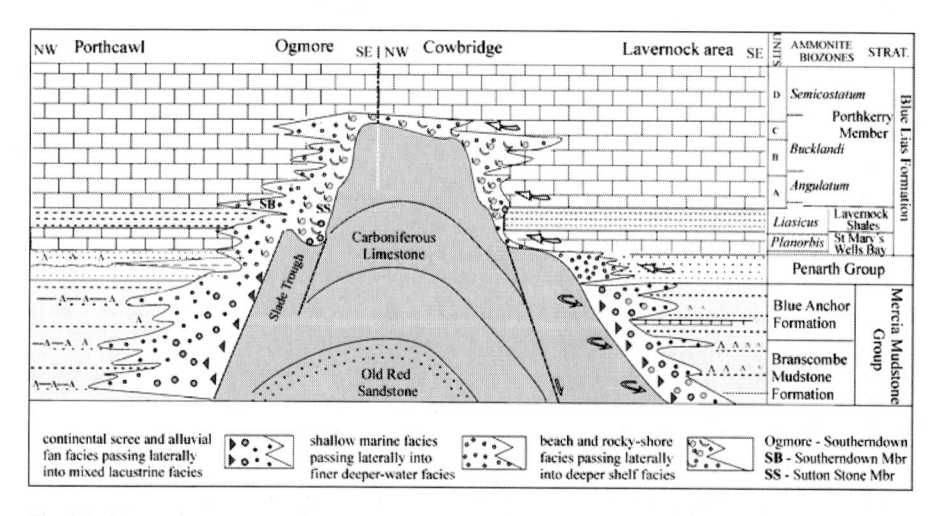

Fig. 3.1 Schematic cross section summarizing the stratigraphy and facies variations seen in Triassic and Lower Liassic sediments deposited on an uplifted block of Carboniferous Limestone in the Vale of Glamorgan (modified from Tucker 1977; Wilson *et al.* 1990).

The geology of the Vale is covered by the 1:50 000 geological maps for Bridgend (Sheet 261) and Cardiff (Sheet 263) and their corresponding memoirs (Wilson *et al.* 1990; Waters & Lawrence 1987) and the 1:25 000 Ordnance Survey map (Explorer 151). Field guides and site descriptions have been published by Cope (1971), Benton *et al.* (1998), Howe (1998), Howe *et al.* (2004) and Simms (2004). Some of the extensive literature covering aspects of the palaeontology and sedimentology of the strata includes, Trueman (1922, 1930), Hallam (1960), Bluck (1965), Wobber (1965), Ager (1974), Tucker (1977), Wu (1982), Fletcher (1988), Johnson & McKerrow (1995), Warrington & Ivimey-Cook (1995), Simms *et al.* (2002), Sheppard (2006, 2007) and Sheppard *et al.* (2006).

3.3 *Itineraries*

The following excursions provide a comprehensive guide to the geology of some of the more accessible coastal outcrops of the Vale. All visits should be planned to optimize the prevailing tidal conditions that are critical for access to Dunraven Bay, Sully and Lavernock. The high cliffs along the coast are unstable, and falls often occur, so it is very important to wear safety helmets and to stay well away from the bases of the cliffs during field studies. No collecting is allowed within the Sites of Special Scientific Interest (SSSI) and, in the interests of geoconservation, all *in situ* fossils and rock specimens should be photographed rather than collected.

3.3.1 *Ogmore-by-Sea*

Travel south from Bridgend on the B4265; turn right at Ewenny onto the B4524 for Ogmore-by-Sea. Just outside the village turn right down the narrow road leading to the main beach car park (Fig. 3.2a). Cars and minibuses should park at the most southerly point near the gate leading to the coast path; coaches can park by arrangement if there is adequate turning space. This coastal section forms part of the Sutton Flats SSSI, which was designated for the exceptional exposures of the unconformity between the Carboniferous Limestone and overlying Triassic conglomerates. The itinerary also begins a field study of the marginal facies of the basal Lias, which rest unconformably on the Carboniferous Limestone at the southwestern end of the coastal exposure. Further studies of the vertical and lateral facies variations that occur in the Lower Lias, are made at the Seamouth and Dunraven Bay localities. The present itinerary, which highlights the most important aspects of the stratigraphy and sedimentology of the area, requires about 4 hours fieldwork and a falling tide, although the localities can be reversed when the tidal conditions are unfavourable.

1a Bwlch y Gro (north). Walk 750 m southeast along the coast path to the corner of the drystone wall (SS 8653 7455), where you can descend onto the Carboniferous Limestone platform. Continue 100 m south along a prominent wave-polished bedding plane, which marks the base of the succession shown in Figure 3.3. This succession, developed in the lower part of the High Tor Limestone (Appendix 1), has a prolific, mainly derived, fossil fauna of rugose corals (*Siphonophyllia,*

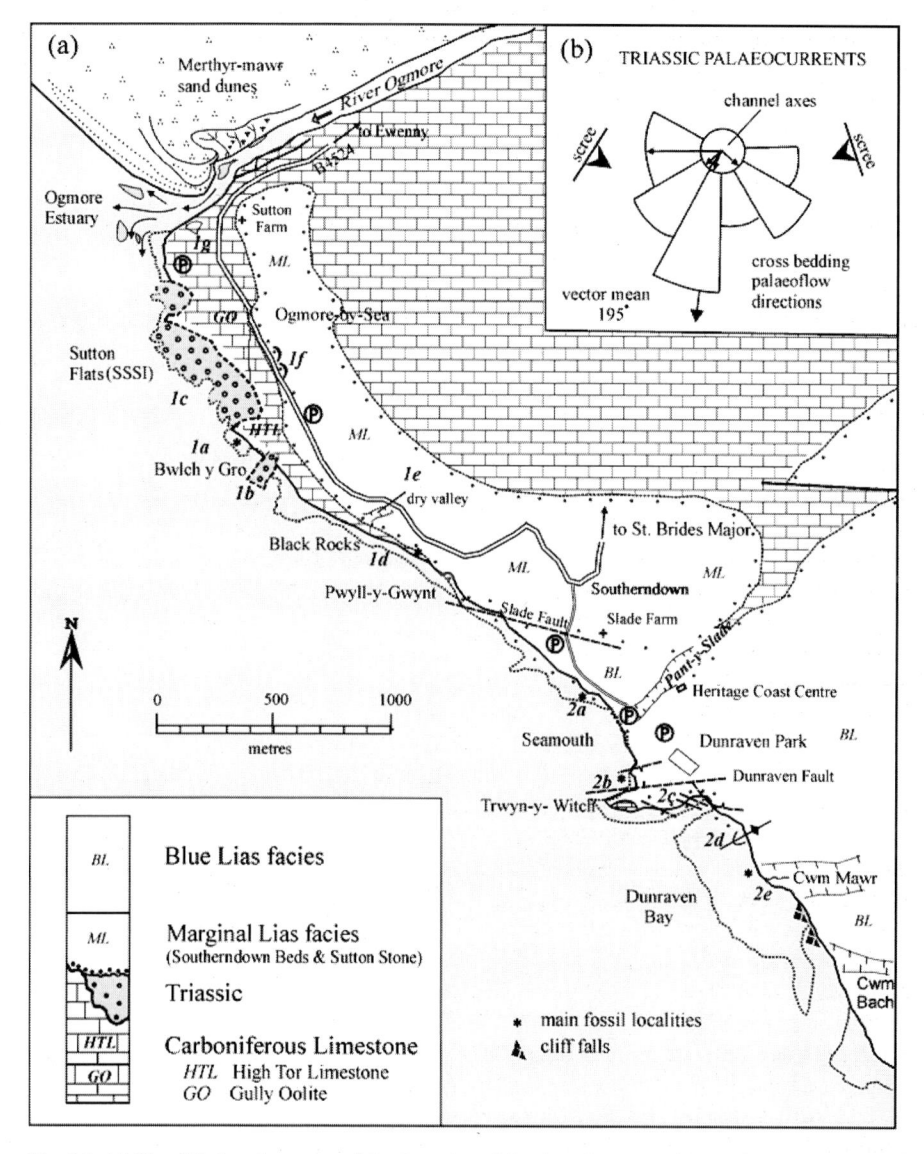

Fig. 3.2 (a) Simplified geology map of the Ogmore and Southerndown coastline showing the localities discussed in the itineraries (based in part on BGS Sheet 261). **(b)** Palaeocurrent vectors from the Triassic breccio-conglomerates.

Palaeosmilia, Michelinia, Lithostrotion and *Syringopora*), brachiopods, crinoids and gastropods. Very large, broken and often bent specimens of *Siphonophyllia* (*Caninia*) are particularly common, as are large valves of the brachiopod *Delepinea*. These derived fossils often form coarse grain lags on the bases of the beds, which grade up into bioclastic packstones with *Thalassinoides* and non-specific horizontal, oblique and vertical burrows belonging to the *Cruziana* ichnofacies. The packstones grade up into wackestones and lime mudstones characterized by the curved feeding traces of *Zoophycos* and tubes of *Planolites*. The intensity of the bioturbation has often completely destroyed the original bedding, producing diffuse mottled textures.

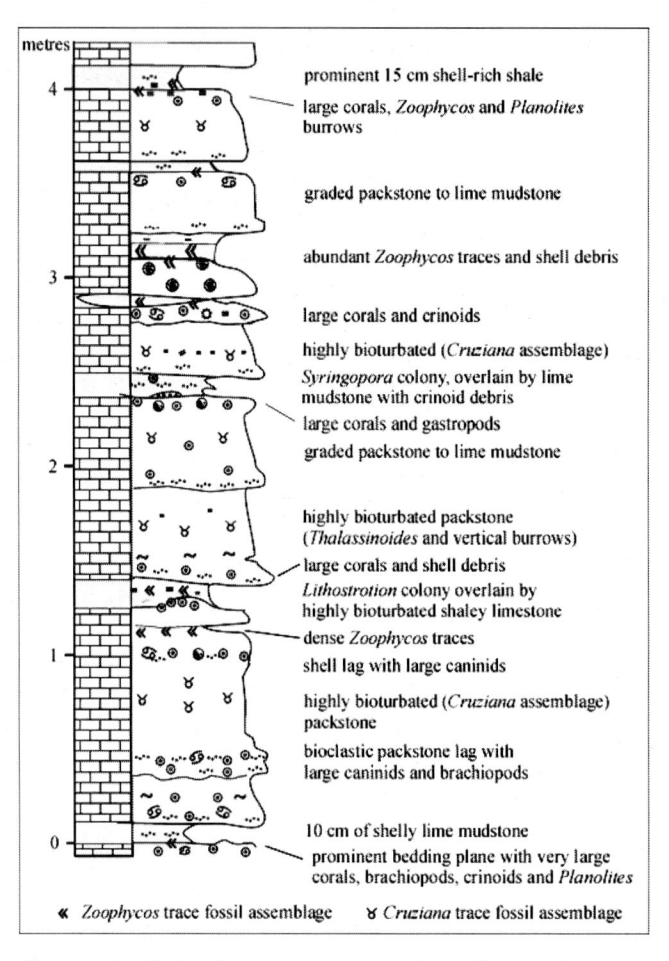

Fig. 3.3 Profile log from the lower part of the High Tor Limestone exposed to the north of Bwlch y Gro (see also Wu 1982).

A number of features described from the above succession suggest that the coarse grain facies were deposited in a storm dominated marine environment. These include beds with the sharp–erosional bases associated with shell lags and graded bedding, broken and displaced fossils, and the presence of two superimposed trace fossil assemblages. The *Cruziana* assemblage, developed in the coarse grain packstones, is typical of nearshore marine environments located between storm and fairweather wave base. Conversely, the *Zoophycos* assemblage, developed in the wackestones and lime mudstones, is more characteristic of deeper water and less oxygenated environments present below storm wave base. One can surmise that the nearshore packstones were remobilised during storms and transported, with their deposit feeding invertebrates, via density flows to a deeper water environment. This assumption is also supported by the absence of wave ripples and HCS in the graded units, the latter structure requiring both wave-generated (oscillatory) currents and storm-induced (unimodal) bottom currents to form. However, it is also possible that the above structures were formed, but then later obliterated as a result of intense bioturbation. The bent rugose corals were possibly deformed by strong directed

Fig. 3.4 View of the High Tor Limestone, at Bwlch y Gro; the limestones are composed of skeletal wackestones and packstone with thin shaly interbeds. Note the vertical Triassic pipe, to the left, filled with marl and speleothem cement.

currents, which deflected their vertical growth prior to them being detached from their colonies by storm currents.

An interesting Triassic pipe cuts through the limestone succession exposed in the cliff-face at the northern corner of the exposure (SS 8656 7452; Fig. 3.4). This feature is about 75 cm in diameter and is filled with red and yellow laminated clays and marls that drape down into sections of the pipe. The fill also displays small desiccation cracks, mud-flake breccias and geopetal structures; many have yielded the fossil skeletons of small mammals. Horizontal offshoots of the main pipe have concentric calcite growths formed by precipitation on the cavity walls (speleothem cement).

1b Bwlch y Gro (south). Return to the coast path and continue southwards to the end of the drystone wall and descend to the bay (SS 8662 7443), where Triassic breccio-conglomerates rest unconformably on the much older Carboniferous Limestone. These very coarse deposits form a southwestwards-extending tongue about 70 m wide (along the coast path), increasing to 110 m (along the beach). In cross-sectional profile the basal erosional surface is prominently stepped, with highly inclined erosional cuts between the steps. Along the northern contact, Triassic facies infill solution pits and fissures (Neptunian dykes) and at one point chaotically bedded breccio-conglomerates form a very large inclined wedge that dips about 28° to the southeast, away from the stepped surface. The uppermost beds of this wedge pass southeastwards into erosionally bedded, finer conglomerates containing outsize boulders. The unconformable contact is exposed again on the southern side of the

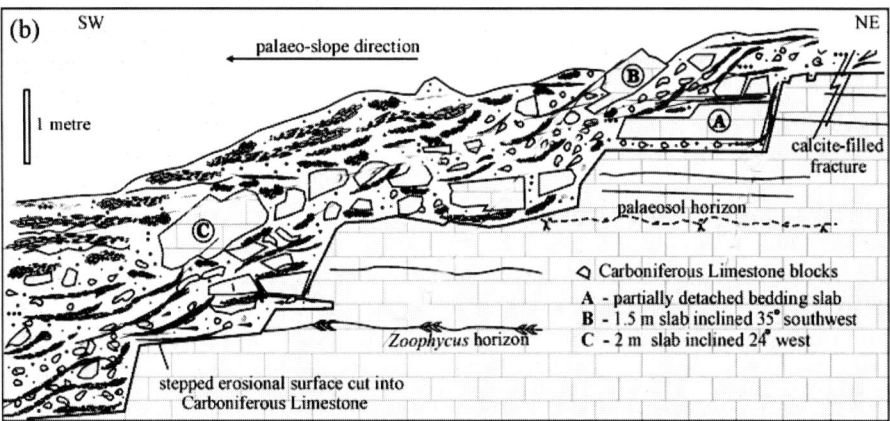

Fig. 3.5 (a) View of the Triassic marginal breccio-conglomerates resting unconformably on the stepped surface of the High Tor Limestone to the south of Bwlch y Gro. (b) Field sketch showing details of the above unconformable surface and the overlying late Triassic facies.

outcrop (SS 8666 7441) where it displays at least five steps cut into the Carboniferous Limestone (Fig. 3.5). The most prominent step occurs above a red–buff-stained bedding plane displaying large fossil caninids, productids, and abundant horizontal *Zoophycos* traces and vertical U-tubes (Fig. 3.5a). A thin variegated clay palaeosol (8–12 cm thick), containing tapering and bifurcating rhizoliths, occurs 1.5 m above the latter horizon. Some of the stepped surfaces have partially detached bedding-plane blocks of limestone, moved as a result of sheet weathering. The overlying Triassic cover (up to 8 m thick) is banked-up against the vertical backwalls of the steps, with many of the larger limestone blocks being inclined up to 35° to the west. Towards the southwest (seawards) the conglomerates become finer, better sorted and display occasional pebble imbrication and low relief erosional bedding.

1c Sutton Flats. At medium to low tide, good wave-polished beds of the Triassic facies can be examined within the main outcrop, where the deposits are up to 20 m thick. Here the limestone fragments are smaller (diameter in the range 2–10 cm), with occasional outsize clasts up to 50 cm diameter. The larger limestone fragments are generally better rounded than the smaller ones. Channels with crude parallel- and cross-bedding can be recognized in the coarser facies; the cross sets giving southerly palaeoflow directions (Fig. 3.2b). In the finer facies the limestone clasts display better sorting and have more closely fitted textures, with some beds displaying normal and inverse grading. The haematite-stained calcareous matrix of these conglomerates is often replaced by pink crystalline barytes and small cubes (2–4 mm) of galena. Stylolitic bed contacts and calcite veins are also common.

The classic unconformity represents a hiatus in the geological record caused by extensive uplift during the Armorican Orogeny, followed by a long period of erosion from the Permian into the late Triassic. Although these Triassic redbeds have not been dated they are generally thought to be of late Norian age (~210–204 Ma). During this time the landscape was subjected to arid and semi-arid weathering, which resulted in limestone blocks becoming detached from the bedrock in a process known as sheeting. The blocks were then, either stabilized by scree deposits, or moved down slope during flash floods. During these periodic flash floods, wadis, extending southwards off the limestone hills, were filled rapidly with poorly sorted conglomeratic facies. In more distal and lower gradient positions, where the wadi channels opened out (intersection point), better-sorted pebbly facies were deposited. Even more distal alluvial fan facies are exposed to the north of Rest Bay (Porthcawl), where the late Triassic succession is composed of finer grain conglomerates and sandstones, interbedded with calcrete palaeosol horizons, and mudstones containing replaced nodular evaporites of playa lake origin (Fig. 3.1).

1d Southeast of Black Rocks. Return to the coast path and walk to the Carboniferous Limestone platform at the southeastern end of the Black Rocks, where another excellent unconformity, between the High Tor Limestone (Dinantian) and the Sutton Stone (Lower Lias), can be examined (Fig. 3.6a). Bedding planes of the former limestone contain abundant cross sections of gastropods (*Euomphalus*) and shell debris. Along the length of the southerly-inclined palaeo-platform the top of the limestone displays southwest-orientated grooves and more irregular cusps and scours, which extend below the unconformable cover. This relationship indicates that the erosional features were formed as a result of wave and tidal scouring prior to the deposition of the overlying Sutton Stone. Small mounds and pockets of serpulid worms encrust the surface, which is also bored by the irregular tubes of *Typanites* and isolated pouches of *Lithophaga*. These trace fossils, belonging to the specialized *Typanites* ichnofacies, are adapted to boring into hardgrounds and fully lithified substrates. Along the palaeo-platform the basal conglomerate of the Sutton Stone infills hollows and channels in the underlying surface. The matrix-supported conglomerate is composed predominantly of rounded to sub-rounded clasts of Carboniferous Limestone, many of which have superficial endolithic algal borings. One very large detached block of limestone measures 5 x 1 m; an even larger mass of fractured limestone, near the end of the coast path, has been interpreted as a collapsed sea stack. Just below the "dangerous rocks and cavities" notice, a small gully, eroded along a northeastwards-trending fault, is filled with Sutton Stone conglomerate, which has been mineralised with coarsely crystalline calcite, radiating

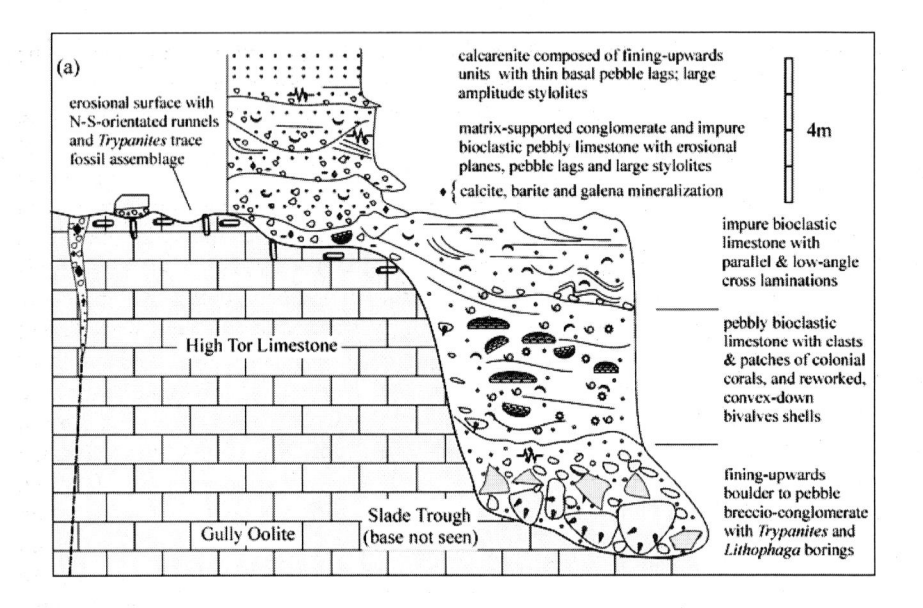

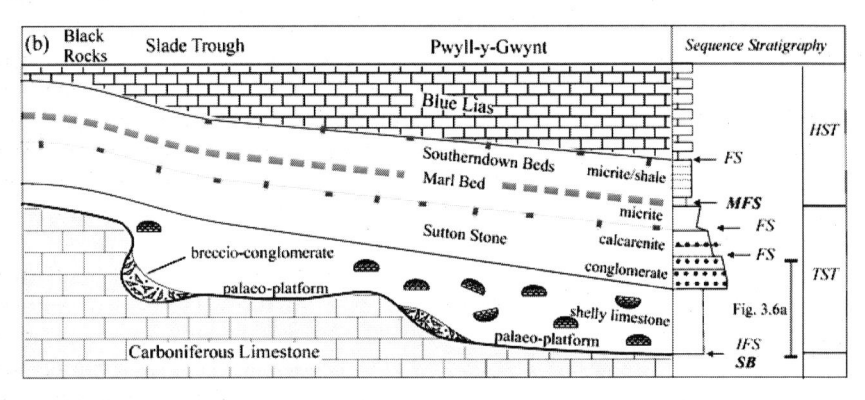

Fig. 3.6 (a) Schematic profile log showing the relationship of the transgressive Sutton Stone marginal facies and the onlapped Carboniferous Limestone palaeo-surface between Black Rocks and the Slade Trough (see Fletcher 1988, for further details). (b) Sequence stratigraphy interpretation of the Lower Lias marginal facies (modified from Sheppard 2006).

crystals of barytes and small cubes of galena. This type of calcite–barite–galena mineralization is also common in the Sutton Stone, immediately above the platform and at lower levels within the Slade Trough. In the latter locality the minerals infill borings in Carboniferous Limestone boulders, and also fossil colonies, coquinas, geopetal structures and vugs in the Sutton Stone.

Higher in the succession the conglomerates pass into calcarenite facies with thin pebble lags marking the bases of small (<1 m) fining-upwards units and a well-defined north–south-tending channel (6 m wide and 2 m deep). Fossil remains in theses calcarenites, include small irregular oyster patches composed mainly of *Liostrea,* corallites and occasional gastropods (for complete faunal lists see Cope 1971; Simms 2004). Another very prominent feature is the development of stylolites with amplitudes up to 15 cm. The contact between the Sutton Stone and more argillaceous and thinly bedded Southerndown Beds occurs at a distinctive dark marl bed, which occurs at a noticeable break of slope in the cliffline. The importance of this bed is discussed further at the next locality (*1e*).

Proceed to the southeastern end of the platform where the unconformable surface steps down (about 1 m) to a lower level. Here the basal shelly calcarenite facies of the Sutton Stone contains buff patches and lenses of highly fossiliferous limestone (coquina) with a rich fauna of colonial corals (*Heterastraea*, *Isastraea* and *Phacelostylophyllum*) and bivalves (*Chlamys*, *Terquemia* and *Lima*). Some of the poorly preserved coral colonies have, in the past, been misidentified as serpulids (Simms *et al.* 2002). All of the coral colonies, seen here and at lower horizons, are not in their life positions, but have been eroded and displaced from their parent reefs. Farther to the southeast, the relief on the unconformable surface increases dramatically, cutting down approximately 8 m from the edge of the platform to a cave at beach level (Fig. 3.6a). This feature (referred to as the Slade Trough) was a palaeo-channel in the Liassic shoreline; consequently it marks the position where the Sutton Stone reaches its maximum thickness of over 16 m. At this locality the Sutton Stone is composed largely of coquina, and pebbles–boulders of Dinantian limestone that become more common towards the margin of the unconformity. An elongated, buff–orange fragment, composed of finely laminated speleothem material, can be seen at one point. This and similar bodies, present at lower levels, have been interpreted as reworked infills of early diagenetic solution cavities. During a falling spring tide it is possible to descend to beach level, down a very jagged and steep limestone surface, to examine the breccio-conglomerates marking the base of the Slade Trough. This descent should only be attempted if you are agile and if the tide has receded far enough to see the sandy beach at the base of the cliffs. The best exposure of the basal facies is seen in a cave, where it is composed of large limestone boulders with borings of *Lithophaga* and *Trypanites*, which occur preferentially on their bases and sides.

The age of the Sutton Stone Member is contentious due to the scarcity of fossils, particularly ammonites of zonal value, and the extreme lateral and vertical diachronaeity displayed by the marginal facies. Most studies suggest that in this area the bulk of the Sutton Stone correlates with the St Mary's Well Bay Member, while the overlying Southerndown Beds correlate with the Lavernock Shales and the lower part of the Porthkerry members (see Fig. 3.1 and later discussion).

Before leaving the locality it is worth discussing the significance of the stratal relationships observed at this important unconformity. The Sutton Stone, which clearly onlaps the highly eroded Carboniferous Limestone surface, is a classic example of a transgressive rocky shoreline–beach succession (Johnson & McKerrow 1995). During this transgression, initiated during the deposition of the Penarth Group, pre-existing late Triassic platforms were re-excavated and any previously deposited marginal facies were removed or reworked. Farther to the northwest (localities *1a–c*), the thicker pockets of late Triassic conglomerate (wadis fills) would have been onlapped, although this stratal relationship is not seen at outcrop. Continued and relatively rapid sea-level rises, punctuated by still-stands, resulted in the complex vertical and lateral facies variations observed in the Sutton Stone and Southerndown Beds, which themselves pass into the alternating limestones and shales of the Porthkerry Member (Blue Lias facies). Regional studies have shown that both the late Triassic and early Liassic marginal facies were deposited on the flanks of fault-defined palaeo-islands, located along the line of the Vale of Glamorgan Axis (Fig. 3.1). This palaeogeographical model is investigated further in later itineraries (i.e. Barry Island and Sully) when other examples of the marginal and normal facies are seen at outcrop.

Sheppard (2006, 2007) has recently presented a sequence stratigraphy model for this classic rocky shoreline environment. This author has recognised four retrogradational parasequences bounded by flooding surfaces, which are marked by decreases in the grain size and facies changes (Fig. 3.6b). Some of the conclusions of this study were:

- (a) The palaeo-platforms in the Carboniferous Limestone were cut during a number of sea-level rises and still-stands, probably over a long period of time (*circa* 2.7 million years).
- (b) The breccio-conglomerates preserved in the Pwyll-y-Gwynt trough and the later Slade trough represent *in situ* foreshore deposits, formed at the base of a sea-cliff, as a result of undercutting and collapse.
- (c) The large clasts in the matrix-supported Sutton Stone conglomerate were derived from storm reworking, and offshore transport of foreshore deposits.

1e Dry valley. Return to the coast path and ascend the small dry valley (erroneously called Pant y Slade by geologists, see Fig. 3.2) that leads up to the main road. Outcrops of the Southerndown Beds occur both sides of the valley, where they comprise irregular and nodular-bedded, bioturbated limestones and lime mudstones. These facies contain granules and small pebbles of Carboniferous Limestone and also debris of bivalves, gastropods and corallites. A distinctive 60 cm bed, composed of dark bituminous marls with limestones nodules, is developed in the lower part of the succession and it can be traced across the southern outcrop. Hallam (1960) referred to this bed as the 'basal marl' and he related its origin to a rapid deepening of the shelf environment. More recently, Sheppard (2006, 2007) used the horizon to mark a major flooding surface in his sequence stratigraphic model (Fig. 3.6b). Towards the top of the valley the facies becomes more argillaceous, but it still retains a large proportion of limestone fragments.

1f Roadside exposure. At the top of the dry valley, turn left on the road and walk back (1500 m) to a small quarry (SS 866 750), just past the entrance to the Sea Lawn Hotel (closed and being redeveloped). In the quarry face, directly opposite the bus shelter, a 20 cm bed of conglomerate with northeast–southwest-orientated scours occurs in calcarenite facies, developed near the top of the Sutton Stone. This bed is composed of angular to sub-rounded flint pebbles up to 10 cm in diameter with the largest pebbles being concentrated at the top of the bed. Fossil bivalves are quite common and many of the limestone beds contain stylolites.

1g Ogmore estuary. Return to the northern end of the car park to view the Ogmore estuary (Fig. 3.2a). At the time of the most recent glaciation (sea-level lowstand) the valley floor was about 8 m lower than it is today, but since then the valley has been filled with a variety of sediments. Between medium and low tide, the mouth of the estuary is seen to be partially restricted by a curved bar, constructed of alternating layers of pebbles and sand. Seawards of this restriction, tidal sandbars divert the flow of the river, while up stream, the river has both mid-channel and side-attached sandbars. A grassy flat, drained by creeks, is developed on the right bank of the river and is succeeded inland by the sand dunes of Merthyr-mawr warren. Here the margin of the dune field shows evidence of fluvial reworking, whereas along Newton Beach the margin of the dunes has been reworked by storm waves. Blowouts can also be observed in many areas of the plant-stabilized dunes.

3.3.2 Seamouth and Dunraven Bay

Travel from Ogmore-by-Sea on the B4524 to Southerndown and turn sharp right, just before the Three Golden Cups inn, onto the minor road that leads to the beach at Seamouth. Cars and mini-buses can be parked either in the car park above the storm beach or in the designated field in Dunraven Park. Unfortunately there is no access to Seamouth for coaches, which should be parked at the top of the hill. There are toilets at the lifeguard station and the Heritage Coast Centre, located 200 m up the dry valley of Pant-y-Slade (Fig. 3.2a). The cliffs along this section are very unstable, consequently safety helmets are essential, and all studies should be confined to the rock platforms located a safe distance from the cliffs. All localities are accessible from mid- to low-tide.

2a Dancing Stones. Walk 40 m down the concrete slip that crosses the storm beach and turn right onto the Dancing Stones (SS 8842 7315; Fig. 3.7), which form a series of gently inclined limestone bedding planes, with conjugate joints trending 110–290° and 20–200°. At this locality the Southerndown Beds contain abundant clasts of very angular to sub-rounded flint up to 10 cm diameter (average 1 cm), occasional oysters, and abundant *Thalassinoides* burrows, which produce the characteristic nodular bedding. Here the marginal facies is considered to be the lateral equivalent of the lower part of Unit B of the Porthkerry Member. What is considered to be the lithological contact with the normal Blue Lias facies occurs 1.75 m higher in the succession, at the base of a second rock platform. This contact is marked by an undulating surface with a thin veneer of oolitic limestone containing large limestone and flint pebbles, with an encrusting fauna of oysters and serpulids, and evidence of bioturbation by a firmground (*Glossifungites*) trace-fossil assemblage. The overlying Unit B of the Porthkerry Member is composed of more pure wackestones, interbedded with dark grey shales and calcareous mudstones. Higher in the cliff a prominent group of limestone beds (4 m thick), which amalgamate upwards, are known locally as the Seamouth Limestone (Unit C; Figs. 3.7, 3.8). This unit has a high limestone: shale ratio, which suggests that it was deposited during a relative sea-level fall (Wilson *et al.* 1990).

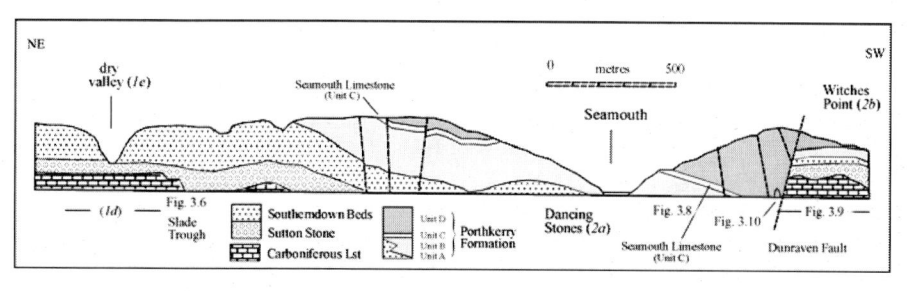

Fig. 3.7 Coastal cross section from the dry valley (locality *1e*) across Seamouth to Witches Point showing the geological structure and Liassic facies (adapted from Trueman 1922).

Fig. 3.8 View of the Blue Lias (Porthkerry Formation) looking northwestwards towards Seamouth.

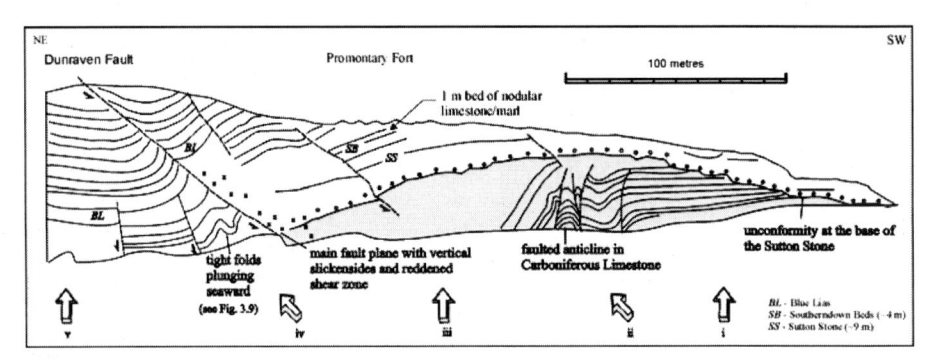

Fig. 3.9 Field sketch of the stratigraphy and structure of the Carboniferous Limestone and Liassic strata exposed at Witches Point, Seamouth. Note that the structural features have been drawn from a number of viewing angles (i – v).

2b Trywn-y-Witch. Viewed from the centre of the bay, the Seamouth Limestone defines a broad anticline with the southern limb dipping gently to the south. Walk southwards across the bay towards the headland known as Trywn-y-Witch (Witches Point). This important geological locality exposes the unconformity between the Carboniferous Limestone and the Sutton Stone and other interesting stratigraphical and structural features (Fig. 3.9). The section, illustrated in Figure 3.9, is drawn from a variety of viewing angles to emphasize the quite complex folding and faulting. From viewpoint (i) the unconformity at the base of the Sutton Stone forms a marked step and then it continues to cut down through the Carboniferous Limestone towards the sea. The stepped morphology of the palaeo-platform is similar to that observed at the Black Rocks locality. Note that from this position the limestone has a very low apparent dip, but viewed from position (ii) the hinge of a steeply dipping and faulted asymmetrical anticline can be seen. Minor faults and parasitic folds occur within the beds in the axial zone of the fold. Farther to the northeast at (iii) a more open and gently dipping anticlinal fold can be seen in the unconformable cover of Sutton Stone (~9 m thick) and Southerndown Beds (~5 m thick), the top of the latter being marked by a nodular limestone and marl bed

Fig. 3.10 Tight seaward-plunging anticline and syncline in the Blue Lias viewed from position (iv) shown n Figure 3.9. Unfortunately the folds are now concealed by a cliff-fall.

(Fig. 3.9). The western corner of the headland is defined by a major east–west-trending reverse fault (Dunraven Fault), which dips 45° southwards and downthrows to the north. The reddened fault plane has slickensides and an associated zone of sheared shales. The displacement on this fault is difficult to calculate accurately, because of the rapid lateral facies variations that occur between the marginal and normal Liassic facies in this area. However, a displacement of 30–40 m is suggested by reference to the Seamouth Limestone, which occurs 100 m to the north at a small normal fault. On the downthrow side of the main fault, the interbedded limestone and shales (normal Blue Lias facies) have been deformed into a very prominent tight anticline and syncline. Their axes plunge seawards, and the fold pair pass upwards into a broader syncline (Fig. 3.10). Unfortunately, this fold has recently been obscured by debris brought down in a cliff fall.

At point (iv), where the cliff-line turns northwards, a normal fault trending approximately east–west and dipping at 70° south can be observed. The fault plane has calcite slickensides and 10 cm of breccia accreted to the surface. Here the rock platform, located well away from the dangerous cliffs, is one of the best localities to examine the fossil assemblage of the Blue Lias. The fossils include nests of *Gryphaea arcuata* in their life positions, many cross sections of *Pinna*, large valves of *Plagiostoma giganteum*, ossicles of *Pentacrinus* and occasional ammonites, including large *Arietites*. Cross sections of lined burrows (*Ophiomorpha*) are also common.

2c Viewpoint above Dunraven Bay. Return to the car park; follow the road into Dunraven Park, and through the walled Victorian garden of the now demolished Dunraven Castle. Exit the garden by the gate in the southern corner, and follow the path through the woods to a viewpoint and information board titled 'Understanding the Landscape' (SS 8895 7270). Here there is a dramatic view of the Blue Lias (Porthkerry Member) forming the cliffs and the wave-cut platform of Dunraven Bay

Fig. 3. 11 Breathtaking view of the Blue Lias cliffs extending southeastwards from Dunraven Bay to Nash Point. Note the contact between Units A and B of the Porthkerry Formation in the centre of the near cliff-face and the axis of a minor anticline extending seaward from the cliffline in the next cove.

(Fig. 3.11). The lower half of the cliff is composed of grey facies with about equal amounts of limestone and mudstone facies (Unit A), whereas the overlying buff-coloured facies have a higher proportion of limestones (Unit B). At medium to low tides the impressive rock platform maps out the strike of the alternating limestone and mudstone beds and the traces of linear cross faults. The top of the Lavernock Shales Member was, until a recent cliff-fall, exposed in a small seaward-plunging anticline seen in the second cove. The relatively flat line of the clifftop represents the 60 m Pliocene platform, which is incised by hanging valleys; in the distance large cliff falls can be seen at Whitmore Stairs.

2d Dunraven Bay Follow the path to the west; turn left through the gate that leads to the clifftop and take the steps down to the beach. At the base of the steps is a rock platform, the backwall of which is composed of limestones with flint impurities interbedded with thin shales. Here a 1 m band exhibits extreme nodular bedding considered to be the result of bioturbation by decapod crustaceans belonging to the ichnogenus *Thalassinoides*. This distinctive bed, which marks the local lithological top of the Sutton Stone, is considered to represent a rapid increase in water depth and is a candidate sequence boundary (Sheppard 2006, his unit D). A prominent bedding plane, below the nodular band, displays large bivalves and *Gryphaea*, small shell mounds with epifauna, and also carbonized veins and fragments of jet, derived from the bark of *Araucaria* (monkey puzzle tree). At this locality the marginal facies is considered to be the lateral equivalent of the Lavernock Shales Member and lowermost beds of the Porthkerry Member (i.e. the *Liasicus* Biozone, Fig. 3.1). To the east the cliffs are cut by three north-westerly-trending, vertical normal faults, which display fault breccia, slickensides, calcite and gypsum mineralization, and also drag folds on their downthrow sides (Fig. 3.12). Walk 200 m to the southeast across the foreshore, to the outcrop of an anticline that plunges at 17° to the

Fig. 3.12 Drag fold located on a northwestwards-trending minor fault exposed on the southern side of Witches Point. The nodular bedding seen in the limestones and marls (Southerndown Beds) is caused by *Thalassinoides* burrows.

southwest. In the past it was possible to see the argillaceous facies belonging to the Lavernock Shales in the core of the structure, but unfortunately these beds are now obscured by a cliff fall.

2e Cwm Mawr. If the tide is receding, it is possible to walk along the rock platform to Cwm Mawr. The floor of this valley is now 10 m above beach level and the stream that runs through it ends at a small waterfall. This misfit valley was formed mainly during the post-glacial rise in sea level when surface runoff was far greater than it is today. Since this time, cliff recession has exceeded the rate of river incision and consequently the present-day stream has been left hanging. Williams *et al.* (1997) have estimated that cliff recession in the Lias occurs at a rate of 1.5–8.0 cm per annum and occurs mainly as a result of cliff falls, caused by toppling or slide failure. Two good examples of cliff falls can be seen 200 m to the southeast of Cwm Mawr and above one of these falls a large stack of rock has slipped away from the cliff face and is now in a very precarious position (Fig. 3.13).

Fig. 3.13 Large tower block of Blue Lias, which has move down the cliff face by sliding and is now in a very unstable condition (summer 2005). In the near future, probably during a storm, the block with slide farther and collapse completely.

3.3.3 Cwm Marcross and Nash Point

Travel on the B4524 from Southerndown to St Brides Major; turn right onto the B4265 and after 2 km turn right again onto a minor road signposted to Monknash, Marcross and St Donat's. At Marcross turn sharp right at the Lighthouse Inn and follow the narrow road that runs parallel to Marcross Brook, to the car park at Nash Point (Fig. 3.14). The narrow valley of Cwm Marcross and the coast around to Nash Point display many interesting geomorphological features; there is also a nature trail along Marcross Brook.

3a Cwm Marcross. The relatively flat seaward-sloping coastal plain, extending from the village of Marcross (60 m OD) to the car park (35 m OD) forms part of the Pliocene platform, cut by marine erosion during the late Neogene sea-level highstand (~6 million years ago). Looking northwest from the car park there is a good profile view of Nash Point, with its distinctive wave-resistant pedestal and softer-weathering overlying notch (Fig. 3.14). Note the fault, extending southwards from the angle of the point across the limestone pavement, and the adjacent cliff-fall debris that merges southwards with storm-beach boulders. Between medium and low tide, waves can be seen breaking on Nash Point Sandbank, located offshore to the northwest. The two lighthouses at Nash Point were built after a passenger paddle-steamer *Frolic* ran aground on the sandbar in 1832 with the loss of 80 lives.

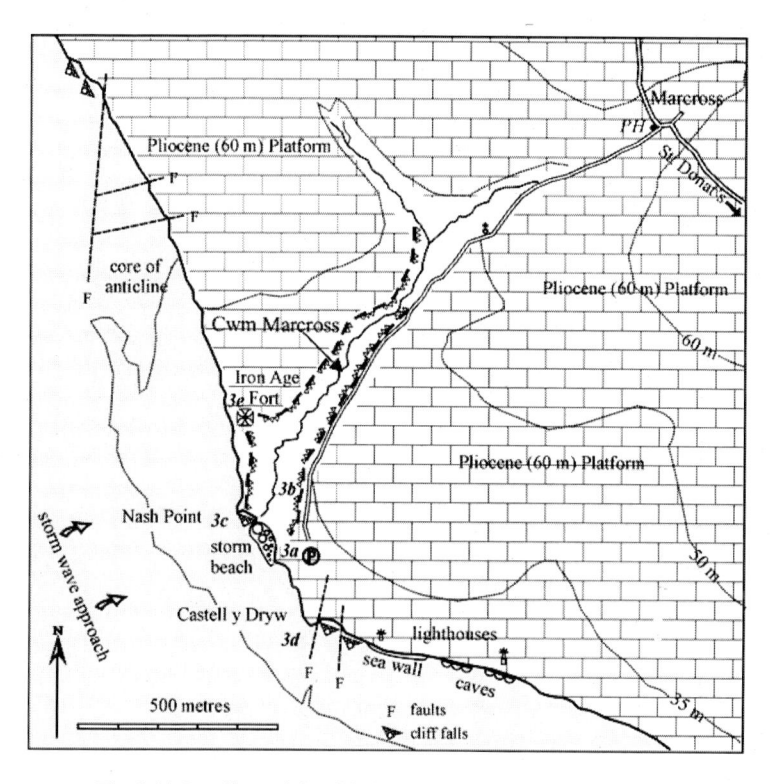

Fig. 3.14 Locality and simplified geological map of Cwm Marcross.

3b Marcross Brook. From the car park take the footpath that leads down into Cwm Marcross and stop at the information board, from where it is possible to examine some of the main geomorphological features of the valley. Note the small dry valley that enters Cwm Marcross from the west. The present-day stream (Marcross Brook) appears far too small to have eroded the relatively deep and wide valley, and it is therefore designated a misfit stream. Also, present-day fluvial processes could not have deposited the well-developed terraces seen on the flat valley bottom. These deposits (over 6 m thick) are incised to a depth of 5 m by the stream channel, but up stream the amount of incision decreases rapidly to about 1 m and small meanders can be seen. The enlarged valley and fluvial-terrace deposits were formed during a period of increased runoff during the melting of ice from the Last Glacial Maximum (Late Devensian). This glacial melt-out resulted in a substantial rise in sea level during the Flandrian interstadial. Later stream incision records a fall in base level probably caused by cliff recession during the Flandrian transgression. The well-developed wave-cut platforms, seen along the Glamorgan Heritage Coast, are a legacy of this process of cliff recession.

3c Nash Point. Descend onto the storm beach that protects the mouth of the valley, noting that the distinctive right-angle cliff face at Nash Point is caused by two small bifurcating north-south trending faults. The adjacent rockfall comprises a jumble of very large blocks of limestone, which become more wave-worn and smaller to the southeast. The older storm-beach pebbles display an overall decrease in average

Fig. 3.15 The impressive headland of Nash Point composed of alternating limestones and thin mudstones belonging to the Porthkerry Member. The Sandwich, Main Limestone and Silicified beds (Bed 39. 49 and 55 respectively of Trueman, 1930) are distinctive marker horizons in Unit B of the member.

diameter from approximately 75 cm in the north, to 25 cm in the centre of the bay, and to less than 1 cm at the southern termination of the beach. This decrease in pebble size is also accompanied by a change in pebble shape from sub-angular to sub-rounded to well rounded. Above the storm beach a wedge of head, composed of angular clasts of limestone, is banked against the northern cliff. This deposit is overlain by white tufa containing abundant planispiral and helically coiled gastropods, with interbeds of alluvium composed of platy limestone pebbles. A very similar succession of head deposits have been recorded farther to the northwest at Cwm Nash (Perkins *et al.* 1979), where the analysis of the gastropod assemblages has indicated overall climate amelioration during postglacial (Flandrian) times.

Nash Point is composed of well-bedded unlaminated limestones and thinner lime mudstones belonging to the Porthkerry Member (*Bucklandi* Biozone). This coastal section is well known from the early studies of Trueman (1922, 1930) of Liassic stratigraphy and the evolution of *Gryphaea*. A number of marker beds, from Unit B of the Porthkerry Member, can be identified (Fig. 3.15). The Sandwich Bed (*39*) comprises a resistant group of interbedded limestones (about 1.8 m thick) forming the top of the erosional notch, with the Main Limestone Bed (*49*) and the Silicified Bed (*55*) occurring 5.5 m and 9 m higher in the section. Note that the bed numbers are those defined by Trueman (1930). The cliff is capped by facies belonging to Unit C of the Porthkerry Member, which is equivalent to the Seamouth Limestone seen at Southerndown (locality *2a*).

On a falling tide it is possible to walk around Nash Point, from where the impressive Liassic cliffs continue 2 km northwestwards to Cwm Nash. These strata are folded into a broad anticline, the core of which is composed of Liassic limestones with a higher proportion of mudstone interbeds belonging to Unit A of the Porthkerry Member. Many faults occur within the middle part of the section and these are often associated with cliff falls.

3d Castell y Dryw. Return to Cwm Marcross and walk 200 m southeastwards to Castell y Dryw, where a number of wave-cut platforms have been eroded into the Blue Lias succession. The middle platform displays excellent clint-and-gryke weathering related to the east–west and north–south conjugate joint system. Continue farther to the southeast to a position on the foreshore where both of the Nash Point lighthouses can be seen. Note that the cliff below the small (disused) lighthouse has been reinforced by stonework, while farther to the southeast many caves occur in the gently inclined Liassic strata. These caves have been excavated along lines of weakness caused by joints, their roofs being supported by the more resistant Sandwich Bed. Along this east-southeastwards-trending part of the coastline cave systems are common and cliff falls are rare. The reason for this is that caves are formed preferentially where northeasterly-directed storm waves strike the coastline at an oblique angle, thus reducing their erosive energy and lowering rates of cliff recession (Williams *et al.* 1997). In this section, cliff fall-debris is present directly after the sea wall, where the coastline swings to a northwestwards orientation; consequently the cliffs are subjected to the full erosive force of storm waves. Most of the cliff falls are also associated with the occurrence of vertical faults.

The gentle southerly dip of the strata results in younger rocks being exposed along the rock platforms to the south. Some of the mudstone beds, below the Sandwich Bed, display intense bioturbation by *Thalassinoides,* while stratigraphically higher in Unit B the top surfaces of some of the limestones display nests of *Gryphaea* and *Liostrea.* Large specimens of *Pinna* and *Lima* are developed particularly above the Silicified Bed in Unit C. All three of the above biofacies are interpreted as omission surfaces formed as a result of pauses in sedimentation. The vertical upward passage from the *Thalassinoides* biofacies to the *Gryphaea–Liostrea* biofacies to the *Pinna–Lima* biofacies records a temporal decrease in water depth (for further details see Sheppard *et al.* 2006).

3e Iron Age fort. Return to the western side of Cwm Marcross and ascend the dry valley, to reach the Iron Age fort. The impressive ramparts of the fort have been truncated as a result of cliff recession. Looking down onto the wave-cut platform, the lines of two northwest–southeast-trending faults can be seen; the conspicuous curvature of the beds between the faults being the result of fault drag.

3.3.4 Whitmore Bay (Barry Island)

Barry can be reached from Cowbridge, via the A48 and A4226 (Waycock Road) or along the coast road from Llantwit Major, via the B4265 and A4226 (Port Road). From the roundabout on the A4226 travel southeast via the B4226, A5050 and A5055 to Barry Island, a popular holiday resort with good rail connections, a railway heritage centre and many leisure amenities. Parking is available in the car park adjacent to Barry Old Harbour or near the seafront at Whitmore Bay, where 2 hours free parking allows enough time to complete the itinerary.

Whitmore Bay is bounded by the headlands of Friars Point and Nell's Point, which together comprise the Barry Island Site of Special Scientific Interest (SSSI) (Fig. 3.16). The site was designated for the exceptional exposures of the unconformity between the Triassic and Carboniferous Limestone, and the interesting

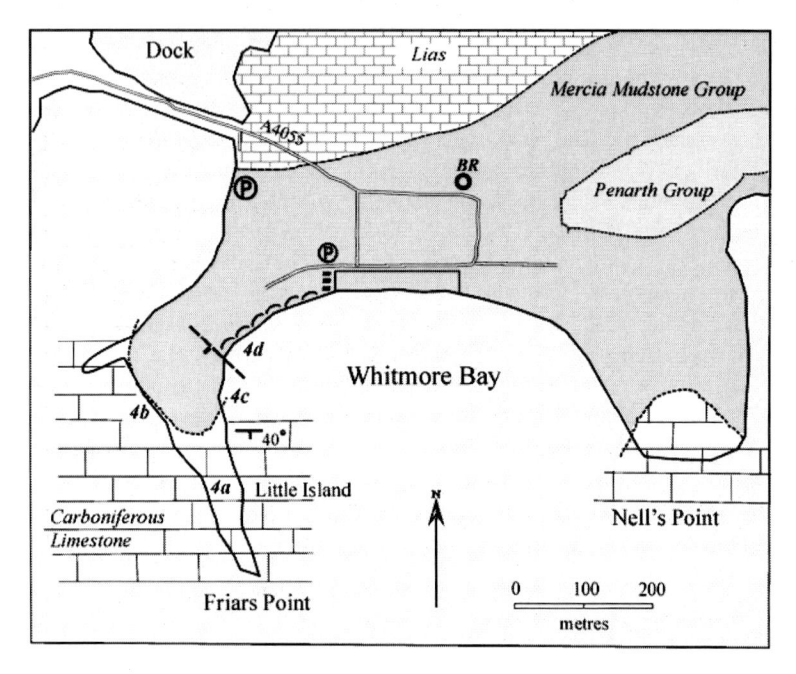

Fig. 3.16 Simplified geology and locality map for Whitmore Bay, Barry Island.

vertical and lateral facies variations developed within the Mercia Mudstone Group. The exposures on Little Island can be examined at all stages of the tide.

4a Little Island peninsula. The peninsula of Little Island (Fig. 3.16) is the type locality for the Friars Point Limestone, which forms the top half of the Black Rock Limestone Subgroup (Appendix 1). Here the limestone is composed of dark grey and thickly bedded crinoidal packstones with thin shaly partings, which dip at 40° due south. The fossil assemblage includes rugose corals (*Zaphrentis* and large *Siphonophyllia*), brachiopods (shells and abundant spines), gastropods, and crinoid debris. Intense bioturbation has often homogenized bedding features. Veins and small vugs (1–2 cm) of crystalline calcite are also common.

4b & 4c Little Island (west and east). These two interesting and geologically similar localities occur on the western (ST 1095 6635) and eastern (ST 1106 6624) foreshores of Little Island. At both outcrops the Carboniferous Limestone is cut by five gently inclined erosional platforms, which are approximately 15 m wide and separated by near vertical, north–south-trending faces up to 3 m high (Fig. 3.17a, b). The unconformable Late Triassic cover is composed of laterally variable breccias and conglomerates, containing boulder- to pebble-size fragments, degraded corals and other fossils derived from the Carboniferous Limestone. These marginal facies display very prominent vertical and lateral facies variations. At some horizons, inclined units of coarse breccia, which are banked against the backwall of the platform (facies A), pass laterally into better-sorted and tabular bedded breccio-conglomerates (facies B). Impure calcarenites to calcirudites (facies C and D respectively) occur on the higher platforms. A very convincing wave-cut notch occurs on the back face of platform 3 (Fig. 3. 18).

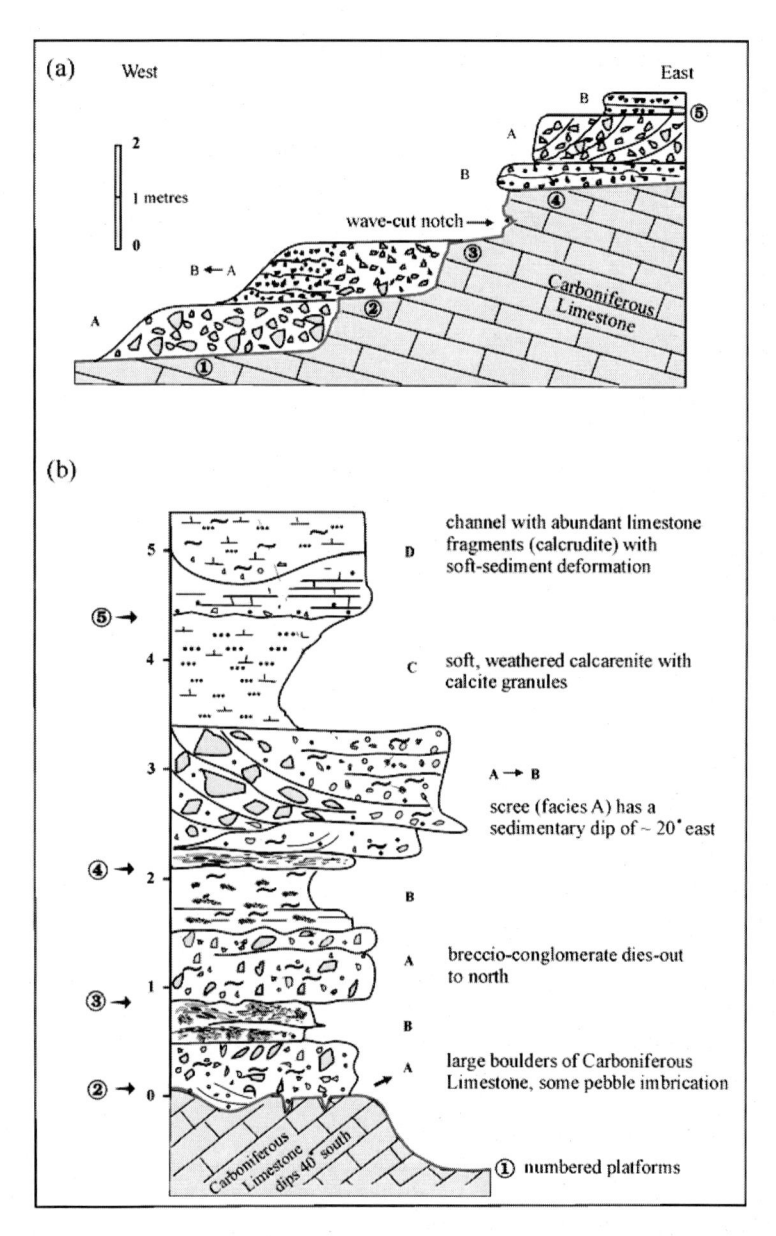

Fig. 3.17 (a) Section illustrating the relationship of the marginal Triassic facies to the Carboniferous Limestone platforms (1–5) on the western side of Little Island (modified from Waters & Lawrence 1987). (b) Profile log of facies developed on the eastern side of Little Island. (See text for further descriptions of facies A-D).

4d Whitmore Bay. Return to the sandy beach at Whitmore Bay where a small normal fault, with a downthrow of 1.5 m, occurs in the low cliff at the bottom of the steps leading down to the beach (ST 1107 6641). The slightly curved fault plane dips 60° south and has a 15 cm shear zone filled with a mixture of clay, calcite and gypsum. To the east of the fault, 10 m of the Branscombe Mudstone Formation

Fig. 3.18 Wave cut notch eroded into the back wall of platform 3, located on the western side of Little Island. Note that the Carboniferous Limestone, which dips 40° to the south, is overlain unconfomably by Triassic breccio-conglomerates. (Metal ruler expanded to 1 metre).

(previously the Keuper Marl) is exposed in the low cliffs that back the beach. The red marls contain abundant white-to-pink gypsum nodules up to 25 cm in diameter. Most of these nodules have suffered extensive dissolution and now appear as voids with marly skins and occasional partial infills of crystalline calcite, referred to locally as potato-stone. The contact of the Branscombe Mudstone and overlying Blue Anchor Formation (previously the Tea Green Marl) is recorded by a change from predominantly red facies to green facies with better-defined bedding. The presence of coarser arenaceous bands in the succession reflects its proximity to the marginal facies described previously.

Analysis of the data from the Barry Island SSSI adds a further dimension to the interpretation of the late Triassic palaeogeography. Compared with the Triassic marginal deposits, examined at Ogmore-by-Sea, the facies seen at this locality are lithologically more variable, and their stratal relationships are more clearly defined. It is evident that the platforms were initially cut into Carboniferous Limestone during a prolonged period of subaerial weathering and erosion. These processes were accompanied by the deposition of thin scree deposits, which were banked-up against the backwalls of the platforms (facies A). Later, some of this scree material was reworked in the shore-zone of an expanding lake, resulting in an increase in the sorting, rounding and bedding of the detritus (facies B). Continued lake expansion resulted in further reworking and the deposition of finer grain calcrudites and calcarenites, which occur on the higher platforms (facies C and D). Regional studies indicate that the lacustrine shore-zone facies also pass very rapidly into lower energy saline lake facies of the Mercia Mudstone Group (locality *4d*).

3.3.5 Sully Island and Sully Bay

From Barry, travel on the A4055 around the outskirts of the docks to the roundabout at the junction with the A4231; continue on the B4267 through Sully village (2.5 km) and turn right, after the sports ground, onto Beach Road and park near the Captains' Wife restaurant at Swanbridge (Fig. 3.19). (Your parking fee is reimbursed if you purchase food or drink in the restaurant or bar). Walk about 50 m west and take the steps down onto the causeway, paying particular attention to the warning sign for dangerous tides. Sully Island can be visited only during a falling tide and the itinerary should be timed to begin when the causeway (a prominent Triassic sandstone bedding plane) is completely uncovered. If the tide is too high to cross the causeway it is suggested that Hayes Point and the Dinosaur trackway locality at Bendrick Rock are visited first.

5a East Point. Cross to the southern side of the island and walk 300 m to East Point (ST 1695 6690), where an angular unconformity, between steeply dipping Carboniferous Limestone and the horizontal Triassic facies, is exposed (Fig. 3.20). A basal Triassic breccia (0–1.5 m thick) can be seen to infill erosional gulleys in the limestone surface, which is cut by a small northerly-trending fault that downthrows 1 m to the east. Cross the fault gully and continue west, where the basal breccio-conglomerates thicken to 4 m, due to the introduction of a 2 m-thick fining-upwards unit. Hummocky structures, developed at the top of this unit, are sand volcanoes, which were formed as the result of the escape of water through vents, causing the internal laminae to be deformed upwards. These de-watering structures are characteristic of beds deposited very rapidly from mass flows. The overlying facies fine upwards into red mudstones with desiccation cracks and calcretes that are succeeded by a 1.5 m bed with densely packed irregular dolomite nodules, which has been interpreted as replaced chicken-wire anhydrite.

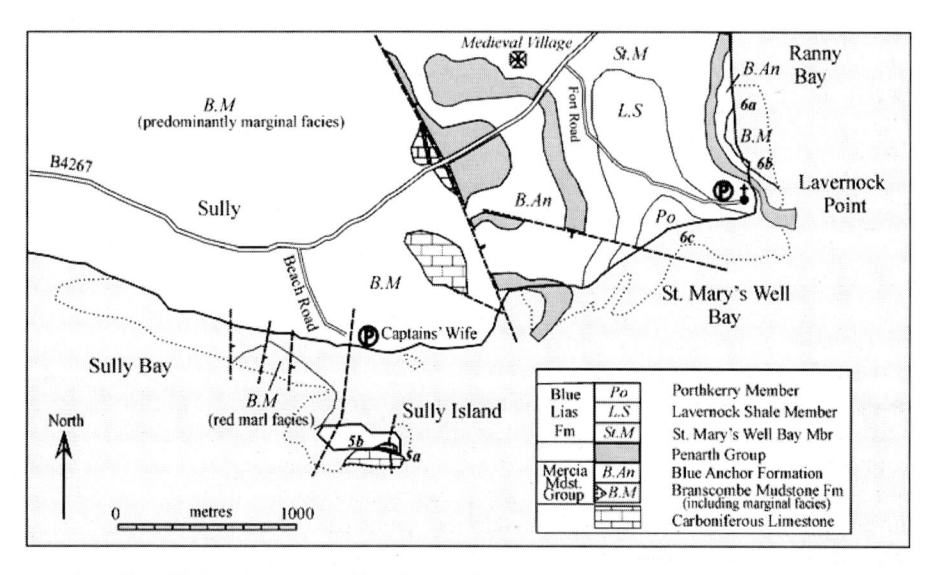

Fig. 3.19 Simplified geology map and locality map for the Sully Island and Lavernock Point itineraries.

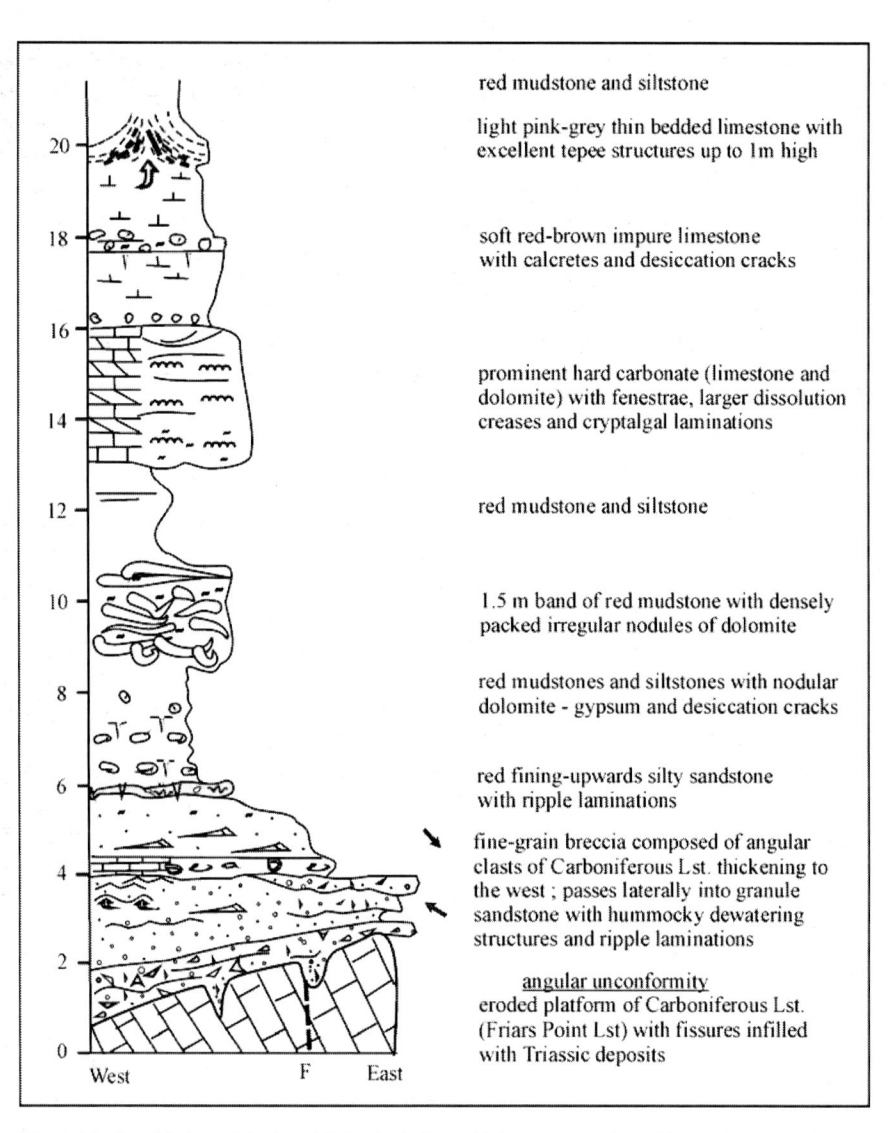

Fig. 3.20 Graphic log of the basal Triassic facies, which rest unconformably on the Carboniferous Limestone at East Point, Sully Island (see also Waters & Lawrence 1987, Plate 5).

5b Central cliffs. Return to the top of the cliff and walk 100 m westward before descending, at a safe point, to examine the very prominent 4 m-high vertical scarp face composed of dolomitic limestone (Fig. 3.21). This limestone displays, cryptalgal laminations, fenestrae and larger dissolution features, with patches of travertine developed at several horizons. The limestone is overlain by soft red mudstones containing desiccation cracks and nodular calcretes. On top of the cliff (ST 1670 6690), the upper beds of the sequence display excellent examples of tepee structures (Fig. 3.22), which form in arid environments where resurging brines break through and upturn surface crusts.

58

Fig. 3.21 View of the Triassic strata forming the cliffs extending to East Point on Sully Island. Note the hard dolomitic limestone forming the scarp cliff-face.

Fig. 3.22 Well defined tepee structure (about 1 m high) preserved in thin-bedded limestones and marls with calcretes, Sully Island.

5c Bendrick Rocks. On returning to the mainland it is possible to walk the 3.5 km along the cliff path west to Bendrick Rock, but it is more time-efficient to drive to a nearer point of access. Return to the B4267; turn left and then left again at the roundabout into Atlantic Trading Estate; park on the roadside 500 m farther west. Walk 300 m southwards along the footpath, adjacent to the security fence for HMS Cambria, to join the coast path. Cliffs and shore platforms between Hayes Point and Bendrick Rock display excellent exposures of the marginal facies of the Mercia Mudstone Group. The unconformable contact, between the Carboniferous Limestone and the overlying Triassic, occurs between the two points (ST 135 670), and the overlying marginal facies can be examined by traversing west to Bendrick Rock. Here the marginal facies are composed of erosionally based trough cross-stratified sandstone and conglomerate units (0.5–2.0 m thick), which provide southerly-directed palaeocurrents. Thinly bedded, graded sheetflood sandstones, which occur towards the top of the main conglomeratic unit, are well known for the preservation of dinosaur footprints. On certain bedding planes it is possible to find the trackways of three-toed dinosaurs, mainly *Grallator* and *Anchisauripus*.

Specimens from this locality are on permanent display in the geology section of the National Museum Cardiff.

The Triassic facies exposed on Sully Island are interpreted as shoreline facies, deposited around the margins of a large hypersaline lake. The basal breccio-conglomerates, seen at East Point, were formed along a narrow beach that was subjected to wave activity (northwest swash), whereas the overlying finer-grain sandstones were deposited in slightly deeper water (southeast backwash). The overlying red mudstones, with desiccation cracks and nodular dolomite, were formed on periodically exposed mudflats similar to sabkha plains. Periods of lake expansion and contraction resulted in several phases of precipitation of anhydrite, whereas more prolonged periods of shallow flooding resulted in the deposition of limestone and dolomite containing fenestrae and algal stromatolites. Upwelling of carbonate-rich groundwater resulted in the precipitation of travertine. A return to sabkha-type conditions is indicated by the calcretes and tepees seen at the top of the sequence. A similar but more fluvially influenced palaeogeographic setting can be interpreted from the strata exposed between Hayes Point and Bendrick Rock. Here the coarse-grain erosionally based cycles were deposited in ephemeral braided streams during periods of flooding. Carnivorous and herbivorous dinosaurs roamed the floodplains and lake margins between flash-flood events.

3.3.6 Lavernock and St. Mary's Well Bay

Return to the B4267 and travel 1.5 km northeast before turning right onto Fort Road, signposted to Lavernock Point. Cars can be parked at the end of the lane, just before St Lawrence Church (ST 1860 6820), or on the side of the road (Fig. 3.19). Note the plaque on the wall of the church commemorating the first radio messages sent across water (Lavernock Point to Flat Holm) by Marconi and Kemp in 1897. Flat Holm is the small flat-top island composed of Carboniferous Limestone, situated in the Bristol Channel 4 km southeast of the point. This island reserve has SSSI status for its gull population, other interesting biological features and for its geology. Members of the Flat Holm Project arrange visits by boat from Barry Docks from April to October. To the northeast the progressive narrowing of the Bristol Channel into the mouth of the Severn Estuary results in an extremely high tidal range (14 m), making the estuary ideal for generating tidal power. One of the more recent feasibility studies for a Severn hydroelectric scheme involves the building of a 16 km concrete barrage, incorporating a dual carriageway, from Lavernock Point to Brean Down near Weston-super-Mare. It has been calculated that turbines, powered by the strong tidal flow, could provide about 6 per cent of Britain's energy needs and would also greatly reduce the carbon emissions produced from burning fossil fuels. However, environmentalists fear that the barrage would destroy the unique ecosystem of the estuary.

This important coastal section is the only one in South Wales where a complete vertical succession from the Penarth Group (Triassic) up into the Blue Lias Formation (Lower Jurassic) is exposed. The Blue Lias is composed of offshore shelf facies, which provide a good comparative study with the marginal facies exposed at Ogmore and Southerndown. The section also has the type localities for the St Mary's Well Bay Member and the Lavernock Shale Member. Thus, the aims and objectives of the itinerary are to examine the fossil and trace-fossil assemblages, sedimentary facies, and modes of origin of the diverse strata beautifully exposed in the cliff section. The itinerary requires about 4 hours field time, more if you intend to examine the fossil assemblages in detail, and a low or falling tide. As the cliffs along the section are unstable, hard hats are required and wherever possible, all observations and collecting should be conducted at a safe distance from the cliffline.

6a Ranny Bay. From the church, walk north for 300 m along the coast path before descending onto the beach. Looking north towards Penarth, the cliffs are composed of gently dipping and faulted red and green marls belonging to the Mercia Mudstone Group. If the tide is receding, walk along the beach to a position where two prominent faults cut the redbeds; small cliff falls occur frequently, so keep well away from the base of the cliffline. Here the red silty marls, belonging to the Branscombe Mudstone Formation, contain a 50 cm bed with bands and pods of pink–white cryptocrystalline gypsum known as Penarth Alabaster (used in the past for carving ornaments). Some of the gypsum nodules display small translucent (recrystallization) patches up to 1 cm in diameter; others have veins of green and red marl and occasional patches of copper sulphate (malachite) on their under surfaces. At the top of the cliff the contacts between and Blue Anchor Formations (formerly the Tea Green Marls) and the overlying Westbury Formation can be seen. A profitable 15 minutes can be spent examining the lithologies, fossils, phosphatic bone remains and trace fossils, seen in the wave-polished fallen blocks of the Westbury Formation, which are located away from the base of the cliff.

6b Lavernock Point. Return south along the beach towards Lavernock Point, noting the broad anticlinal structure of the bay, and the contact between the red and green marls at a small promontory. Nodular bands of highly weathered pink gypsum occur 3.75 m above the contact and these are overlain by a 1 m-band composed of four redbeds (the 'Pink Band'). Large desiccation cracks and small synaeresis cracks can be seen in fallen blocks of green marl. Upwards the facies changes in colour from predominantly greenish-grey to grey. Continue south for 20 m past the sea wall (ST 1875 6824) to a position where a group of resistant beds extend seawards from the base of the cliff. Here the erosional disconformable contact between the Blue Anchor Formation and overlying Westbury Formation, previously referred to as the *Rhaetavicula contorta* Shales, is exposed (Fig. 3.23). This erosional surface is recognized throughout northern Europe and is referred to as the Kimmerian II unconformity. Six depositional cycles have been recognized in the Westbury Formation, each cycle beginning with a hard sandstone or impure limestone and continuing with dark shales and bivalve-rich horizons. Cycle 1 is composed of 70 cm of shales and limestones with U-tube burrows of *Diplocraterion*. At the base of cycle 2 the beds are truncated by a conglomeratic bone bed facies, which pinches out laterally and represents a vestige of Storrie's Fish Bed (see Waters & Lawrence 1987). This type of conglomeratic facies, called a tempestite, was deposited during a single storm event. All of the above facies display rapid vertical and lateral facies changes, but unfortunately beach pebbles frequently cover the best wave-polished exposures. The base of cycle 3 comprises interbedded graded sandstones, shales and beds containing abundant phosphatic bone fragments of reptiles, fish scales and coprolites. Pyritized burrows are common on the bases the sandstone units. A diverse fossil assemblage has been identified including

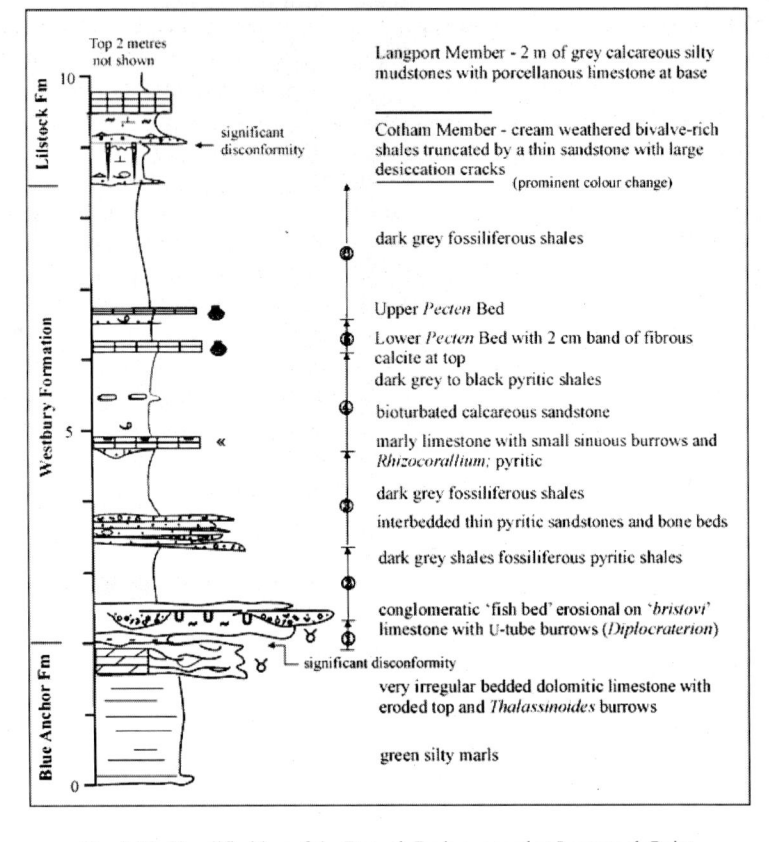

Fig. 3.23 Simplified log of the Penarth Beds exposed at Lavernock Point.

abundant shells of the bivalve *Eotrapezium*. Samples of the bone beds and the overlying limestones (including the two *Pecten* Beds) can be found as fallen blocks at the base of the cliff. These limestones contain good fossil bivalve assemblages including *Chlamys* (*Pecten*), *Eotrapezium*, *Placunopsis*, *Protocardia* and *Rhaetavicula* and the small gastropod *Natica* (for a complete fossil list see Fig. 30 in Waters & Lawrence 1987). The overlying Lilstock Formation is composed of calcareous shales, which weather to a cream colour and contain a fauna of shallow marine sessile bivalves, including *Liostrea*, *Modiolus* and *Dimyopsis*. These beds are truncated by a rippled sandstone with large desiccation cracks, which represents a significant disconformity and lowstand of sea level. A bed with soft-sediment deformation is believed to have formed as a result of seismic shocking from a distant meteor impact (see Hounslow & Ruffell 2006).

6c St Mary's Well Bay. Walking southwards to Lavernock Point and then westwards to St Mary's Well Bay a complete conformable transgressive sequence from the top of the Penarth Beds into the Blue Lias can be examined. The Blue Lias has been subdivided into three members based on the ratios of limestone to mudstone beds; their nomenclature and the biostratigraphy are summarized in Fig. 3.24. Note that the Triassic–Jurassic boundary (199.6 ± 0.6 Ma) is placed at the base of the *Planorbis* Subzone, which occurs about 5 m above the Paper Shales marking the lithological base of the Blue Lias. Just north of a concrete-protected sewage pipe (ST 1873 6816) the lower half of the succession can be clearly seen in the cliff face (Fig. 3.25). The alternating limestone and mudstone facies and marker beds of the Bull Cliff Member are most easily studied along the wave-polished platform where they contain a good fossil assemblage of bivalves (*Liostrea* and *Modiolus*), echinoderms (*Diademopsis* and *Eodiadema*), detached spines and ossicles of *Pentacrinus*. Fossil vertebrae of marine reptiles (ichthyosaurs and plesiosaurs) can sometimes be found. Bioturbation is particularly clear on the bases of the limestones, and some of the nodular bedding can be attributed to large *Thalassinoides* burrows. About 7 m above the contact, within the St Mary's Well Bay Member, fissile dark shales contain poorly preserved examples of the ammonite *Psiloceras planorbis* (*Planorbis* Mudstones). The fossil fauna also includes pectinids such as *Camptonectes*, *Oxytoma* and *Terquemia* and shallow burrowing bivalves such as *Plagiostroma*, and *Pinna*. Farther to the west the overlying transitional contact with the Lavernock Shales and the broad synclinal structure of the bay are also clearly visible. These calcareous shales with limestone nodules,

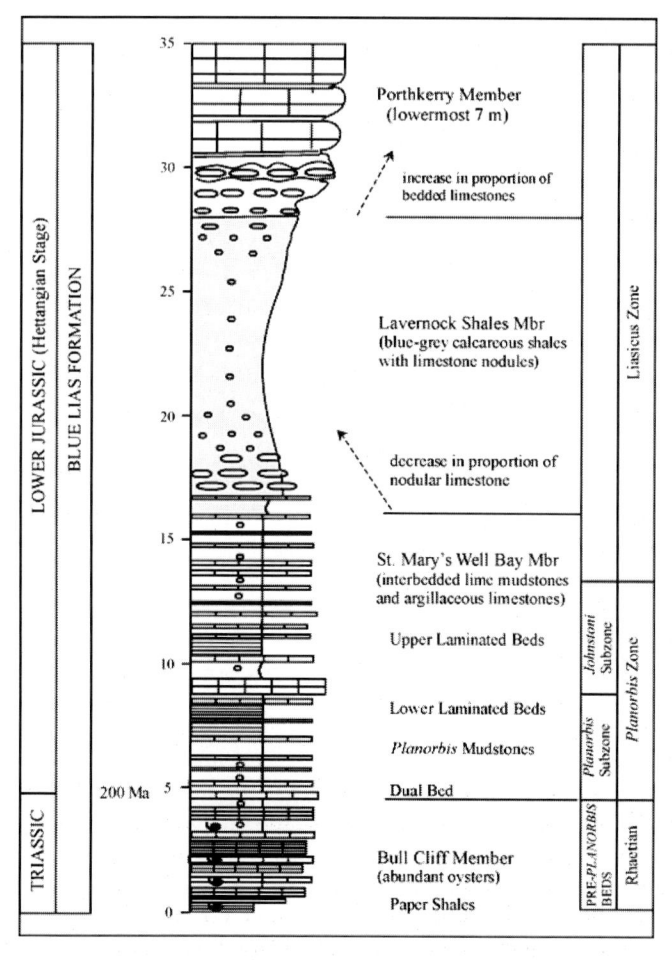

Fig. 3.24 Profile log showing the facies and biostratigraphy of the late Triassic and lower part of the Blue Lias exposed between Lavernock Point and St Mary's Well Bay (modified from Waters & Lawrence 1987).

contain deeper burrowing bivalves such as *Cardinia*. The upper 10 m of the cliff section is composed of alternating limestones and lime mudstones belonging to the Porthkerry Member. At the western end of St Mary's Well Bay (ST 176 677), the strata between the top of the Blue Anchor Formation and the base of the Westbury Formation are faulted against Triassic marginal facies. To the east of the fault the base of the Westbury Formation is marked by an erosional surface filled with a mudstone conglomerate (> 0.25 m thick). This unit is overlain by a limestone coquina packed with *Liostrea bristovi*, which becomes more argillaceous upwards; its top being cut by small scours filled with mudstone pellets.

The complete succession records a progressive sea level rise, which was accompanied by a change in climatic conditions. The red argillaceous facies of the Mercia Mudstone Group were deposited in a shallow hypersaline lake, which often experienced extended periods of desiccation resulting in the precipitation of sabkha-type sulphates. Alternating lake expansion and contraction resulted in the deposition of the green facies with occasional units of red marls and sulphate horizons. Studies of microfossil assemblages suggest that marine influences occurred towards the top of the Blue Anchor Formation. Marine conditions were further established during the deposition of the Westbury Beds where transgressive events are recorded by the bone-bed lags and deepening-upwards cycles. A significant regression, resulting in lagoonal conditions and regional emergence (desiccation cracks), is recorded in the Cotham Member, which was followed by a return to restricted marine conditions in the Langport Member. The above facies variations also record a change in climate from arid through semi-arid to humid. The Blue Lias facies marks the onset of a regional marine transgression, which covered most of southern Britain in a relatively shallow epeiric sea. The fauna of the Bull Cliff Member indicates slow sedimentation and a firm substrate, whereas the fossils in the succeeding St Mary's Wells Bay Member point to increased sedimentation rates, gentle currents and well-oxygenated bottom conditions (Simms 2004). The succeeding Lavernock Shale Member records a rise in sea level from the nodular limestone facies to the shale-dominated facies. This was followed by an extended period of offshore shelf sedimentation, recorded by the Porthkerry Member, with a notable shallowing event occurring in late *Bucklandi* Biozone (Unit C) times.

At present there is no consensus view on the origin of the cyclic sedimentation in Jurassic successions, including the decimetre-scale rhythmic alternation of the limestones and shales, which characterizes the Blue Lias. In the past the rhythmites have been interpreted in terms of primary (depositional), secondary (diagenetic), and by combinations of both processes. Many of the depositional theories rely heavily on changes in eustatic sea level and climate, operating on a relatively short time scale (third-order cycles). Such high frequency changes are considered to be the result of variations in solar radiation produced by variations in the Earth's orbit i.e. eccentricity (~100 and 400 ka cycles), obliquity (~40 ka cycles) and precession (~20 ka cycles). Climatic and sea-level changes, brought about by this type of Milankovitch cyclicity, would influence ocean circulation, oxygen levels, carbonate productivity, freshwater runoff and the amount of detrital clay entering the basin; and consequently the deposition of alternating lime/mud lithofacies. Most post-depositional models invoke an initial homogeneous aragonite mud and create the limestone alternations during early diagenesis by, (a) the dissolution of the unstable aragonite, followed by the precipitation of calcite to form a micritic limestone bed, (b) diagenetic segregation or (c) authigenic production of the limestone units by

Fig. 3.25 Cliff section at Lavernock Point showing the vertical transition form the late Triassic (Rhaetian) into the basal Lias (*Planorbis* Subzone). (See Fig. 3.24 for stratigraphy).

microbial processes. From field studies, it is clear that pauses in sedimentation and bioturbation have modified many of the limestone–shale bedding planes present in the Blue Lias. For example, pauses in sedimentation have resulted in the formation of firmgrounds and hardgrounds (omission surfaces), while bioturbation by *Thalassinoides* has frequently resulted in the formation of nodular bedding (Fig. 3.12). Recently, Sheppard *et al.* (2006) have interpreted the majority of the limestone/shale alternations in the Blue Lias at Nash Point as pseudo-bedding, formed as a result of diagenetic enhancement of the limestone units by microbial processes, although they relate the cyclic bundling of omission surfaces to Milankovitch-scale climatic cycles.

Chapter 4

Gower

4.1 Introduction

The Gower peninsula was the first 'Area of Outstanding Natural Beauty' designated in the United Kingdom. Ridge (1999) described the area as "a happy accident of geological forces, climatic conditions and human enterprise; a harmonious blend of landscape, nature and history". Unquestionably, the beautiful landscape of the peninsula is a direct result of its geological setting (Fig. 4.1). Cefn Bryn, an anticlinal ridge composed of resistant red conglomerates and sandstone, divides the peninsula along its length. To the south, high limestone cliffs are interrupted by the wide bays at Oxwich and Port Eynon, which are eroded along synclines preserving soft shales. Gower's western coastline is dominated by Rhossili Bay, which extends northwards for 5 km between Worms Head and Burry Holms. Along the northern coastline of the peninsula, sand dunes, low-lying saltmarshes, mudflats and sandbanks of the Loughor Estuary now front the stranded cliffline.

Gower is rich in archaeological sites including Iron Age hill and promontory forts (The Bulwark, Cilifor Top and The Knave); burial chambers (Parc le Broes, Arthur's Stone, Paviland Cave and Sweyne's Howes) and bone caves (Bacon Hole, Minchin Hole and Catshole). Paviland Cave (Goat's Hole), located 5 km southeast of Worms Head, is one of the most important prehistoric archaeological sites in Britain. The burial was first excavated by William Buckland in 1823, and later by William Sollas in 1912, both professors of Geology at Oxford University. Buckland identified the ochre-stained human bones as a female buried during Roman times (the 'Red Lady' of Paviland), but later work, and modern radiocarbon dating, proved the body to be a male of Palaeolithic age (*circa* 26 000 BP).

About 65 per cent of the coastline is owned and administered by the National Trust, who together with the Gower Society, the Countryside Council for Wales and the South Wales Wildlife Trust, support and encourage conservation and tourism. Road links from Swansea, to all parts of the peninsula, are generally good, but the main routes are often very busy during the summer. All of the principle beaches and villages can be reached using the 'Gower Explorer' and other bus services. Swansea has good shopping facilities and there are many hotels in the city, the newly developed Maritime Quarter, and along the coast road to the Mumbles. There are many well-organized and unobtrusive camping and caravan sites. Swansea market is renowned for its fresh produce from Gower, including the gastronomic delights of saltmarsh lamb, lava bread and Penclawdd cockles. A guide to the events and attractions in and around the city are published in the free magazine *What's on in Swansea*.

4.2 Geological history

Gower is composed almost entirely of Upper Palaeozoic strata of Devonian and Carboniferous age, with only small remnants of a once-extensive Mesozoic cover, and patches of drift (Fig. 4.1a). The Devonian rocks, usually referred to as the Old Red Sandstone (ORS), comprise redbed facies, which were deposited in arid to semi-arid continental environments. These rocks are divided informally into the Red Marls and Brownstones (Lower ORS) and the Quartz Conglomerate (late Upper ORS). A major unconformity, separating the two groups, is considered to represent the Middle to early Upper ORS time interval. The oldest ORS rocks exposed in the area are red mudstones, siltstones and sandstones belonging to the Brownstones. These facies are often organized into fining-upwards cycles, which were deposited in a variety of low- to medium-energy alluvial channel environments. The Quartz Conglomerate represents higher-energy deposits of alluvial fans and ephemeral braided rivers.

Approximately 50 per cent of Gower is composed of Carboniferous Limestone (Dinantian). Most of the pioneer work on this limestone was carried out in the first half of the 20th century, when the succession was subdivided into zones based on the distribution of coral and brachiopod faunas. More recently the lithostratigraphical succession has been revised and renamed by reference to type localities from the many excellent outcrops in South Wales (Fig. 4.2; Appendix 1). This long episode (about 35 million years) of carbonate deposition was initiated by a major rise in sea

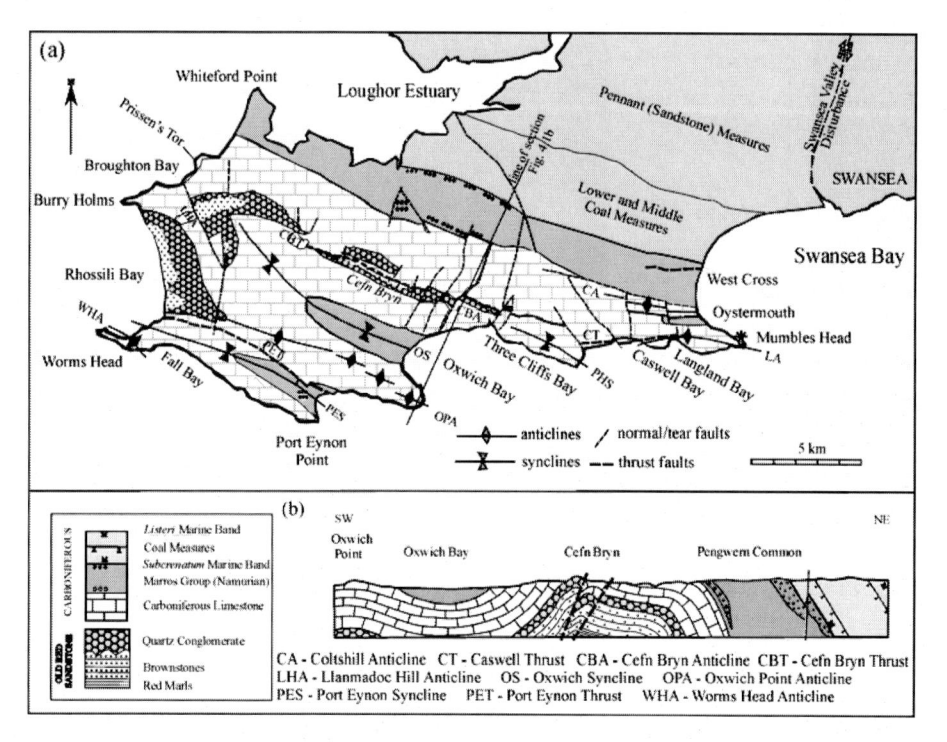

Fig. 4.1 (a) Map of the geology and structure of Gower (modified from BGS 1:50 000 maps for Swansea (Sheet 247) and Worms Head (Sheet 246). **(b)** Simplified cross section to illustrate the structure of Gower (modified from BGS Sheet 247).

level, which resulted in the flooding of the southern margin of the ORS continent and the development of a southward-dipping ramp environment. In this setting about 1000 m of limestones and subordinate shales were deposited in a warm tropical sea. Periodic glacio-eustatic sea-level changes resulted in the deposition of alternating transgressive and regressive sequences, with the more significant sea-level falls resulting in the subaerial emergence of the ramp and the formation of palaeokarst surfaces (see Fig. 2.11). A more sustained rise in sea level began during the late Dinantian, which resulted in the deposition of the Oystermouth Formation (Upper Limestone Shales), and the succeeding deeper water Namurian facies.

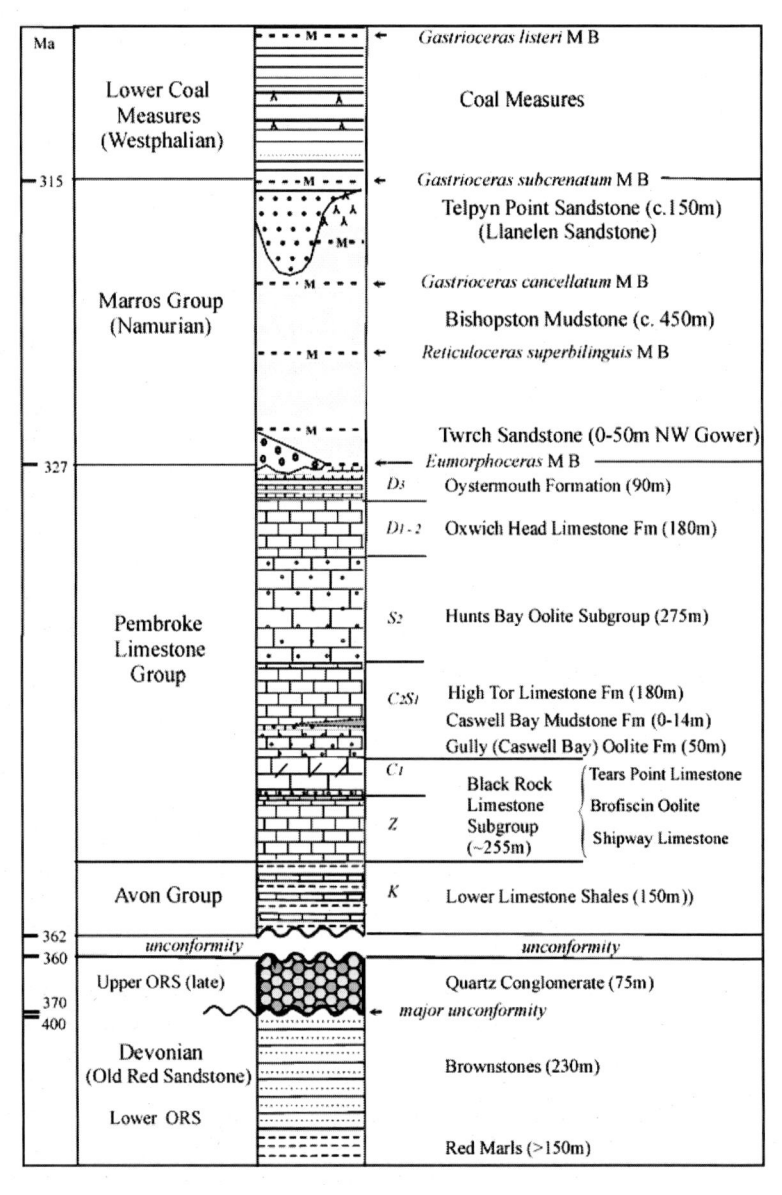

Fig. 4.2 Upper Palaeozoic stratigraphy of Gower.

During the succeeding Upper Carboniferous (Silesian) the northward propagation of the Variscan deformation front resulted in the uplift of terrigenous source areas, particularly along the northern margin (foreland bulge) of the basin. This developing orogeny, resulted in an extended period (20 million years) of clastic sedimentation, recorded by the Marros Group (Namurian) and Coal Measure (Westphalian) strata. The rise in sea level, which had started in the late Dinantian, continued into the Namurian (Bishopston Mudstone Formation) with the deposition of deepwater basinal shales containing many goniatite-bearing marine bands. A thin development of pebbly quartz arenites, belonging to the Twrch Sandstone (formerly the Basal Grit), occurs at the base of the Marros Group to the east of Llanrhidian, where they form the prominent hill of Cilifor Top. Later Namurian sequences contain sandstone facies of shallow-marine and fluvio-deltaic origin, and thin marine shales (e.g. the *Gastrioceras cancellatum* and *G. cumbriense* marine bands) that record eustatic sea-level rises. A major fall in sea level occurred during the late Namurian, resulting in the erosion of a large incised valley, which was filled with fluvio-deltaic sands (Telpyn Point Sandstone) during a subsequent sea-level rise. Continued sea-level rise led to the deposition of the regionally developed *G. subcrenatum* Marine Band, which defines the base of the Westphalian. The succeeding, but very poorly exposed, Coal Measures of north Gower are composed of fluvio-lacustrine facies containing freshwater bivalves (mussel bands), plant remains, coal seams and occasional thin marine bands. These coal-bearing strata supported an important coal mining industry until the early 1930s.

At Port Eynon an isolated patch of red breccio-conglomerate represents the remains of a once-extensive cover of Triassic strata. Other traces of this cover are recorded by redbed deposits preserved along fault planes, in fractures (Neptunian dykes) and caves, and also by the secondary reddening of the Carboniferous Limestone that is seen at some localities. Much of Gower was also covered by Mesozoic strata, which were removed during the extended period of denudation that followed the Variscan Orogeny. The presence of this cover is supported by extensive Permo-Triassic and Jurassic deposits, which occur offshore, to the south of the Gower Peninsula.

The final episode of sediment accumulation in Gower is recorded by a variety of Pleistocene and Holocene deposits. Some of the most easily recognizable Pleistocene sediments are the raised beach and cave deposits that occur along the coast. The *Patella* beach (Hunts Bay Beach, Hoxnian–Ipswichian) is an important horizon that occurs on a narrow platform at the foot of the limestone cliffs 7–10 m above high-water mark. At Minchin Hole the *Patella* beach is overlain by bone bearing cave deposits and a further raised beach (the slightly younger *Neritoides* beach). At other localities the *Patella* beach is overlain by solifluxion (periglacial) deposits and a variety of glacial drift and outwash sediments formed prior to, and during the Late Devensian glaciation (Dimlington Stadial) (Fig. 4.3).

The principal post-glacial (Holocene) sediments found in Gower are the drifts of aeolian sand, which form the burrows at Oxwich, Pennard, Llangennith–Broughton and Whiteford. A large proportion of this sand was blown onshore between 700–300 years ago. In some areas drifts of blown sand and storm beach deposits have impeded drainage, resulting in the development of localized freshwater marshes such as those at Bishopston Pill, Pennard Pill, Oxwich, and Llangennith. Extensive salt marsh and tidal-flat deposits extend along the north coast of the peninsula from Whiteford Point to Gowerton. At Broughton Bay a variety of recent deposits, including peat from the submerged forest and wind-blown sand, are exposed (Figs 4.4, 4.5). Here the peat contains *in situ* stumps and fallen branches of alder, birch, hazel and oak, and in some areas it has been altered to a ferruginous hardpan. The broad-leaved forest that produced this peat began growing about 8000 years ago as a result of an increase in temperature and rainfall after the last Devensian glaciation; growth continuing until the coastal plain was flooded during the later part of the Flandrian transgression.

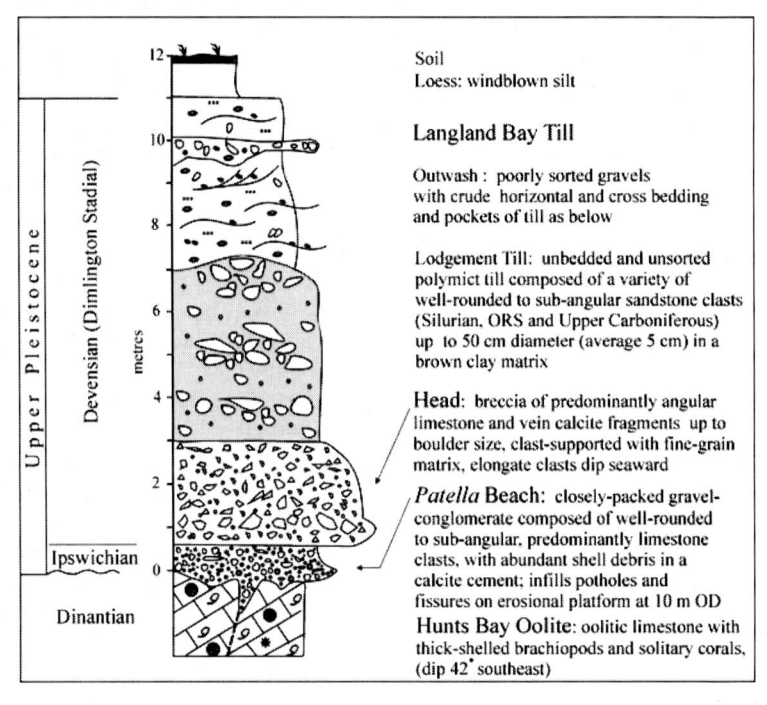

Fig. 4.3 Graphic log illustrating the Upper Pleistocene succession at Rotherslade (SS 613 871) (stratigraphy after Bowen 1971).

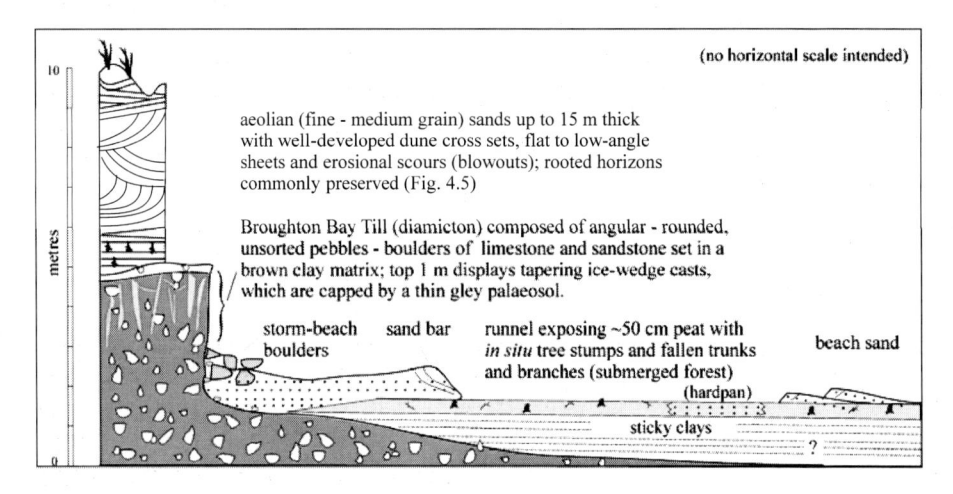

Fig. 4.4 Generalized cross section of the superficial deposits seen at Broughton Beach (2007–8).

Fig. 4.5 Large-scale dune sets and sand sheets (4 m thick) exposed at Broughton Burrows (March 2008).

Professor T. Neville George published the seminal work on the structure of Gower in 1940, and many of his maps and cross sections have been simplified for this guide. The main Variscan structural elements of Gower are a series of ENE-trending asymmetrical folds that are cut by many cross faults and a few important thrusts (Fig. 4.1). The major anticlines expose the oldest rocks belonging to the Old Red Sandstone, whereas the most notable synclines preserve concealed Namurian shales. Thrust faults run approximately parallel to the fold axes and dip steeply to the south, although the Caswell Thrust dips to the north. George (1940) demonstrated that both the cross faults and thrusts occurred more or less contemporaneously with the folding. Some of the tectonic structures of Gower, which were reactivated during the northward propagation of the Variscan front, had an important influence on Dinantian and Namurian sedimentation.

The Gower is covered by the 1:50 000 geology maps of Swansea (Sheet 247) and Worms Head (Sheet 246) and the 1:25 000 Ordnance Survey map (Explorer 164). Some of the main sources of geological information are George (1940, 1970), Bowen (1971, 1999), Owen (1971), Ramsay (1987), Cossey *et al.* (2004) and Howe *et al.* (2004).

4.3 *Itineraries*

Gower has a wealth of excellent geological sites and it was difficult to decide which ones to include in this guide. The final selection is intended to provide a reasonable coverage of the stratigraphy and lithofacies of the Old Red Sandstone and Carboniferous Limestone, and also the little-known, but important, outcrop of the Marros Group (Namurian) at West Cross in Swansea Bay. Rhossili and Three Cliffs

Bay requires full days; Caswell Bay and Oystermouth to West Cross each require 3 to 4 hours field study. Many of the localities are accessible only from medium to low tides, but alternative routes are suggested when tides are high.

4.3.1 Rhossili

Enter Rhossili via the B4247 road and continue to the large car park (with adjacent toilets), situated at the western end of the village (Fig. 4.6). The village, which is on the Gower Explorer bus route from Swansea, has a hotel, a restaurant and a couple of small shops and cafés. St Mary's Church contains a memorial to Edgar Evans, who was born in the village and died in 1912 on the ill-fated South Pole expedition led by Captain Scott. Rhossili Bay is one of the finest stretches of sand on the Welsh coastline. The bay curves gently northwards for 5 km to the headland of Burry Holms, and is exposed to the prevailing southwest winds and Atlantic storms. Consequently, it has been the resting place for many ships wrecked along the coast. Bough-timbers, belonging to the *Helvetia* (grounded on November 1st 1887), are still visible on the beach (see Fig. 1.1). A more famous undiscovered wreck was that of a Spanish galleon (known locally as the 'dollar ship'), which came to grief during the 17th century. Silver coins from its cargo were uncovered during exceptionally low tides in 1807 and 1833, with some of the booty financing the building of a cottage (Dollar Cot) and a farm at Llangennith. Spanish gold coins (doubloons) from another wreck have also been found along the coast to the north of Burry Holms.

The beach is backed by a narrow platform of recent solifluxion deposits (the Warren) that drape the steep western flank of Rhossili Down up to the 40 m topographic contour. The National Trust owns a large part of the area, including the beach, cliffs, the Warren, Rhossili Down and Worms Head. Before beginning the itinerary it is necessary to check the times of the tides, which are posted outside the National Trust visitors centre and the lookout station. This is important because the causeway leading onto Worms Head is accessible only for about 2.5 hours each side of low tide. This time slot will determine whether the inland localities (*1a–1c*) or Worms Head (*1d–e*) are visited first.

1a The Beacon. If the tide is high and the weather is good the climb to the Beacon (193 m OD), the highest point on Gower, is invigorating and offers excellent panoramic views. From the car park follow the road to the church and then turn left onto the track; note the fine monolith of Quartz Conglomerate erected by the people of the village to commemorate the Millennium. Pass the gate (which leads to the beach) and continue through another gate signposted to Rhossili Down. Directly ahead is the former Rocket House where the lifesaving apparatus (bosun's chair) was kept and used to save the crew of the *Helvetia* and many other shipwrecked sailors. Follow the wooden steps that ascend the spur leading up to The Beacon, but before reaching the summit stop to admire the views of the Rhossili Peninsula, terminating at the spectacular Worms Head (Fig. 4.7b). 'The Worm', as it is known locally, derives its name from its apparent similarity to a humpback sea serpent, especially when the tide is full and a blowhole, towards its tip, is active. On a clear day it is also possible to see the north Devon coast and Hartland Point, and from the summit there are fine views of the coast of south Pembrokeshire as far as St Govan's Head.

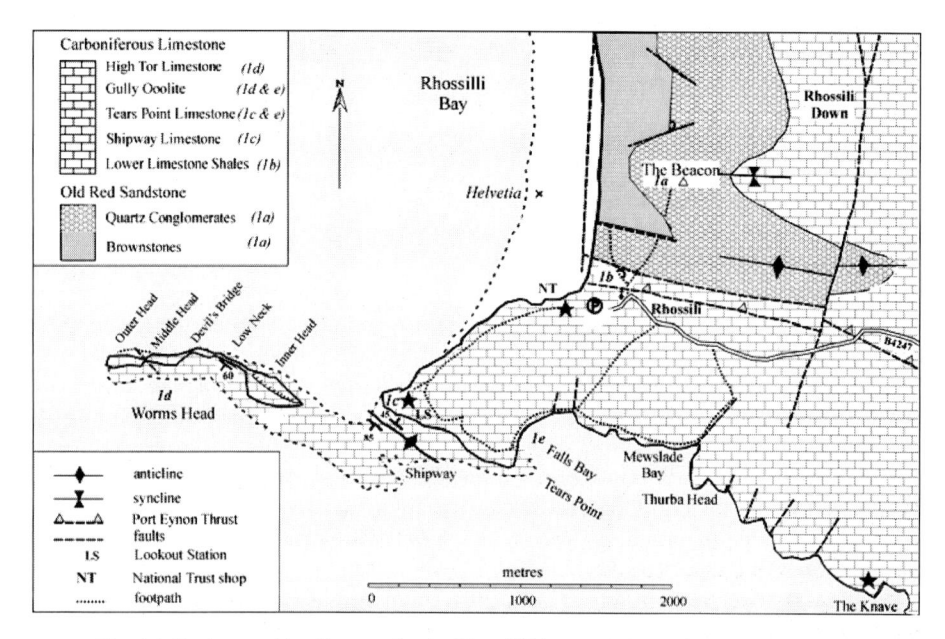

Fig. 4.6 Geology and locality map for the Rhossili itinerary (adapted from George, 1940).

Rhossili Down is a ridge of resistant Quartz Conglomerate (Upper ORS), which is cut by faults and deformed into an anticline at its southern termination (Fig. 4.6). Along the summit, just below the triangulation point (SS 4200 8887), conglomerates crop out in low north–south trending crags. These conglomerates are composed of sub-rounded to rounded white and pink vein quartz pebbles (up to 10 cm diameter), with minor amounts of jasper, quartzite and lithic fragments. Occasional isolated sets of cross stratification and graded bedding are present. At a prominent crag, located about 500 m to the north (SS 4199 8908), a bed of conglomerate, dipping 60° south, is seen to erode into a red silty-sandstone. This locality is very close to the mapped contact between the Quartz Conglomerate and the underlying Brownstones (Fig. 4.6). The same contact can also be located within the intermittent outcrops on the side of the track leading down the flank of the ridge from the Rocket House. All along the ridge blocks of the conglomerate have been used in the construction of Bronze Age burial mounds and also the earlier Neolithic tombs, known as Sweyne's Howes, located 1 km to the north of The Beacon.

1b Old quarry. Descend from The Beacon and walk 100 m along the track that leads to Rhossili beach. A small quarry on the right exposes weathered and deformed calcareous shales and thin harder limestones, containing fossils of the brachiopod *Camarotoechia mitcheldeanensis* and crinoidal debris. These beds belong to the Avon Group (Lower Limestone Shales), which at this locality have been compressed and faulted by the Port Eynon Thrust.

1c Lookout station. Return to the village and take the road that passes the National Trust visitors centre and leads to Worms Head. The cliffs along the route were extensively quarried until the end of the 19th century, the limestone being exported to Devon where it was burned to produce agricultural lime (Fig. 4.7b). Continue along the footpath to the lookout station where there are good views of the

Worm and the eroded axis of the Worms Head Anticline, exposed on the causeway. Descend to beach level, noting the large warning sign of the dangerous currents and giving other safety information. Walk about 100 m onto the tidal causeway where both limbs of the asymmetrical anticline are exposed. The axis of the fold is cut by north–south trending faults filled with coarsely crystalline rhombic and dogtooth calcite. This is the type locality of the Shipway Limestone, which comprises thin-bedded crinoidal limestones, containing silicified brachiopods (*Spirifer*, S*yringothyris* etc.) and nodules and bands of chert.

1d Worms Head. Although the geology of Worms Head is not unique, the scenery is spectacular. The round trip to the Outer Head and back takes at least 2 hours, but beware that the steeply dipping limestones of the Causeway and Long Neck are very sharp and require good footwear. The walk to the Inner Head takes about 15 minutes and crosses limestones that young westwards from the Shipway to Tears Point Limestones. The Long Neck is a narrow strike-orientated ledge of the Tears Point Limestone that dips 65° southwards, and is cut by many cross faults and fractures. A large natural arch, known as the Devil's Bridge, occurs at the western termination of the Inner Head.

Follow the footpath to the Middle Island, composed of oolitic limestone (Gully Oolite), noting the tectonic crumpling associated with a fault, which cuts the bare rocks connecting it with the Outer Island. In this part of west Gower, located to the south of the Port Eynon Thrust, the Caswell Bay Mudstone is absent and consequently the High Tor Limestone, which forms most of the Outer Island, lies directly on the Gully Oolite. However, to the north of the thrust the distinctive lagoonal lithofacies reappears at Burry Holms where it is 2 m thick, while at the northern end of Broughton Bay (Prissen's Tor) it reaches its maximum thickness of 14 m. At the latter locality the Caswell Bay Mudstone is composed of shallowing-upwards limestone, dolomite and mudstone cycles containing algal laminites, desiccation cracks and tepee structures; its top being erosionally truncated at the conglomeratic base of the High Tor Limestone.

To complete the Worms Head itinerary you can climb to the top of the Outer Head, following a fault plane filled with red sandstone and blocks of mudstone of Triassic age. From this point there are good views looking back along the Worm towards the flat top of the 60 m platform of south Gower.

1e Fall Bay. Return to the mainland and walk 400 m along the low-level path, up a small dry valley to the clifftop; then continue eastwards around Tears Point before descending to Fall Bay (Fig. 4.7c). The foreshore along the western margin of the bay is composed of highly bioturbated and dolomitized limestone (Tears Point Limestone), which contains abundant crinoid and shell debris, ribbed brachiopods and occasional solitary rugose corals (*Zaphrentis* and *Siphonophyllia*). The dolomitized limestone facies also contains many vugs, which are either completely filled with calcite or have partial fills of drusy calcite. Dolomitization probably occurred during burial diagenesis when magnesium-rich fluids were expelled from the underlying Lower Limestone Shales. The vugs were formed as a result of dissolution caused by migrating acidic (carbon dioxide-charged) fluids, at a later stage of diagenesis.

Fig. 4.7 (a) View of Rhossili Bay looking north to Burry Holms; note the solifluxion terrace. (b) View of Worms Head from the cliffs to the west of the information centre. The very irregular cliffline is the result of old quarries cut into the limestone. (c) Fall Bay looking eastward to Mewslade Bay and Thurba Head; note the irregular erosional base of late Pleistocene *Patella* beach to the left of the 1 m-tape.

Fig. 4.8 (a) View of Three Cliffs Bay and the limestone pinnacles forming the eastern headland. (b) View of Pennard Pill from the castle showing the river floodplain with creeks, ponds and a large meander loop; note the flat top of the 60 m platform. (c) Peat-rich floodplain deposits exposed seaward of the storm-beach; the ravinement surface is being covered by beach sand (see also Fig. 4.11b).

A prominent exposure of the late Pleistocene (Ipswichian) *Patella* Beach (30–75 cm thick) occurs on the eroded top of the limestone (SS 4102 8725; Fig. 4.7c). This deposit consists of rounded pebbles of limestone in a shelly sandy matrix containing complete shells of the limpet *Patella* and broken columellas of gastropods. The raised beach is overlain by Devensian sediments consisting of a basal unit (0–1.5 m thick) of sandy to gritty orange clay with occasional rounded boulders of limestone, and a further 5 m of typical head composed of well cemented angular limestone fragments.

Farther along the beach the limestones are cut by numerous northeast–southwest-trending faults and fractures, which are filled with iron-stained coarsely-crystalline calcite. The calcite often encloses pods and thin veins of brick-red clay and sandstone, which represent remnants of the Triassic cover. A more substantial remnant of this cover is seen at the bottom of the steps leading down into the cove, where an elongate pod of red sandstone and draped clays enclose large angular blocks of limestone. Thick head deposits occur at the back of the bay and many large displaced blocks can be examined on the beach. The transitional contact between the dolomitized limestones and the succeeding Gully Oolite can be located in the centre of the bay, while the top of the oolite forms a distinct notch in the vertical cliff at the eastern end of the bay (Fig. 4.7c). A stepped path, located in the northeast corner of Fall Bay, joins the coastal path at a small limekiln, from where you can cross the strip fields to Middleton and walk back to Rhossili on the B4247. It is also possible to walk along the cliff path to Mewslade Bay and then up the dry valley to reach the road. From Mewslade the High Tor Limestone forms magnificent cliffs extending southeastwards via the promontory fort known as The Knave (SS 432 864), Paviland Cave (SS 437 859) and on to Port Eynon Point.

4.3.2 Three Cliffs Bay

Travel westwards on the A4118 through Parkmill and at the top of the hill turn right at Penmaen Church; drive through the gate, over the cattle grid, and into the National Trust parking area for Cefn Bryn (SS 532 887). This is the nearest and most convenient access point for Three Cliffs Bay (Fig. 4.8a), which is a 1 km walk (Fig. 4.9). The bay can also be reached by walking 2 km westwards along the clifftop from the National Trust car park at Southgate. As there are no facilities at Three Cliffs, a packed lunch is recommended. If the tide is high, the walk westward along Cefn Bryn provides excellent views of Gower, especially from the impressive Neolithic burial chamber known as Arthur's Stone, located at the northwestern termination of the ridge near Reynoldston.

From the car park return to the A4118, noting the exposure (opposite the Old School House) of the Quartz Conglomerate (Upper ORS) with many south-eastward-tending vertical slickensides. Cross the busy road and continue along the minor road for 200 m before turning right at the National Trust signpost for Notthill, which forms the southeastern termination of the Cefn Bryn Anticline (Fig. 4.9). Continue to the end of the track following the signs for the beach. From the southern flank of Notthill (SS 534 884) there are fine views of the bay and the three limestone pinnacles, which define the eastern margin of the cliffline (Fig. 4.8a). The bay owes its origin to preferential erosion, along a northeast–southwest-trending tear fault with a sinistral displacement of up to 200 m. The fault runs just west of the limestone

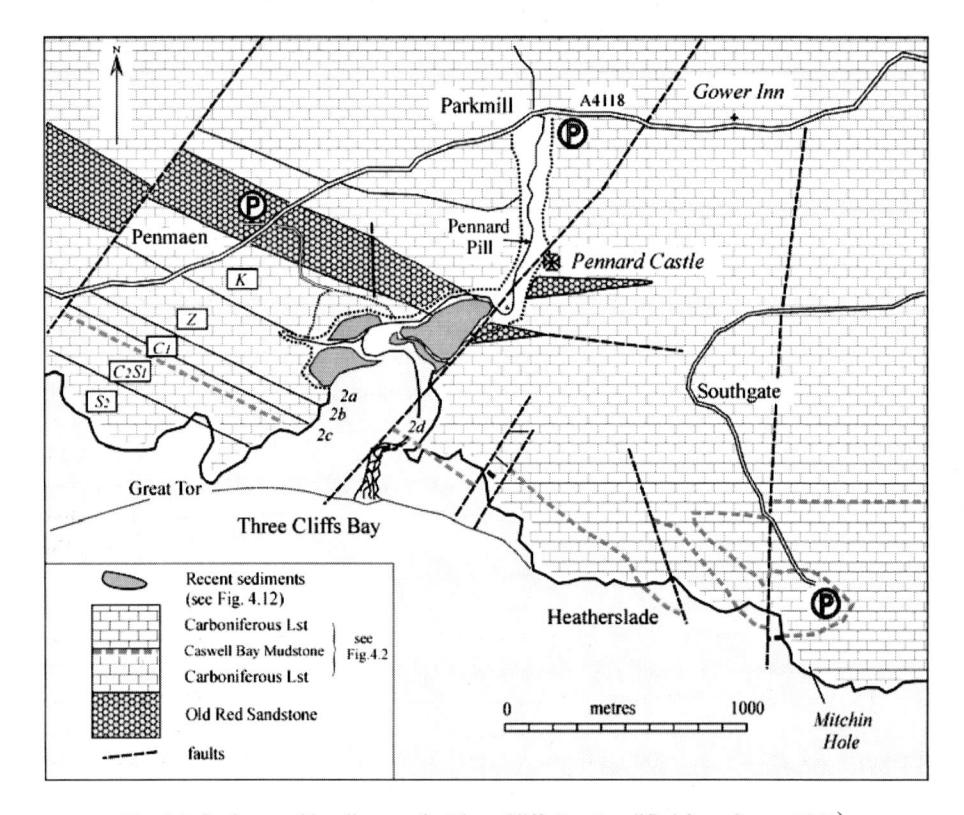

Fig. 4.9 Geology and locality map for Three Cliffs Bay (modified from George 1940).

crags supporting the ruined ramparts of the 13th century (Norman) Pennard Castle. Take the right-hand footpath that leads down the steep southern flank of Notthill, noting that the steps are composed of fractured and slickensided blocks of Quartz Conglomerate. On reaching the beach there are two options depending on the state of the tide. The solid geology localities (*2a–d*) are accessible from medium to low tide; if the tide is high a couple of hours can be allocated to a study of the recent sedimentary environments and geomorphology of the area (localities *2e–i*; Figs 4.8a, b, c and 4.12).

2a Western cliffs (north). The oldest Carboniferous Limestone strata belonging to the Lower Limestone Shales (Avon Group) are mostly concealed and form the westward-extending hollow known as Stonesfield. To the south, steeply dipping (60–80° south) fractured limestones assigned to the Shipway Limestone (lower part of the Black Rock Limestone Subgroup) are exposed at the northeast termination of the cliffs against the dunes (SS 535 881). Here they comprise alternating bioclastic limestones and bioturbated lime mudstones and shales, which are succeeded by thicker-bedded bioclastic packstones with graded bedding and sets of hummocky cross-stratification (HCS). These are followed by about 4 m of cross-laminated bioclastic facies displaying a thickening-upwards trend. All of the above facies contain an abundant fauna of chonetids and other brachiopods, crinoid debris, bryozoans and a variety of trace fossils, including *Zoophycos, Chondrites, Rhizocorallium,* small *Thalassinoides,* escape burrows and occasional U-tubes.

Silicified lenses and stylolitic seams are common. Higher in the succession graded beds, up to 30 cm thick with sets of hummocky bedding, are interpreted as storm-event beds. The complete succession was deposited on a storm-dominated outer to mid-ramp environment.

2b Western cliffs. Walk 300 m southwestwards, along the foot of the cliff, to a small recess where the strata are deformed into a low-amplitude fold and are cut by four faults (Fig. 4.10). The core of the fold is composed of about 6 m of cross laminated grey dolomitic limestone, which passes up transitionally into a light-grey oolitic and crinoidal limestone (3.75 m thick) with large-scale cross beds. The latter correlates with the Brofiscin Oolite, which occurs towards the middle part of the Black Rock Limestone in other parts of South Wales. Detailed diagenetic studies of the Brofiscin Oolite indicate that the top of the shoal was emergent prior to the deposition of the overlying Tears Point Limestone. Evidence for this period of emergence (i.e. a palaeokarst surface) was removed as a result of erosion (ravinement) at the base of the transgressive Tears Point Limestone. This transgressive mid-ramp facies is composed of highly bioturbated and impure crinoidal dolomitic limestones.

2c Western cliffs (south). About 80 m higher in the succession (SS 5340 8788) a further shoaling-upwards sequence involves the Langland Dolomite and the lower half (*circa* 24 m) of the Gully (Caswell Bay) Oolite. A prominent set of HCS with an amplitude of 5 cm and a wavelength of 50 cm occurs 4 m above the base of the Gully Oolite; the cycle continuing with oolitic grainstones (~20 m thick) displaying erosional surfaces and sets of cross strata.

2d Eastern cliffs. On the eastern side of the bay (SS 5390 8790) the previously described sequence is overlain by a further regressive cycle (~20 m thick), which has an irregular erosional base and is succeeded by cross-bedded oolitic limestone with stylolitic seams. About 4 m above the base of the cycle a bed with large brachiopods and recrystallized gastropods occurs below another erosional surface. Large-scale sets of cross strata occur towards the top of the oolitic facies and these are capped by a prominent palaeokarst surface, draped by yellow–buff claystones containing rhizocretions, calcretes and scattered limestone blocks. The succeeding Caswell Bay Mudstone (5.5 m thick) is composed of alternating, laminated to homogeneous, micritic limestones, dolomites and softer-weathering lime mudstones.

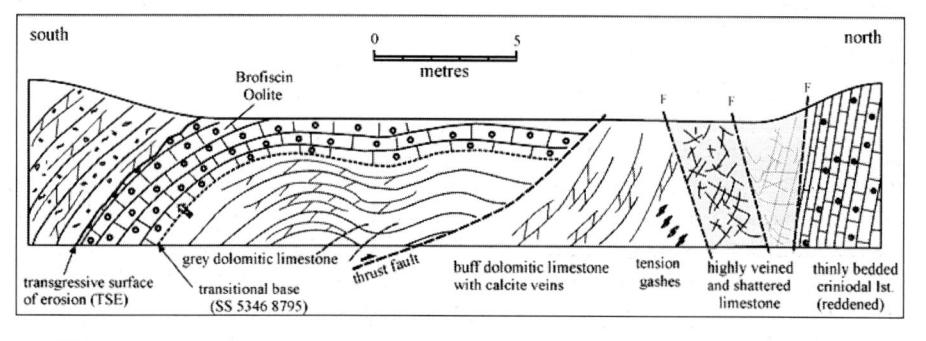

Fig. 4.10 Field sketch illustrating the structure and succession in a small cove in the western cliffs of Three Cliffs Bay. Note that the Brofiscin Oolite occurs in the middle part of the Black Rock Limestone.

These facies are devoid of macrofossils, but they are bioturbated and contain laminated sheets and mounds, some of which are interpreted as algal stromatolites. Desiccation cracks are also common on the tops of some of the minor coarsening- and shallowing-upwards units.

The coarsening-upwards cycles, developed in the grainstone facies of the Brofiscin Oolite and the Caswell Bay Oolite, are interpreted as seaward-prograding (regressive) shoals (Fig. 4.11a). Ooid shoals are most readily formed in warm (tropical), shallow (<2 m depth), current-agitated marine environments, where the water is saturated with dissolved calcium carbonate. Under these conditions thick oolitic sandbodies can be deposited, especially during still-stands or slow falls of sea level. In this setting active progradation results in the build-up of an emergent strand-plain, displaying ridge-and-swale topography and karstic weathering features. The preservation potential of this type of falling-stage lowstand sequence is dependent on the dynamics of the subsequent transgressive event. For example, during a rapid high-energy transgression the karstic surface and a variable amount of the oolitic facies could be removed by erosion (e.g. the Brofiscin cycle); conversely, during a longer duration, lower energy transgression the succession would have greater preservation potential and would be overlain by back-barrier peritidal deposits (e.g. the upper Gully Oolite cycle). This depositional model is discussed further when the successions at Caswell Bay are examined.

2e Pennard Pill. A variety of interesting recent sedimentary environments and related geomorphological features occur within the confines of Pennard Pill and Three Cliffs Bay (Fig. 4.12). Two relatively large meanders occur along the course of the small river that flows through Pennard Pill (Fig. 4.8a, b). The up stream-located meander loop has been, and still is migrating to the south. Looking down from the crags below Pennard Castle, it is possible to see a former course of the river, which cuts across the neck of the meander. Evidence for the lateral migration

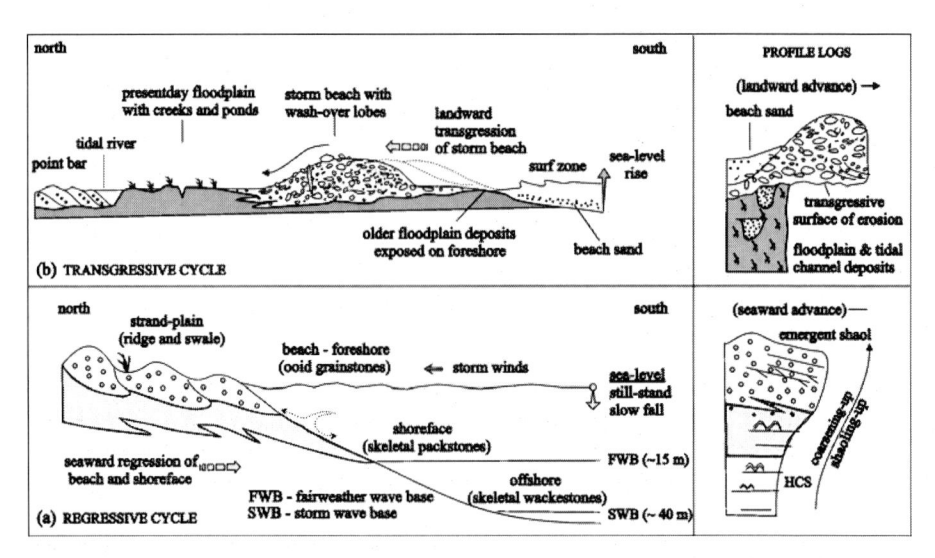

Fig. 4.11 (a) Depositional model for regressive oolite shoals such as the Brofiscin Oolite and Gully Oolite (Caswell Bay Oolite). (b) Depositional model to illustrate the formation of the present-day transgressive storm beach at Three Cliffs Bay.

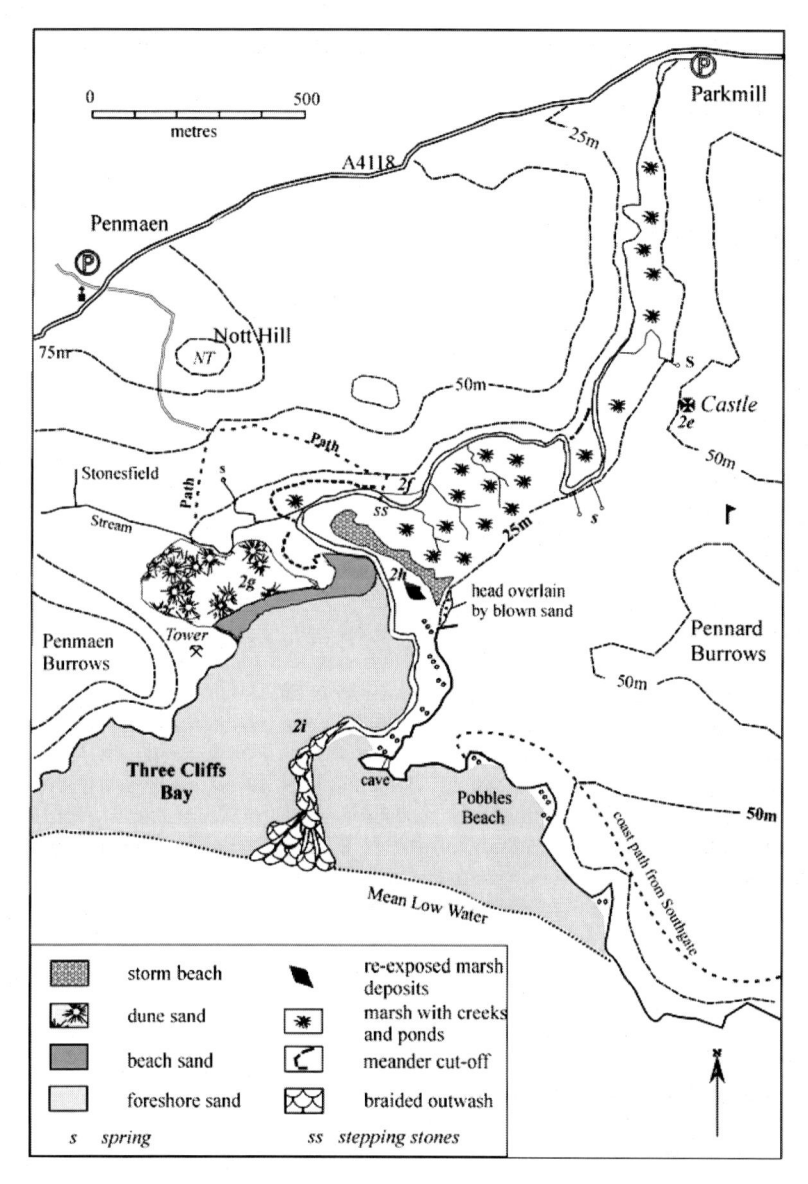

Fig. 4.12 Geomorphology and locality map for the recent sediments at Three Cliffs Bay and Pennard Pill.

of the meander can sometimes be detected by colour variations in the vegetation. The tight bend in the meander, adjacent to the boardwalk, usually displays undercutting and bank-collapse features. Thin patches of blown sand extend onto the marshy floodplain, which is dotted with small ponds and drained by creeks.

2f Stepping-stones. Just up stream of the stepping-stones the tidal river has deposited a sandbar on the convex bank of another meander (Figs 4.8a, 4.12). This sandbar usually displays large linguoid ripples with multiple crests that pass laterally into sinuous-crested ripples. Both types of ripples give downstream flow directions.

A shallow stream, which enters the pill 200 m to the west of the stepping-stones, is a good place to study the mode of formation of linguoid ripples and other small-scale bedforms.

2g Dunes. At the mouth of the pill, the tidal river is diverted as a large meander around an area of aeolian dunes, which are up to 17 m thick and are being stabilized by marram grass, attractive blue-grey sea holly, yellow flowering evening primrose, and many other plants adapted to colonize mobile dunes. Wind ripples can usually be observed on the bare sand, and occasional blowouts reveal remnants of bedding in the well-sorted, fine-grain sands. Waves have reworked the seaward side of the drift to form a relatively straight margin with the beach, while the landward side of the dunes has been incised as a result of the lateral migration of the river meander.

2h Storm beach. A well-defined cuspate storm beach, composed of cobbles and boulders of limestone, occurs inside the previously mentioned meander. It reaches a height of 3 m above high-water mark and has a maximum width of 40 m. On the seaward side of the storm beach a patch of organically rich saltmarsh deposit is being re-exposed as a result of present-day sea-level rise and erosion (Fig. 4.8c). During the spring of 2005 an area of 20 m² was exposed, the top surface being ramified by *in situ* roots and matted stem debris. Internally the deposit was composed of thin alternations of grey organic rich sediment, buff sands and clayey sands. The vertical and lateral facies relationships seen here are important for the interpretation of sequences deposited during transgressive events (Fig. 4.11b).

2i Foreshore. Seawards of the three limestone pinnacles the river subdivides to form a shallow braided delta that can be seen at low tide. This part of the river is excellent for studying the formation of a variety of sandy bedforms such as current ripples and other higher-velocity structures related to standing waves and anti-dunes.

4.3.3 Caswell Bay

Travel to Caswell Bay on the B4593; park at the bottom of the hill at Bishop's Wood, adjacent to the beach (Fig. 4.13). Bishop's Wood is a nature reserve with a visitor centre nearby. The itinerary takes about 3 hours to complete and requires a low or receding tide, although localities *3a* and *3b* can be examined at high tide.

3a Western cliffs. From the car park cross the road and walk down the ramp, to the left of the beach cafés, to a concrete platform behind the building used for storing deckchairs. The low cliffs at this locality (SS 5925 8765) expose a section from the top of the Gully (Caswell Bay) Oolite to the base of the High Tor Limestone. The palaeokarst, at the top of the oolitic packstone, displays many of the characteristic features of this key surface, including pot-holes filled with a buff–red clay crust, isolated calcretes and angular clasts, and also vague *in situ* root traces and fractures with iron-oxide rims. The overlying Caswell Bay Mudstone consists of alternating thin limestones, dolomites and variegated claystones. Many of the limestone beds have nodular, sharp or erosional bases and they fine upwards into algal-laminated facies. Some of the claystones are completely homogenized as a result of bioturbation by roots and invertebrates. A very hard porcellaneous limestone, marking the top of the sequence, is truncated by a prominent erosional surface at the base of the overlying coarse-grain crinoidal limestone (High Tor Limestone). This surface has an erosional relief of 25 cm and it contains limestone intraclasts up to 15 cm in diameter. A small *Syringopora* colony and cross sections of caninids occur in the overlying thick-bedded limestones. Looking to the east across the beach the Caswell Bay Mudstone is seen to crop out in the cliffs just below the Surfside café, its anomalous position being caused by a sinistral displacement (~40 m) along the Caswell Valley Fault (Fig. 4.13). The mudstone succession on the east side of the bay is identical to that described previously.

3b Eastern cliffs. Walk 175 m to the southeast to a well-known locality where the Caswell Thrust intersects the cliffs. Here the Langland Dolomite has been pushed up and over a syncline defined by the Caswell Bay Mudstone (Fig. 4.14). The thrust dips 50° northwards, and the dolomite beds, towards the top of the cliff, display fault drag. At the northeast termination of the syncline the palaeokarst surface, defining the base of the Caswell Bay Mudstone, displays irregular tongue-shape pot-holes up to 45 cm deep, which are filled by variegated clays containing scattered angular to rounded limestone clasts (Figs 4.15, 4.16). Within the weathered zone below the pot-holes the host limestone is discoloured and displays nodular patches and small haematite-filled veins. Walk 50 m along the lip of the syncline to a small indentation in the platform, where the basal deposit is up to 75 cm thick and contains a disorganized array of rounded to angular pebbles and boulders, cemented by greenish-yellow clay. The clay matrix has been forcibly injected into some of the limestone fragments. At this locality the Caswell Bay Mudstone is 6.5 m thick, the increase in thickness from localities *3a* and *3b* being attributable to younger beds

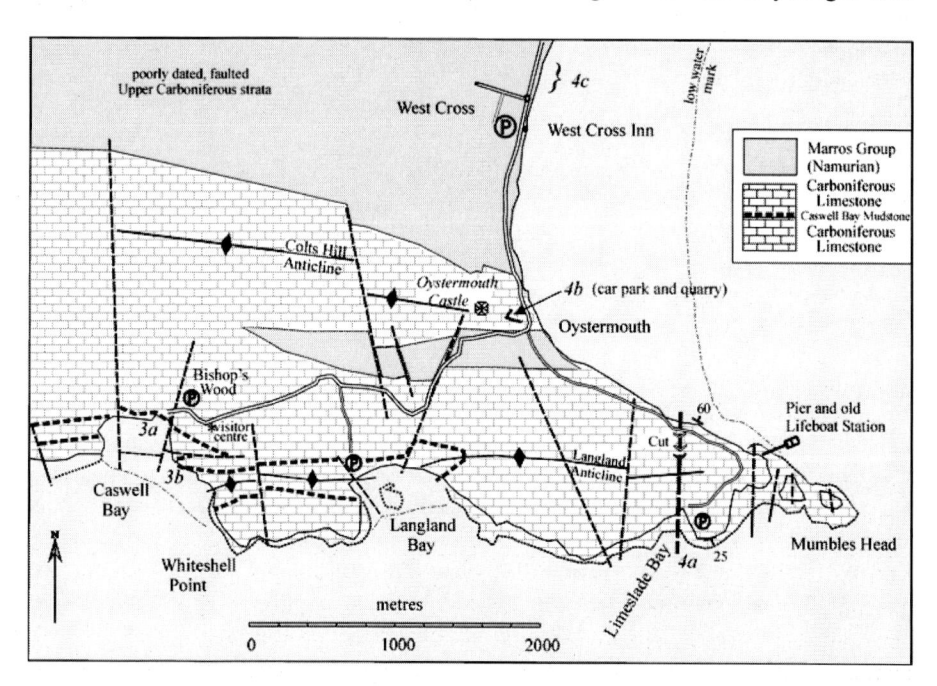

Fig. 4.13 Geology and location map for the Caswell Bay and Mumbles (adapted from George 1940).

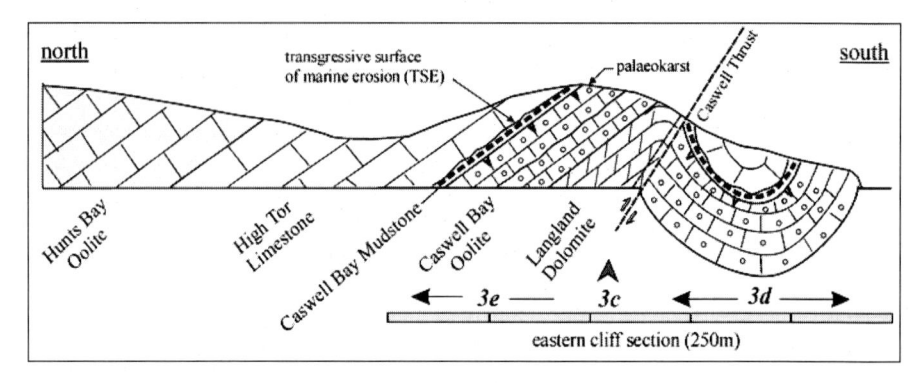

Fig. 4.14 Cross section illustrating the stratigraphy and structure of the Carboniferous Limestone across the eastern margin of Caswell Bay (adapted from George 1940).

Fig. 4.15. View of the type locality of the Caswell Bay Mudstone (CBM) showing its stratigraphical relationships with the underlying Caswell Bay Oolite (CBO) and overlying High Tor Limestone (HTL). The base of the latter limestone is a transgressive surface of marine erosion (TSE) or ravinement surface.

(porcellaneous limestone and calcareous mudstone) being preserved below the erosional base of the High Tor Limestone. These younger facies display small synsedimentary faults and some evidence of contemporaneous slumping and minor folding. The erosional surface at the base of the High Tor Limestone is marked by a lag composed of shell debris and intraclasts, and it has north–south orientated scours and small gutter casts. Occasional vertical lined burrows extend across the erosional surface. Return to the succession located to the north of the Caswell Thrust and make a traverse through the Caswell Bay Oolite, which displays many of the classic features of grainstone shoals. Working up the succession the oolitic facies display an overall increase in bed thickness, with large cross sets appearing towards the top of the first cycle. The second cycle begins with two thin units of erosional storm

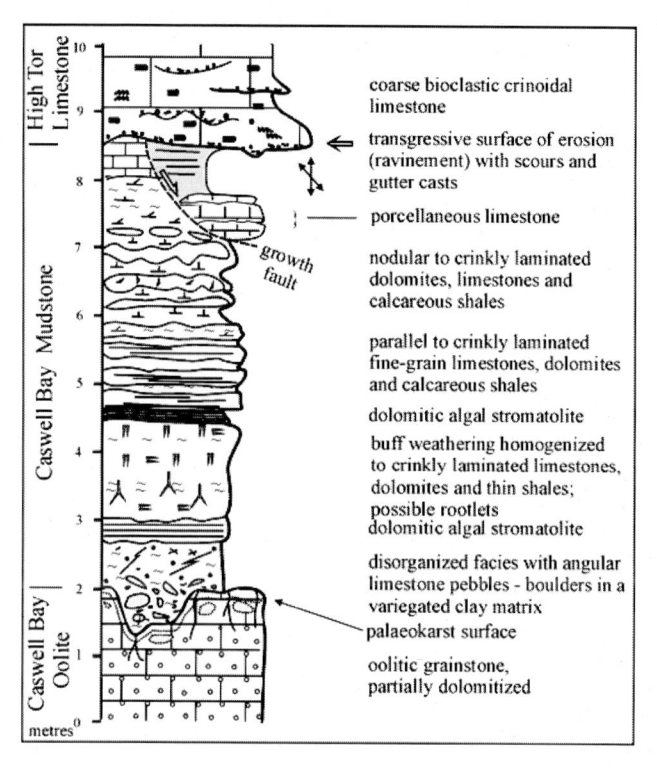

Fig. 4.16 Graphic log of the Caswell Bay Mudstone showing the sequence stratigraphy interpretation of the key surfaces.

beds displaying coarse-grain ripples and small- to large-scale hummocky bedforms, which are succeeded by a thickening-upwards cross-stratified facies sequence.

The succession described above illustrates the control that sea-level changes exerted on depositional processes. In warm tropical seas ooids would be generated very rapidly in high-energy upper shoreface and beach environments, resulting in the seaward progradation of a shoal. Further progradation would be promoted during periods of sea-level still-stand or slow falls, leading to the preservation of extensive beach-ridge strand plains. These abandoned beach ridges would be subjected to sub-aerial weathering, leading to the processes of karstification and pedogenesis. The partial dolomitization of the oolitic facies could also have occurred at this time as a result of the mixing of fresh meteoric and marine pore waters. The very mature palaeokarst surface at the top of the Caswell Bay Oolite, records an extended-period of sub-aerial weathering, which occurred during a significant relative fall in sea level (Fig. 4.17a). In terms of sequence stratigraphy, this major palaeokarst surface would equate with the lowstand sequence boundary. A slow rise in sea level is recorded by the overlying Caswell Bay Mudstone, which was deposited in a low-energy lagoon to emergent tidal flat environment. As wave activity and the rate of sea-level rise increased, the barrier bar that protected the lagoon transgressed landwards. Erosion in the high-energy surf zone resulted in the removal of the contemporaneous barrier deposits and the upper part of the earlier lagoonal deposits (Fig. 4.17b). Storm-surge ebb currents transported the coarser products of this erosional phase offshore to form a coarse-grain lag, located

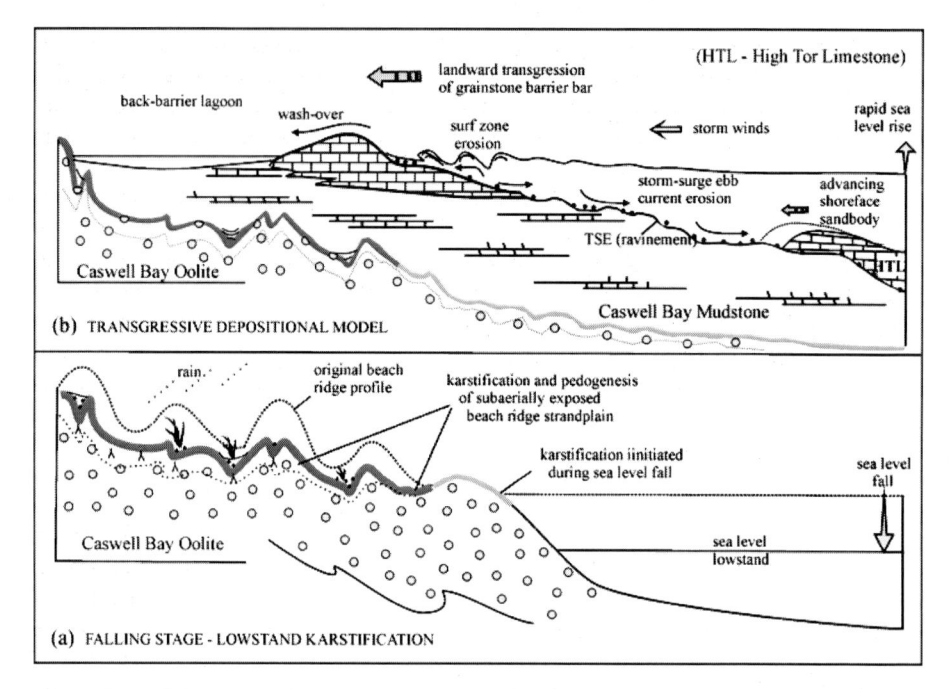

Fig. 4.17 (a) Schematic model illustrating the formation of the palaeokarst surface at the top of the Caswell Bay Oolite. (b) Transgressive deposition model illustrating the formation of the Caswell Bay Mudstone and the overlying High Tor Limestone.

above the ravinement surface. A further sea-level rise resulted in the ravinement surface being preserved below the offshore carbonate sandbody (High Tor Limestone). These depositional sequences record a significant period of slow sea-level fall, succeeded by a more rapid rise. This type of asymmetrical sea-level signature is typical of the depth changes caused by glaciations (slow falls) and interglacial melt-outs (rapid rises).

4.3.4 The Mumbles and Swansea Bay

The Mumbles is the name given to the southeastern peninsula of Swansea Bay, extending from Oystermouth around Mumbles Head to Limeslade Bay. Today it is a popular holiday destination with a variety of water sports and entertainments, small hotels and guesthouses, many good restaurants, bistros and public houses. The original village was an important oyster fishing centre, which flourished until the 1920s. There are good views of Mumbles and Swansea Bay from the Norman keep of Oystermouth Castle. The low-lying ground located to the south of the castle is composed of Namurian shales preserved along the axis of an east–west syncline, whereas the remainder of the peninsula is composed of folded and faulted Carboniferous Limestone (Fig. 4.13).

4a Limeslade Bay. Park at the western end of the large car park at Bracelet Bay. Mumbles Head, to the east, is composed of limestone belonging to the Hunts Bay Oolite that forms the axis of the fault-dissected Langland Anticline. From the car park walk 300 m east to Limeslade Bay; take the steps down to the concrete platform and continue to the low limestone cliff (Oxwich Head Limestone) 20 m to the east. Here a north–south-trending fault zone (2–4 m wide) is filled with concentrically zoned crystalline calcite and blood-red haematite (Fig. 4.18). The haematite occurs as a replacement of the concentric growths of calcite and also as more irregular pods and stringers. The iron oxides within this fracture zone, which extends from Limeslade Bay across the Mumbles Hill nature reserve, have been worked periodically from Roman times until the end of the 19th century. Evidence of past mining can be seen in the walled-up adit on the beach and another behind the shack on the road. Workings along the fault on the northern flank of the hill have created a 15 m-wide gully known as the Cut. This feature (now overgrown) is best seen from the Knab Rock car park near the slip road to Mumbles pier (SS 6258 8761). The mineralization is considered to be the result of the leaching of iron from a previous cover of Triassic deposits, followed by its precipitation (secondary enrichment) in the fault cavity, and by the replacement of existing vein calcite.

4b Clements' Quarry. The abandoned Clements' Quarry (SS 6145 8834), now the long-term car-park for Oystermouth and Mumbles, is located 200 m to the east of Oystermouth Castle and 100 m north of the junction of the B4593 and the A4067 road from Swansea. Although the quarry is partly overgrown, it is still one of the best outcrops of the Oystermouth Formation (previously the Upper Limestone Shales) in the area. At this locality the facies comprise interbedded impure limestones (20–50 cm thick) and calcareous shales (5–15 cm thick), containing a fauna of brachiopods, including *Spirifera oystermouthensis, Martinia, Schellwienella* and *Eomarginifera,* and rare corals and trilobite fragments. The nose of the Colts Hill Anticline (Fig. 4.13) is exposed behind the electricity terminal. An information board in the car park gives details (compiled by Ron Austin) of the Mumbles Marble, which was quarried locally from this stratigraphical horizon.

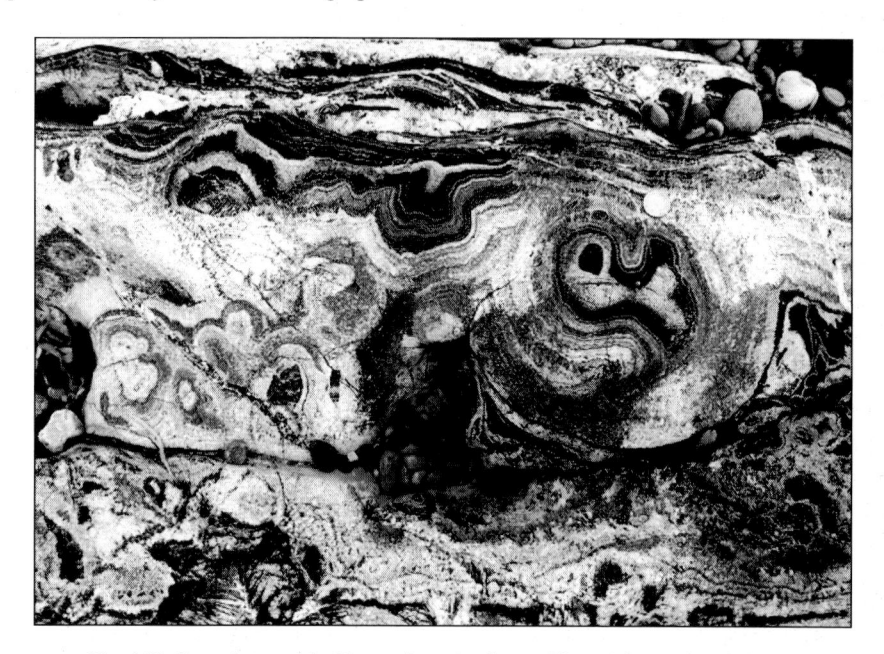

Fig. 4.18 Secondary enriched haematite ore and crystalline calcite at Limeslade Bay

The increase in the amount of argillaceous facies in the Oystermouth Formation marks the final demise of the generally clear-water carbonate ramp that had existed throughout the Dinantian. This influx of terrigenous mud marked the onset of the clastic sedimentation that characterizes the Upper Carboniferous succession. In southern Gower the Dinantian passes up transitionally into the Namurian, the passage beds being marked by black shales with radiolarian cherts. This transition has been documented in a poorly exposed stream section at Barland Common, where the Bishopston Mudstone Formation is composed of about 350 m of black shales with goniatite-bearing marine bands and one thin band of distal turbidites. This deepwater (basinal) succession records a significant increase in water depth from the Dinantian carbonate ramp setting. During later Namurian times the water depth gradually decreased as coarser sediments were brought into the basin. Evidence for this shallowing is seen in the late Namurian strata, which outcrop along the coast at West Cross.

4c West Cross. From Oystermouth travel 1 km northwards on the A4067 to West Cross; turn left at the mini-roundabout, left after the parade of shops and left again to reach a small car park located directly opposite the West Cross Inn. Cross the A4067 and walk 100 m northwards along the cycle path to low-lying outcrops at the top of the beach. The first rocks exposed comprise interbedded shales, siltstones and very fine sandstones, which form a prominent asymmetrical east–west trending anticline. Fossils are rare, although *Gastrioceras cumbriense* and *Lingula* have been recorded. Farther northwards along the foreshore four thin coarsening-upwards cycles are seen in intermittent exposures (Fig. 4.19). The tops of the sandstone members are often bioturbated and contain rootlets. After another gap in the succession, a mature silica-rich palaeosol (ganister) with *in situ* stumps of club mosses (*Stigmaria*), and a thin coaly-shale are exposed. A sharp-based fine-grain sandstone with hummocky cross sets and small gutter casts overlies the palaeosol.

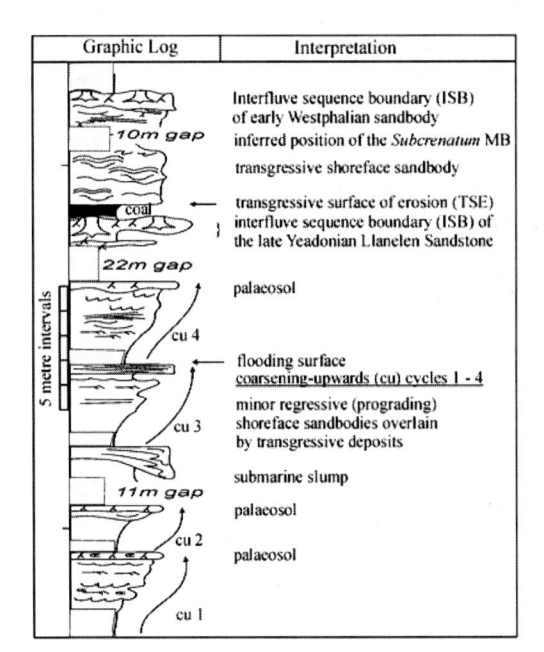

Fig. 4.19 Log and interpretation of the Namurian strata partly exposed on the foreshore at West Cross (modified from George 2001).

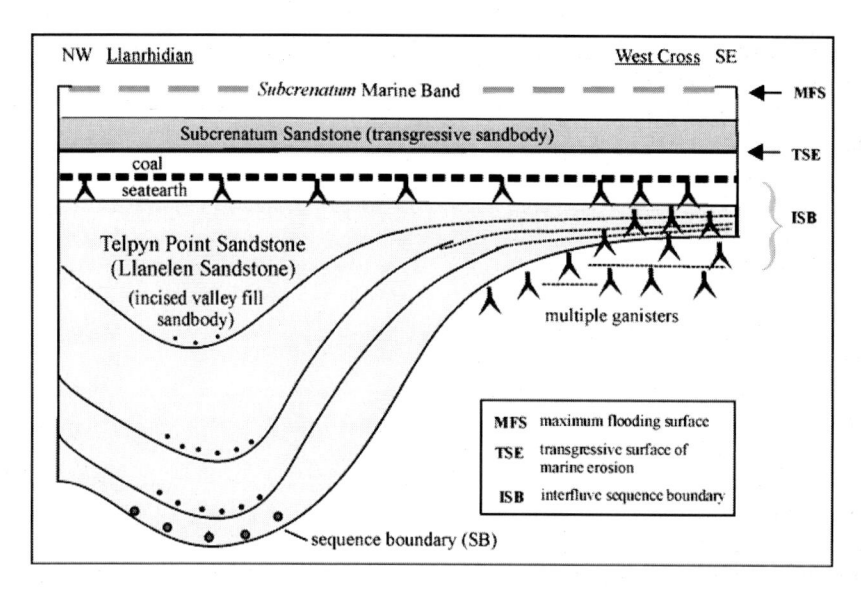

Fig. 4.20 Conceptual sequence stratigraphy model for the late Namurian strata developed between West Cross and Llanrhidian (modified from George 2001).

The interpretation of these interesting strata is speculative in the absence of stratigraphical control. The coarsening-upwards sandstones, capped by palaeosols, appear to correlate with a thick multi-storey sandbody (Llanelen Sandstone), mapped to the north and northeast of Llanrhidian (Figs 4.1a, 4.2), which has been interpreted as an incised valley fill (George 2001). In this scenario, the West Cross cycles could represent the interfluve facies genetically related to the incised valley (Fig. 4.20). The sharp-based sandstone unit with HCS is interpreted as a transgressive sandbody, and it is tentatively correlated with the Subcrenatum Sandstone seen in the headwaters of the Neath and Swansea valleys (see Frontispiece A). If this correlation is correct, the *Gastrioceras* s*ubcrenatum* Marine Band should overlie the sandstone, but unfortunately the overlying shales are not exposed. The regional implications of this interpretation are addressed further in the Chapter 5 itineraries.

Chapter 5

Headwaters of the River Neath and the River Tawe

5.1 Introduction

The headwater regions of both the River Neath and the River Tawe are located within the southern margin of the Brecon Beacons National Park, in an area noted for its beautiful wooded valleys, impressive waterfalls and large cave systems (Dan-yr-Ogof, Porth-yr-Ogof and Ogof Fynnon Ddu). This region is easily reached from the M4 (junction 45) at Neath; then continuing on the A465 (T) (Heads of the Valleys road) to Glynneath. From the termination of the M50 at Ross-on Wye, you can travel on the scenic route via the A40 (T) to Abergavenny and then westwards on the A465 (T) to Glynneath. Hotels are available in Neath, Swansea and Merthyr Tydfil, and numerous inns also provide inexpensive accommodation. There is a Youth Hostel at Ystradfellte, and many guesthouses and farmhouses provide bed & breakfast.

The Fforest Fawr UNESCO Geopark has a wealth of attractions, including the Welsh National Showcaves (Dan yr Ogof), the Black Mountain, Pen y Fan, Carreg Cennen Castle and the Craig-y-nos Country Park. Other, nearby, geology-related attractions include the Big Pit and Ironworks at the Blaenavon World Heritage Site. The Big Pit is a very popular mining museum, which gives visitors an authentic underground tour of the coalmine. Other sites of interest are the Aberdulais Falls (National Trust), which is mainly an industrial archaeology site, but also has interesting outcrops of the Pennant Sandstone, and the nearby Melincourt Falls (Glamorgan Wildlife Trust).

5.2 Geological history

The area is located along the northern rim (North Crop) of the South Wales Coalfield, and the itineraries consider the strata from the top of the Carboniferous Limestone (Viséan), through the Marros Group (Namurian) to the base of the Lower Coal Measures (Westphalian A) (Fig. 5.1, Appendix 1). The lithostratigraphy, biostratigraphy and sequence stratigraphy of the successions are summarized in Figure. 5.2 (see also Howells 2007, Fig. 42 N.B. adapted from George 2000).

The age and lithology of the uppermost Carboniferous Limestone strata varies across the area as a result of erosion at the base of the Namurian, the amount of which increases to the east. Thus, at Penwyllt, the base of the Namurian rests on the lowermost beds of the Oystermouth Formation (Upper Limestone Shales), whereas farther east in the Pontneddfechan area it rests on the older Oxwich Head Limestone.

The succeeding Namurian strata, recently designated the Marros Group, has been subdivided into the Twrch Sandstone Formation (formerly the Basal Grit Group) and the Bishopston Mudstone Formation (formerly the Shale Group). The Twrch Sandstone is composed of pebbly to fine-grain quartz arenites and interbedded argillaceous facies, whereas the succeeding Bishopston Mudstone is dominated by argillaceous facies containing minor coarsening-upwards or fining-upwards sandstone bodies. The above facies were deposited in mixed deltaic, marginal marine and shelf environments. Goniatite-bearing marine bands are extremely important for the biostratigraphical subdivision and correlation of Namurian strata, but unfortunately, many of those in the Twrch Sandstone are absent as a result of contemporaneous erosion, which occurred during falls in sea level. Within the Bishopston Mudstone Group, six informally named sandbodies (allostratigraphical units) have proved important for lithostratigraphical correlation and sequence stratigraphy studies (Fig. 5.2). The base of the Westphalian is marked by the *Gastrioceras subcrenatum* Marine Band, which is overlain by a thick incised fluvial channel sandstone known as the Farewell Rock.

The Namurian and Westphalian rocks of the South Wales Coalfield were deposited in a peripheral foreland basin, which evolved during the Variscan Orogeny. Consequently, sedimentation, especially during Namurian times, was strongly influenced by tectonic structures activated during the northwards propagation of the Variscan front (see Kelling 1988; Leveridge & Hartley 2006). In the field study area the most important of these structures were the northeastwards-trending Swansea Valley and Vale of Neath disturbances and a group of northwestwards-trending faults developed between Glynneath and Ebbw Vale (Fig. 5.1).

The area is covered by the Explorer (1:25 000) map for the Brecon Beacons National Park (165); the geology of the district is summarized in Barclay *et al.* (1988) and the accompanying geological map (Sheet 231). Jones & Owen (1957)

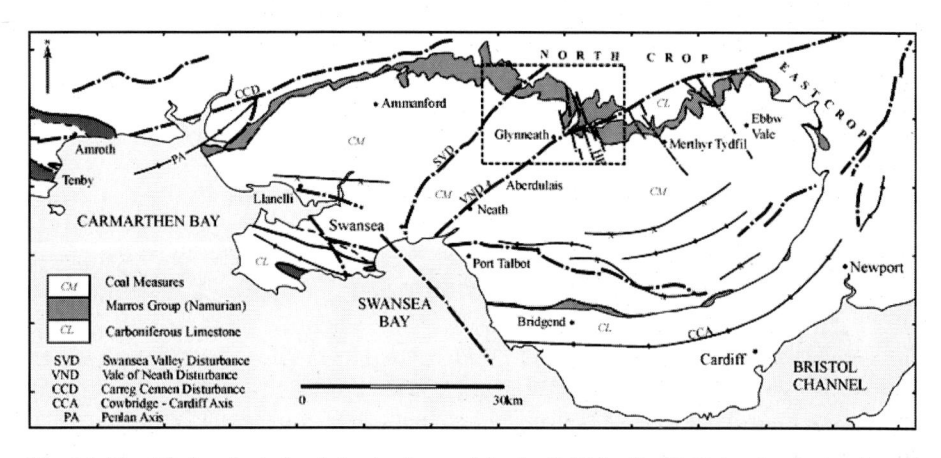

Fig. 5.1 Simplified geological and structural map of the South Wales Coalfield showing the location of the headwater region of the River Neath and River Tawe.

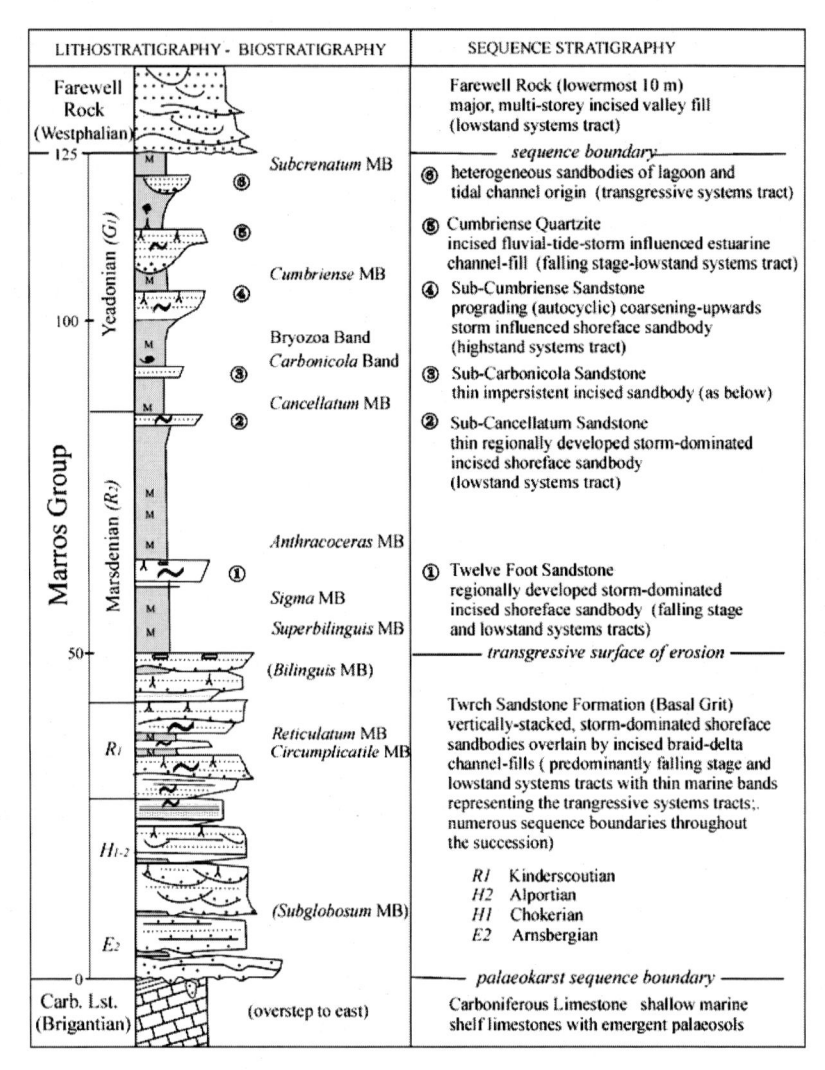

Fig. 5.2 Lithostratigraphy and sequence stratigraphy of the Marros Group (Namurian) exposed in the upper Neath and Swansea valleys (modified from George 2001).

accomplished much of the early research on the palaeontology and biostratigraphy of the Carboniferous strata; field guides of the area include those by Kelling (1971), Owen (1971), Thomas (1971) and Davies *et al.* (1978).

5.3 Some aspects of sequence stratigraphy

The main aims and objectives of the itineraries included in this Chapter are to introduce some of the sequence-stratigraphy concepts, which can be used to describe and interpret sedimentary successions. This approach is particularly suited to the study of the Namurian and lowermost Westphalian successions of South Wales because they were deposited when relative sea level was being strongly influenced by repeated glaciations and deglaciations on the southern continent of Gondwana. Furthermore, processes such as basin subsidence and uplift, syn-sedimentary faulting, differential sedimentation and the compaction of argillaceous sediments, were also exerting important, if more localized, controls on relative sea level. Distinguishing between the global and basinal processes responsible for sea-level change is often difficult and can only be resolved by detailed regional studies, which are beyond the scope of this book. Fortunately, the evidence for sea-level changes on all scales can be analysed by reference to concepts developed for the relatively new discipline of sequence stratigraphy. Some of these concepts are discussed below; more detailed accounts of the sequence stratigraphy of the Namurian strata of the area have been published by Hampson (1998) and George (2001, 2002).

5.3.1 Eustatic sea-level changes

Global changes in sea level, occurring on a range of time scales, have been recorded throughout the stratigraphical record. Long-duration changes, measured on a timescale of hundreds of millions of years, are considered to be the result of variations in the volume and spreading rates of the mid-ocean ridge systems. In this scenario, highstands of global sea level correspond to episodes of continental fragmentation and drifting, when the volumes of buoyant oceanic ridges were at their maximum, resulting in the displacement of ocean water onto the continents. Lowstands, on the other hand, equate to periods when the continents united to form supercontinents. For example, during the Palaeozoic Era (*circa* 291 million years duration), sea level rose, rather erratically, from the Cambrian to reach a peak in the late Ordovician, which was a time when the continents were dispersed and spreading rates were presumably relatively high. At this time the Iapetus Ocean, separating southern Britain from Scotland and northwestern Ireland, was at its widest extent (see Fig. 2.4a). Thereafter, sea level fell slowly to a lowstand in the Permo-Triassic, which corresponded with the consolidation of the supercontinent Pangaea (see Fig. 2.4c). The conflicting position of the continental redbed facies of the ORS within this sea-level curve was a result of the Caledonian Orogeny, which elevated Wales and northern Britain above the level of the Devonian sea.

Many of the successions described in this field guide were deposited during much shorter duration sea-level changes. For example the Carboniferous sequences of South Wales were deposited during glacio-eustatic sea-level cycles, which are estimated to have had periodicities of 65–100 ka. These cycles were caused by repeated glaciations (slow sea-level falls) and melt-outs (more rapid sea-level rises), which occurred on the southern continent of Gondwana. A good example of a similar glacio-eustatic sea-level cycle, lasting about 120 ka, occurred during the Last Glacial Maximum and the succeeding Flandrian transgression in Britain. During this period the volume of continental ice fluctuated, but reached its maximum extent in the late Devensian (about 20 000 years ago). This caused a major fall in sea level and created a land connection between Wales and Ireland, which was rapidly flooded by the Irish Sea during the succeeding Flandrian transgression (see Fig. 2.17a). It has been estimated that sea level rose at a rate of up to 10 mm per year, but later the rate decreased to 1 mm per year (see Fig. 2.17b). Interestingly, the latest research into global warming and climate change suggests that sea level rises, of 8–10 mm per year, could return before the end of the century.

5.3.2 Sequence stratigraphy terminology and concepts

In sequence stratigraphy, sedimentary successions are divided into unconformity-bound sequences, which were deposited during a single sea-level cycle. These sequences can be subdivided further into four systems tracts, which correspond to the falling, lowstand, transgressive and highstand stages, shown on the sea-level curve in Figure 5.3a. Sedimentary facies deposited within these stages can be recognized in the field by their three-dimensional geometries and a variety of key stratal surfaces; including sequence boundaries (SB), interfluve sequence boundaries (ISB), regressive surfaces of erosion (RSE), transgressive surfaces of marine erosion (TSE), marine-flooding surfaces (FS) and maximum flooding surfaces (MFS) (Fig. 5.3b). Many of these features can be illustrated on a relatively simple sequence-stratigraphy model proposed for a clastic ramp environment (Fig. 5.4).

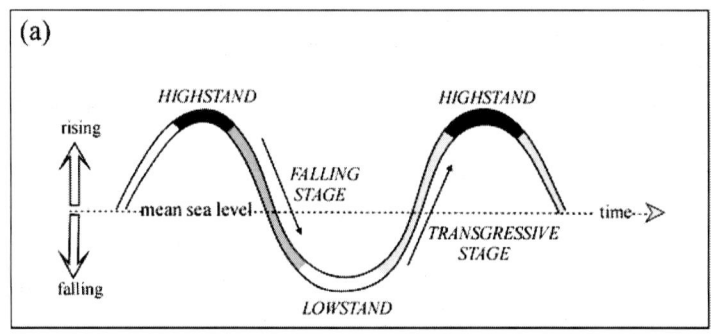

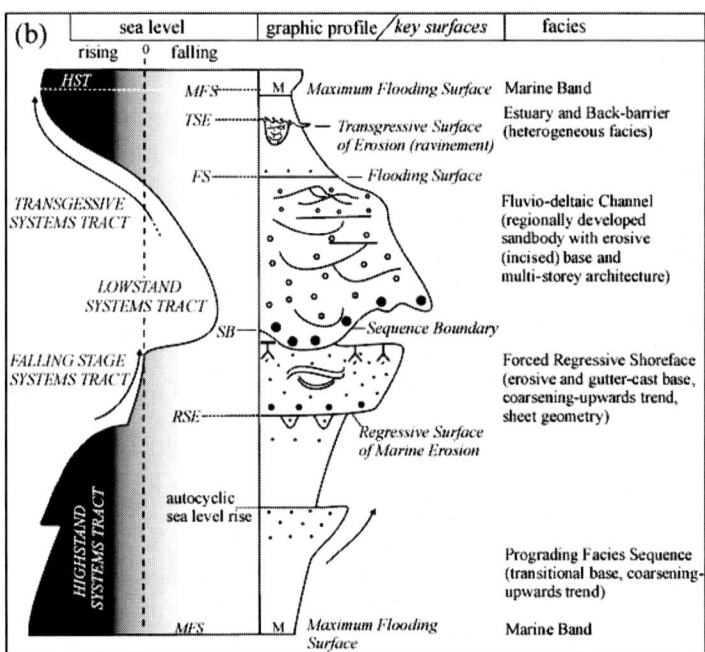

Fig. 5.3 (a) Simplified sea-level curve and stages (b) Generalized log and sequence stratigraphy of a typical Upper Carboniferous succession from South Wales.

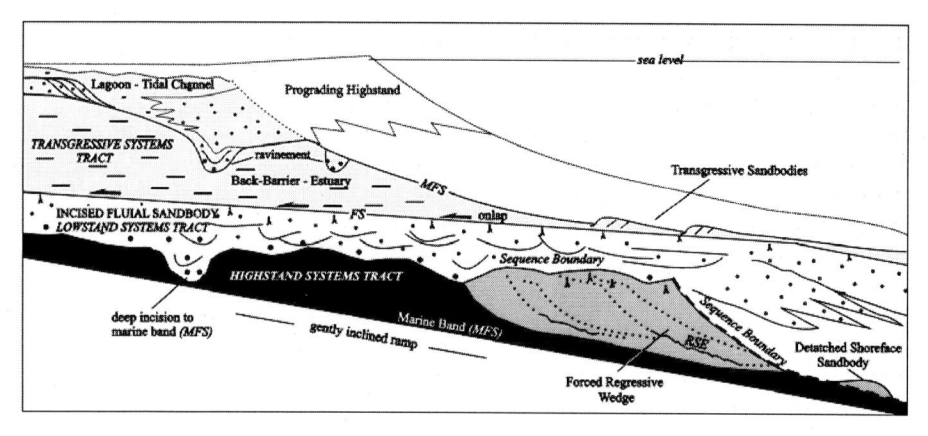

Fig. 5.4 Generalized sequence-stratigraphy model for a clastic ramp environment (modified from Plint & Nummedal, 2000).

The Namurian and Westphalian strata discussed in the following itineraries provide good field examples of falling stage, lowstand stage, transgressive stage and highstand stage sequences. Some of the main attributes of these sequences are summarized below:

- Falling-stage sequences are particularly well represented by forced regressive sandbodies, which have sharp–erosional bases, usually marked by gutter casts, and their component facies display coarsening- and thickening-upwards trends and sets of hummocky and swaley cross stratification (HCS and SCS respectively). These important structures comprise gently undulating and crosscutting laminae that are arranged into convex-upwards hummocks and concave-upwards swales. Such bedforms are believed to have been deposited from combined wave-generated oscillatory currents and seawards-flowing density currents, which are commonly generated on storm-dominated shelf and ramp environments.

- Incised fluvio-deltaic systems that formed during sea-level lowstands are well represented in the Namurian Twrch Sandstone and Cumbriense Quartzite, and in the early Westphalian Farewell Rock. These sandbodies have prominent erosional bases (sequence boundaries), which erode deeply into older deposits and often remove marine bands. On a regional scale they display multi-storey and multi-lateral channel geometries and many also exhibit fining-upwards trends. They are usually capped by palaeosol facies, which merge laterally with more mature and well-drained palaeosols (interfluve sequence boundaries). During major eustatic sea-level falls, fluvio-deltaic systems can erode through emergent shelves to reach the shelf-edge and supply sediment directly to deepwater submarine fans (lowstand fans). This type of incision is believed to have occurred during the late Namurian lowstand, when fluvio-deltaic systems, sourced from the forebulge of the South Wales Basin, supplied large volumes of sediment to lowstand submarine fans in the Culm Basin of Devon and Cornwall (see Fig. 2.13).

- Facies belonging to the transgressive systems tract are characterized by shales, which contain marine fossils (marine bands) and trace-fossil assemblages. These facies sequences, either begin with low energy flooding surfaces, or higher energy transgressive surfaces of erosion, marked by thin bioturbated lags containing quartz granules, reworked fossils and phosphate fragments. Within these marine facies the maximum flooding surfaces occur within condensed beds characterized by high densities of goniatites and other fossils.

- Highstand deposits are characterized by a variety of transitionally-based coarsening-upwards cycles of deltaic and shoreface origin. These parasequences were deposited as a result of autocyclic processes such as delta and shoreface progradation. Their vertical–offset stacking patterns were produced as a result of a combination of basin subsidence and processes such as the lateral migration or the rapid switching of depositional systems (Appendix 5). The compaction of argillaceous facies may also have generated accommodation space for further highstand sedimentation.

The Carboniferous Limestone of South Wales also displays many fine examples of sequences deposited during fluctuating sea levels on a carbonate ramp–shelf environment. In carbonate sequences it is important to remember that the sediments are derived almost entirely from biogenic products formed within in the photic zone of the contemporary marine environment. As a result, sediment production rates are highest during transgressive and highstand stages, lower during falling stages, and virtually zero during lowstands. During lowstands of sea level the shelf was often exposed to the elements, resulting in the formation of karstic surfaces with thin soil profiles. Succeeding transgressive systems tracts are commonly composed of back-barrier lagoonal deposits, the actual barrier bar usually being truncated by a ravinement surface. Falling stage forced regressive deposits are often composed of a large proportion of reworked carbonate sand, which display sets of hummocky and swaley bedforms. Highstand deposits are composed of a variety of bioclastic and oolitic facies that frequently display coarsening-upwards (shoaling-upwards) trends.

5.4   Itineraries

Three itineraries have been selected to illustrate the relationship between sedimentology and sequence stratigraphy in the Carboniferous successions exposed across the northern margin (North Crop) of the South Wales Coalfield Basin. To minimize the amount of lithological description, a summary of Namurian and lower Westphalian facies is given in Appendix 4, and in the text the facies are referred to by their codes. Note that the river and waterfall localities should not be visited during or after heavy rainfall, and that hard hats are necessary for visits to the disused quarries at Penwyllt.

5.4.1 Upper Neath Valley

The small village of Pontneddfechan, located 2 km northeast of Glynneath, is the confluence point of the three tributaries, the Mellte, Nedd Fechan and Sychryd, which form the River Neath. These three river valleys expose classic sections of Carboniferous strata (Carboniferous Limestone, Marros Group and Coal Measures), many of which are located at waterfalls formed where north-northeastwards-trending faults bring quartz-rich sandstones into contact with much softer shales (Fig. 5.5).

1a Porth yr Ogof. From Pontneddfechan take the minor road to Ystradfellte, and about 200 m past the youth hostel (1 km outside the village) turn right down the narrow lane to Cwm Porth car park. Coaches can stop to let passengers alight at the junction and the driver should then continue to Ystradfellte to park. From the car park take the path adjacent to the toilets; cross the stile and follow the rough path down to the River Mellte, where you turn left and continue to the impressive main entrance to Porth yr Ogof (SN 9276 1232). At this point the river enters the cave system, estimated to have about 2.2 km of passages, and re-emerges 250 m down stream. The rockface above the cave entrance is approximately 30 m high and is composed of dark bioclastic facies (lower 5 m) and lighter oolitic facies belonging to the upper part of the Hunts Bay Oolite (formerly the Dowlais Limestone Formation), which also forms the host rock for the caves systems in the Tawe Valley at Dan-yr-Ogof and Ogof Ffynon Ddu. A prominent erosional surface occurs about 10 m above the river bed, and the limestone face is cut by two sets of vertical joints orientated north–south and east–west, many of which have associated calcite veins. After a period of dry weather it is possible to walk into the mouth of the cave to see the dripstone features and white tufa coatings on its roof. Another cave entrance, from which a strong waterflow can be detected, occurs at the northeast termination of the path leading back to the car park.

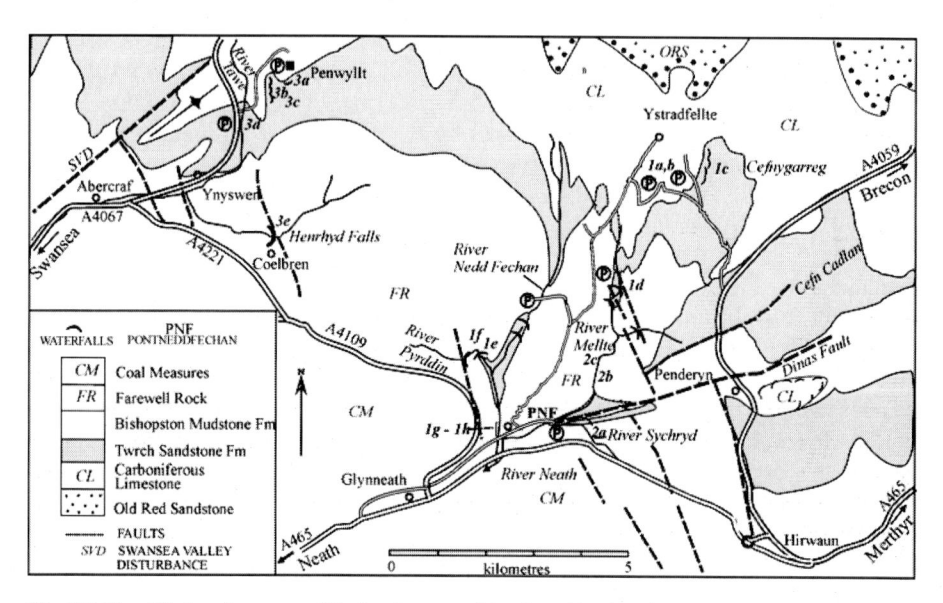

Fig. 5.5 Simplified geology map of the headwaters of the rivers Neath and Tawe showing the localities described in the itineraries.

1b Blue pool. Return to the car park; cross the road and follow the track signposted "Blue Pool and waterfalls", and then turn right down the path marked "access for cavers". This path leads down a dry valley, which displays good karstic features, including clints and grykes, and swallow holes marking the vertical entrances into the cave system. At the narrow resurgence point (SN 9273 1223) the limestone has been worn smooth and displays small pot-hole features. Farther down stream there is a picnic site and the path continues 1.5 km down the left bank of the River Mellte to the Clun Gwyn Falls.

1c Cefnygarreg. From Cwm Porth car park travel to the pretty village of Ystradfellte, which has an inn and good parking facilities. Cefnygarreg is not accessible to large vehicles so coach parties should proceed to Upper Clun Gwyn Falls. Cars and mini-buses should turn right at the inn; cross the river and continue for 1.5 km up the hill and through the forest. At the T-junction turn sharp left and continue northwards for 1 km, through the forest, and park on the side of the road near a stile and information board. Looking to the east the Twrch Sandstone Formation is exposed along the north–south-trending crags of Cefnygarreg. Below the gritstone crags, the underlying Carboniferous Limestone gives rise to a smoother weathering profile, with intermittent outcrops producing a second scarp face. Small caves, with gritstone overhangs, mark the position of the unconformable contact between the two lithologies. At this locality all of the Oystermouth Formation (Upper Limestone Shales) and the upper part of the Oxwich Head Limestone have been removed by erosion at the base of the grits. Some asymmetrical channel features, and well-defined sets of large-scale cross-strata, can be observed in the gritstone crags. Note that the cross sets are inclined to the south (southerly palaeocurrent flow) and that the channels trend at oblique angles to the ridge.

Ascend the slope via the footpath, which runs parallel to the dry-stone wall, to the point where the wall terminates against the base of the crag (SN 940 130). Here the lowermost facies of the grit forms a very impressive channel feature (approximately 50 m wide and over 10 m deep), which cuts down into the concealed limestone. The channel is filled with very pebbly quartz arenite facies, displaying multiple sets of trough cross strata. The sub-angular to well rounded pebbles (max. diameter 10 cm) are composed predominantly of vein quartz, with some darker chert and red jasper fragments.

Proceed to the top of the crags and walk about 100 m northwards to a small cave (SN 941 132), where the unconformable contact between the Carboniferous Limestone and the Twrch Sandstone can be examined. Here the base of the grit displays very irregular, steep-sided minor channels and scours with pebble lags, which cut into the limestone surface. Another pebble lag, about 2 m above the unconformity, defines the base of a further channel storey. Palaeocurrent readings, taken from cross sets on the flanks and axis of the channel, indicate a southwesterly palaeoflow direction. Return to the top of the crags and walk a further 100 m to the north, where a large channel (30 m wide by 12 m deep) is exposed. At the northern end of the outcrop it is possible to follow a rough path along the surface of unconformity, which has an erosional relief of about 7 m and is marked by a basal conglomerate containing large lenses and weathered boulders of limestone enclosed by pebbly grits. Here the grits are composed of a variety of vein-quartz, chert, silicifed limestone and limestone debris, and they display irregular erosional scours, disorganized bedding and crude cross sets.

The Twrch Sandstone at this locality is interpreted as a proximal multi-storey and multi-lateral fluvial channel complex, which formed a southwesterly-flowing supply system for a more distal braided fan delta (Fig. 5.6a, c; see also Fig. 2.12). In terms of sequence stratigraphy, the unconformity at the base of the grits represents a major sequence boundary, caused by a combination of sea-level fall, tectonic uplift and karstic weathering on the northern margin of the basin. The quartz arenite facies were deposited during a period of time when high sedimentation supply rates and slow basin subsidence enhanced the effects of sea-level falls, resulting in the deposition of vertically stacked lowstand channels (lowstand sequence sets).

1d Clun Gwyn Falls. Return to Ystradfellte, travel 1 km towards Pontneddfechan and park in the layby at SN 9180 1080, from which point a gravel track leads to the Clun Gwyn Falls. About 500 m along the track take the right-hand footpath, which leads to a viewpoint above the lower falls (SN 9239 1066). The upper and lower falls both occur where northeastwards-trending faults upthrow resistant quartz arenites (Twrch Sandstone) against softer shales developed below the Twelve Foot Sandstone (Fig. 5.2). Looking due south the impressive gorge, eroded by the River Mellte, is over 70 m high and cuts through the lower half of the Namurian succession. The lowermost 55 m of the gorge, including the falls, is composed of about seven quartz arenite units separated by more thinly bedded siltstones and soft black shales. A rather precarious path, above the river, follows the thickest of these shale units. The quartz arenite units display sheet geometries with occasional evidence for erosional truncations, cross bedding and thinning-upwards trends. These features, and the finer grain size of the facies, suggest that they were deposited in a more distal fan-delta setting than the grits seen at the previous locality (Fig. 5.6a, b). The top 15 m of the cliff is composed of overgrown shales and a poorly exposed sandbody (Twelve Foot Sandstone).

Return to the main path, which descends to the upper falls, where the top 10 m of the Twrch Sandstone is exposed (Fig. 5.7a). Here the uppermost surface of the falls is defined by a highly siliceous undulating bedding plane, containing *in situ* bleached stumps and roots of *Stigmaria*, detached log casts and a thin plant-bearing carbonaceous shale (10–20 cm). This palaeosol marks the top of a 2-m thick coarsening- and thickening-upwards shoreface sandbody. The carbonaceous shale is truncated by a thinner (1 m) coarsening- and thickening-upwards unit, which is capped by a very poorly sorted bioturbated lag (18 cm thick), containing angular to sub-rounded pebbles and granules of quartz. This unit is interpreted as a minor transgressive sandbody, deposited during a relative rise in sea level.

Down stream, the lower part of the Bishopston Mudstone Formation is downfaulted into the succession, where it consists of about 8 m of black shales with thin calcareous siltstone beds displaying hummocky cross stratification (HCS). One of these units (2 m above the base) is 35 cm thick and displays small basal gutter casts, swaley cross laminations and a hummocky top. This predominantly shale sequence contains both the *Superbilinguis* (maximum flooding surface) and the *Sigma* Marine Bands, but neither are accessible. The overlying Twelve Foot Sandstone can be observed high in the cliffs farther down stream, and also on the rough path, adjacent to the falls, on the right bank of the river. Here the sandbody displays its very characteristic two-tier architecture. The lower tier (75 cm thick), which has basal gutter casts and displays a thinning- and fining-upwards profile (facies SF 2→1), is interpreted as a minor incised shoreface sand, the development

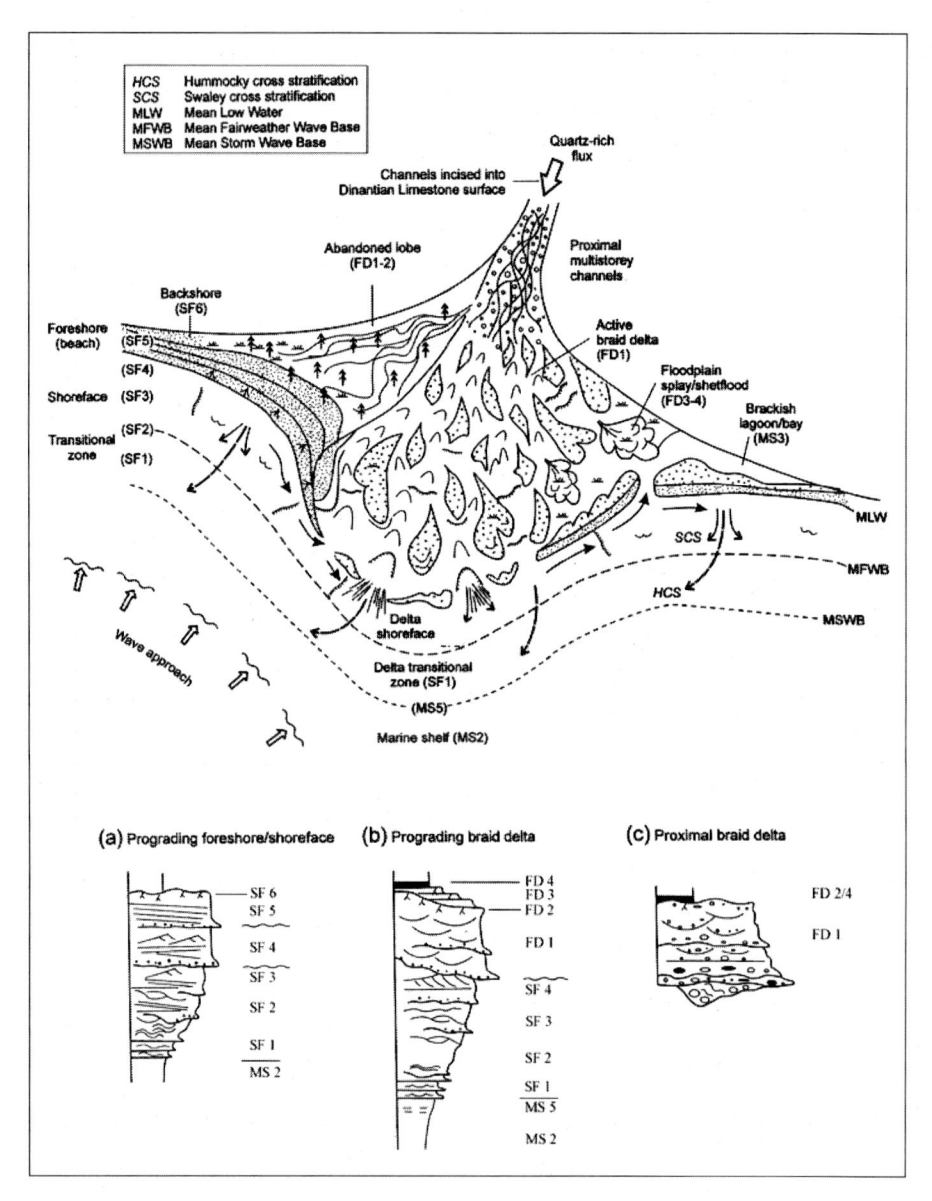

Fig. 5.6 Depositional model for the Twrch Sandstone Formation (Basal Grit) of the North Crop (see Appendix 4 for facies codes; modified from George 2000).

of which was interrupted by a minor rise in sea level. The upper tier has a pebbly base, with very pronounced flutes and gutter casts that are overlain by 4.25 m of erosively bedded, very coarse-grain and poorly sorted arenites (facies FD 1; Fig. 5.8d). This sandbody is interpreted as an incised braided channel, the erosional base of which records a significant basinward shift in facies and is therefore designated a sequence boundary (Fig. 5.4).

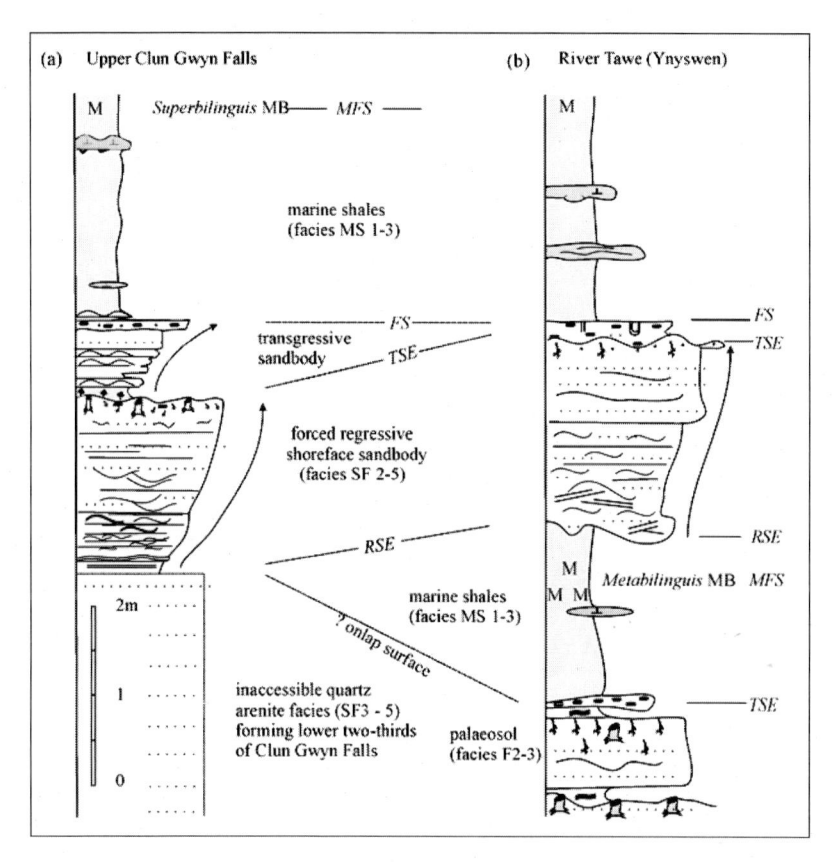

Fig. 5.7 Sedimentology and sequence stratigraphy summaries of the facies exposed at the top of the Twrch Sandstone; **(a)** the upper Clun Gwyn Falls, **(b)** the River Tawe.

1e Sgwd Gwladys. Travel back to Pontneddfechan, where parking is available adjacent to the Angel Inn or near the post office. Take the footpath up the right bank of the River Nedd Fechan to its confluence with the River Pyrddin (a 2.5 km walk). Cross the footbridge; turn sharp left and continue 300 m up the left bank of the River Pyrddin to the waterfall known as Sgwd Gwladys (SN 8970 0930). This waterfall is located where the quartz arenite facies of the Twelve Foot Sandstone crosses the river. Above the plunge pool about 7 m of shales are exposed containing the *Superbilinguis* and *Sigma* marine bands, and these are succeeded by the Twelve Foot Sandstone, which consists of two sharp-based tiers separated by 75 cm of black shales (Fig. 5.8b). Both tiers have basal gutter casts and are composed of siliceous siltstones and sandstones with coarsening- and thickening-upwards profiles. Follow the rough path to the top of the falls, noting the swaley and low-angle laminations in the coarse-grain arenites developed towards the top of the upper tier. The undulating top of the sandstone displays *in situ* roots and bleached *Stigmaria* stumps (facies SF 6), and erosional pockets of pebbly sandstone containing silicified ribbed brachiopods, horizontal burrows and vertical to sub-vertical U-tubes (facies MS 1). It is overlain directly by fossiliferous shales containing *Lingula* and *Anthracoceratites* (facies MS 2–3), which are exposed at the base of the cliff on the right bank of the river, and are accessible only during very low water.

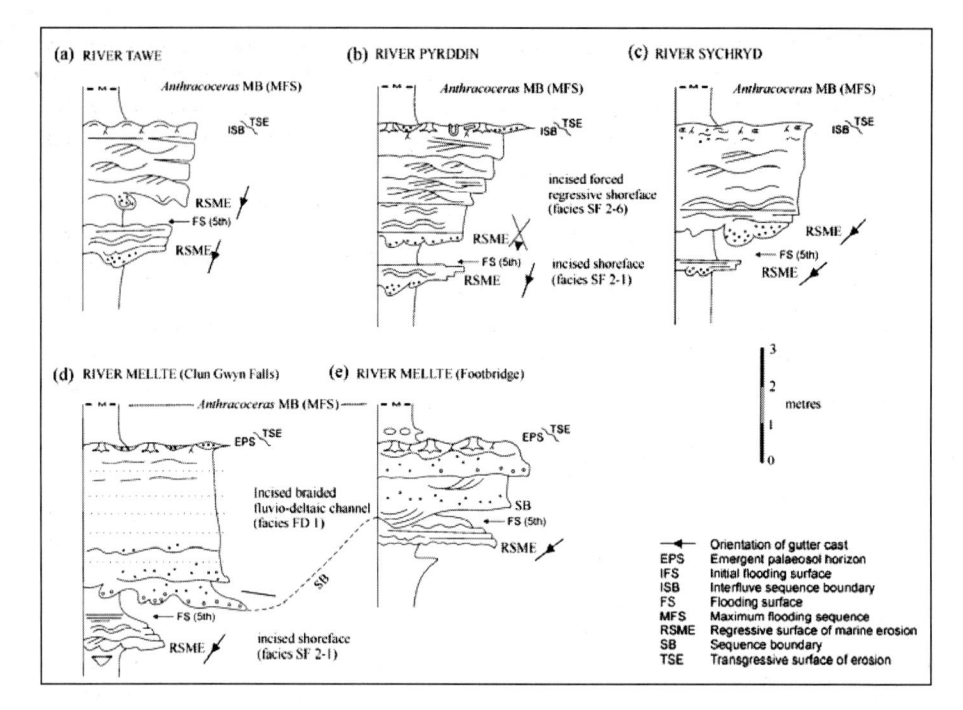

Fig. 5.8 Profile logs and sequence stratigraphy of the Twelve Foot Sandstone exposed along the banks of **(a)** the River Tawe, **(b)** River Pyrddin, **(c)** River Sychryd, **(d)** and **(e)** the River Mellte.

The abundance of HCS and SCS in both tiers of the sandbody indicates that deposition occurred on a storm-dominated shoreface environment. Their sharp, gutter-cast bases record facies dislocations, considered to be the result of relative falls in sea level. Coarsening- and thickening-upwards profiles, best seen in the upper tier, indicate that shoreface progradation occurred during these sea-level falls. All of the above features are diagnostic of forced regressive shoreface deposits in which successive sandbodies prograde further into the basin (Fig. 5.4). Emergence of this sandbody is recorded by the ganister and an unknown thickness of floodplain facies, which were removed during the formation of the overlying transgressive surface of erosion. The succeeding marine shales were deposited during a transgressive phase, which culminated with a maximum flooding surface represented by the *Anthracoceras* Marine Band (Fig. 5.8b).

1f Upper reaches of the River Pyrddin. Farther up stream from the falls, the Bishopston Mudstone Formation is exposed in a vertical cliff face, on the left bank of the river. At this locality the Sub-Cancellatum Sandstone (1 m thick) has an erosive base with loaded gutter casts, which is succeeded by quartz arenites (facies SF2–3) displaying low-angle swaley sets and some hummocky surfaces. Pockets of coarse bioturbated sandstone (facies MS 1) occur on the sharp, slightly undulatory, top of the sandbody. The *Cancellatum* Marine Band is developed 1 m above the sandstone, and about 5.5 m higher in the succession the Sub-Carbonicola Sandstone is represented by 0.5 m of siltstone that is overlain by shales, containing very large specimens of *Carbonicola*. Specimens of the freshwater mussel are best collected from fallen blocks at the base of the cliff. At the top of the cliff a prominent coarsening- and thickening-upwards unit, composed of flaggy and hummocky-bedded facies, represents the transitionally based Sub-Cumbriense Sandstone.

The Sub-Cancellatum Sandstone is interpreted as another forced regressive shoreface sandbody, deposited during the falling stage of the sea-level cycle (Figs 5.3, 5.4). Its limited thickness could be the result of; (i) a rapid fall in relative sea level, (ii) post-depositional erosion (ravinement) caused during the succeeding transgression, or (iii) a combination of both processes. A rapid sea-level fall would have the effect of limiting the creation of accommodation space, while reducing the length of time available for shoreface progradation. Evidence for post-depositional erosion is recorded by the undulating surface and granule lag, marking the top of the sandbody. As this surface separates falling stage deposits from transgressive deposits it must also represent a sequence boundary. The overlying *Cancellatum* Marine Band represents the maximum flooding surface of the transgressive systems tract.

About 500 m farther up stream the downfaulted Farewell Rock forms an impressive waterfall, known as Sgwd Einion-gam (SN 8910 0935). However, this locality is not recommended because access is quite difficult and involves crossing the river several times. It is more practical to return down stream to examine further sections of Namurian strata exposed in the River Nedd Fechan.

1g River Nedd Fechan. This well-known exposure of the Cumbriense Quartzite occurs approximately 500 m up stream from the Angel Inn (SN 900 081). Here the main sandstone has a prominent scoured base marked by a granule lag, which erodes into slivers of a much finer hummocky facies preserved in gutter casts (facies EC 4; Fig. 5.9). This surface is overlain by fine- to medium-grain quartz arenite facies (facies EC 2) with minor internal lags, mud drapes and mixed swaley and hummocky sets. A prominent internal channel, with a more massive fill, is developed in the upper part of the sandbody as it is traced up stream. Both of these facies are overlain by a complex palaeosol, comprising ganisters and a bleached fireclay (facies EC 3).

A regional study of the Cumbriense Quartzite indicates that at this locality the sandbody can be interpreted as a bay-head delta distributary channel, the erosive base of which represents a combined transgressive surface of erosion and sequence boundary. This interpretation is considered further when the much thicker sandbody, exposed in the River Mellte, is examined.

Farther down stream, the Shale Group succession is quite highly faulted and is difficult to correlate. A marine band containing a rich fauna of goniatites and brachiopod debris, exposed on the left bank of the river, is considered to represent the *Cancellatum* Marine Band. A little farther down stream, it is possible to locate the Sub-Carbonicola Sandstone and the overlying *Carbonicola* freshwater mussel band (Fig. 5.2), on both banks of the river. Here the erosively based silty sandstone is just over 1 m thick and contains sets of HCS. Abundant *Carbonicola* can be found 75–90 cm above the top of the sandstone.

1h Ravine. About 100 m farther down stream the erosive base of the Farewell Rock can be traced from exposures west of the footpath to the river-bed, where both the channel sandstone and underlying *Subcrenatum* Marine Band are exposed. Here the major sandbody has a prominent erosional base (relief of 1–2 m) marked by a lensoid lag. The *Subcrenatum* Marine Band occurs less than 1 m below this

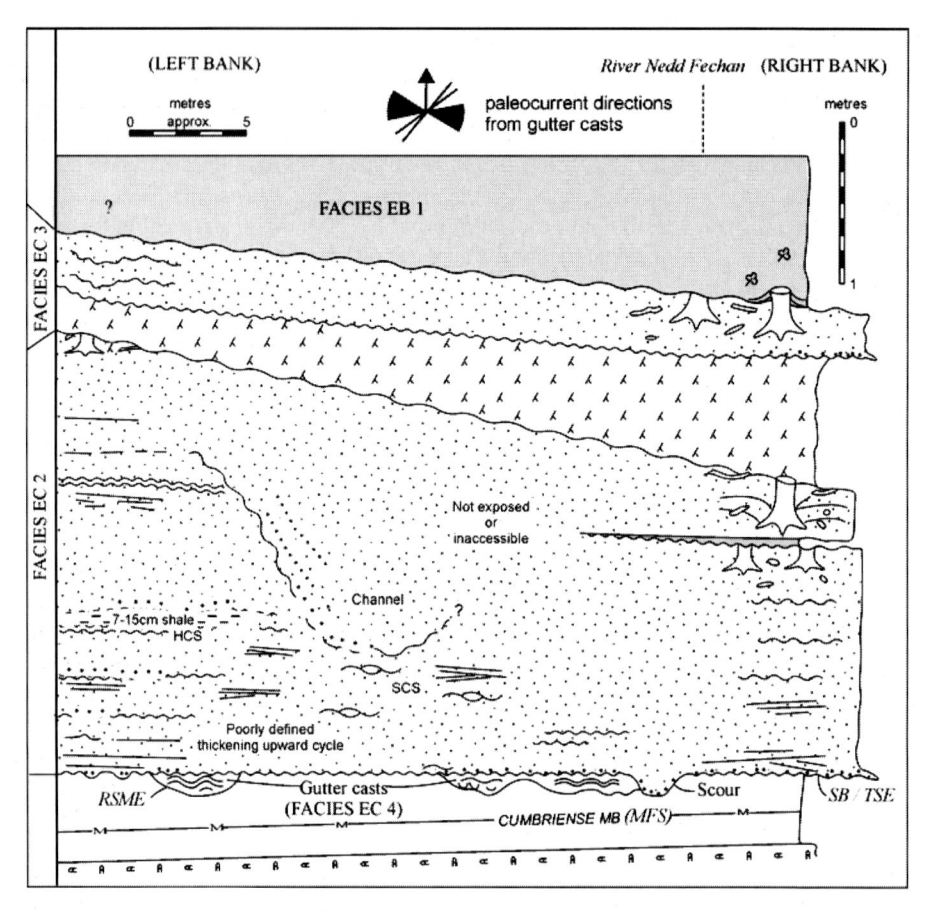

Fig. 5.9 Vertical and lateral facies relationships and key stratal surfaces present in the Cumbriense Quartzite exposed in the River Nedd Fechan (modified from George 2001).

erosional surface, and comprises two bands, containing abundant goniatites and a varied shelly fauna. Lower in the marine band the goniatites are replaced by fossil debris and pyritized burrows, which occur above a very highly bioturbated siltstone preserving the curved traces of *Zoophycos*.

The Farewell Rock crops out in an impressive ravine from this point down stream to the bridge at Pontneddfechan. This thick sandstone displays many internal erosional planes marked by pebble lags, log casts and soft-sediment deformation structures; large sets of cross-strata indicate southerly palaeoflow directions. A convenient exposure occurs by a millstone, just inside the metal gate, at the beginning of the footpath. At Pontneddfechan the Farewell Rock is estimated to be about 75 m thick, but the sandbody thins and dies-out about 3 km to the east. The pronounced multi-storey channel geometry of the Farewell Rock is typical of an incised valley, which was cut and filled during the falling and lowstand stages of sea level. The erosive base of the fluvial channel is interpreted as a sequence boundary, which presumably passes into an interfluve sequence boundary 3 km to the east.

5.4.2 Craig y Dinas

Cars and minibuses can be parked in the Craig y Dinas car park (SN 911 079), located at the eastern end of the village of Pontneddfechan. Coaches cannot cross the Mellte Bridge and these should be parked on the approach road or in the village. The car park is part of the old quarries, cut into vertical beds near the top of the Carboniferous Limestone. The ridge of limestone follows the east-northeast-trending line of the Vale of Neath Disturbance, the main splay of which (the Dinas Fault) is located in the River Sychryd. Unfortunately, the path leading from the car park through the gorge between Craig y Dinas and another impressive folded block of limestone, known as Bwa Maen ('bow of stone'), has been closed for some time because of the high risk of rock falls.

2a Old silica mines. Take the footpath to the north of the car park; ascend the old tramway to the top the ridge, and continue along the path that descends down to the mines and the River Sychryd. The three adits to the left of the path (SN 9167 0799) were the main entrances to the very extensive underground workings of the Dinas silica mine. The mine workings followed a 4 m band of very pure and fine-grain quart arenite (facies SF 5; Fig. 5.10) that was used to make high-quality silica bricks. This section should only be viewed from the path and on no account should you attempt to enter the adits. Details of the facies and interpretation of the succession, exposed in the quarry face, are summarized in Figure 5.11. In this log, the sequence-stratigraphy interpretation has been aided by reference to a gamma-ray profile, measured with a hand-held spectrometer. Note that the lowstand and falling-stage quartz arenite facies produce low readings of 25–50 API (American Petroleum Institute units), whereas the transgressive facies record higher values, with the marine bands giving readings well above 125 API.

Cross the footbridge to the left bank of the River Sychryd, where the silica horizon has again been mined and the lower 3 m of the section illustrated in Figure 5.11 can be examined. Down stream the lower 20 m of the Twrch Sandstone and its unconformable/palaeokarst contact (sequence boundary) with the Oxwich Head Limestone Formation, are exposed in the bed and banks of the river, but this section is unsuitable for field-trip visits.

2b River Mellte. Return to the car park; cross the bridge and follow the track up the right bank of the River Mellte. The flat area of land, adjacent to Rose Cottage, was the site of the demolished gunpowder factory (1857–1931), and farther up stream there are remains of buildings and the weirs and leats that were used to provide water power for the processing of the black powder. Continue up the track (old tram line) to the footbridge; cross the river and walk 150 m down stream (SN 9189 0840) to a locality where the Twelve Foot Sandstone is exposed. Here the quartz arenite sandbody is 2.75 m thick and displays its distinctive two-tier architecture (Fig. 5.8e). The upper tier (2 m thick) is composed of three beds of poorly sorted, coarse-grain and granule-rich quartz arenite facies (FD 1), capped by a ganister (facies FD 2) with stumps of *Stigmaria*, horsetail roots and plant debris. It is overlain by slightly calcareous silty mudstones with pyrites and occasional fossils representing the *Anthracoceras* Marine Band. The complete sandbody can also be seen (but is inaccessible) on the right bank of the river.

Fig. 5.10 Sharp-based forced regressive sandbodies exposed adjacent to the old adits of the Dinas Silica Mine (see Fig. 5.11, 22 – 38 m).

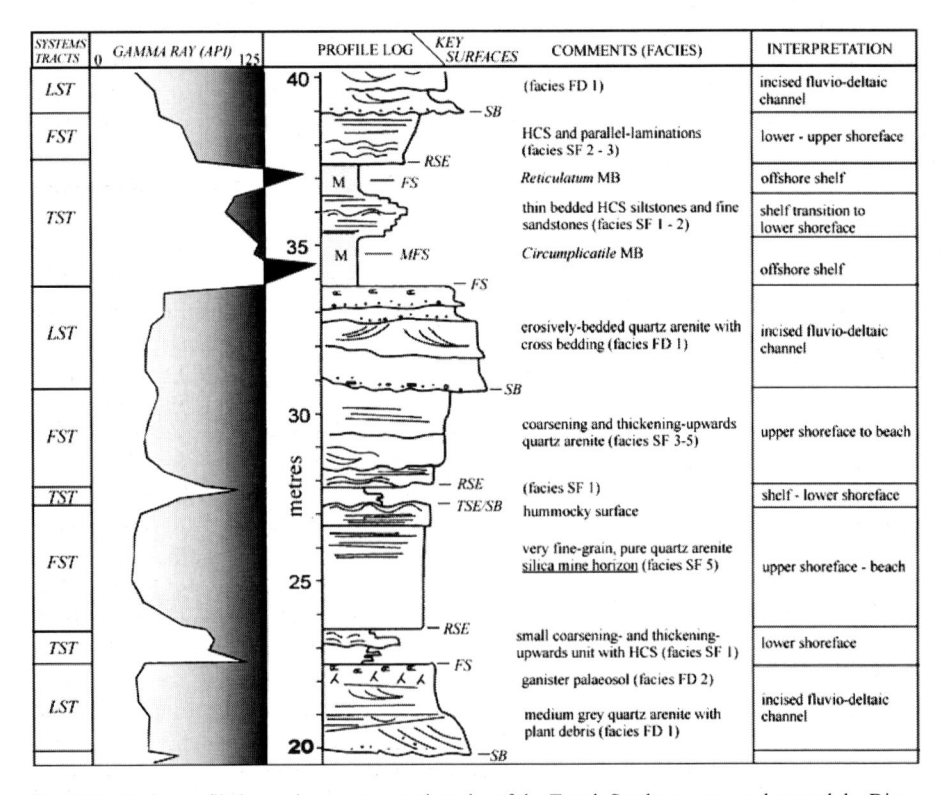

SYSTEMS TRACTS	GAMMA RAY (API) 0–125	PROFILE LOG	KEY SURFACES	COMMENTS (FACIES)	INTERPRETATION
LST		40	– SB	(facies FD 1)	incised fluvio-deltaic channel
FST			– RSE	HCS and parallel-laminations (facies SF 2 - 3)	lower - upper shoreface
		M – FS		*Reticulatum* MB	offshore shelf
TST				thin bedded HCS siltstones and fine sandstones (facies SF 1 - 2)	shelf transition to lower shoreface
		35 M – MFS		*Circumplicatile* MB	offshore shelf
		– FS			
LST				erosively-bedded quartz arenite with cross bedding (facies FD 1)	incised fluvio-deltaic channel
		– SB			
FST		30		coarsening and thickening-upwards quartz arenite (facies SF 3-5)	upper shoreface to beach
TST			– RSE – TSE/SB	(facies SF 1) hummocky surface	shelf - lower shoreface
FST		25		very fine-grain, pure quartz arenite silica mine horizon (facies SF 5)	upper shoreface - beach
TST			– RSE – FS	small coarsening- and thickening-upwards unit with HCS (facies SF 1)	lower shoreface
LST		20	– SB	ganister palaeosol (facies FD 2) medium grey quartz arenite with plant debris (facies FD 1)	incised fluvio-deltaic channel

Fig. 5.11 Facies profile log and sequence stratigraphy of the Twrch Sandstone exposed around the Dinas Silica Mine and on the banks of the River Sychryd.

The lower tier of the sandbody is interpreted as a lower shoreface facies, deposited on a storm-dominated shelf. Its sharp base and thinning- and fining-upwards trend, indicates that the sandbody was deposited as a result of a sharp fall in relative sea level, followed by a more gradual rise. The overlying coarser-grain tier is interpreted as the fill of an incised low-sinuosity (braided) channel. Its erosional base records a significant basinward shift in facies, which is one of the most diagnostic features of a sequence boundary.

2b Gunpowder works. Return via the footbridge to the right bank of the river. Note that the Sub-Cancellatum Sandstone (75–100 cm thick) developed about 20 m above the Twelve Foot Sandstone, outcrops intermittently on the slopes above the footpath, just down stream of the bridge. Follow the footpath a further 750 m up stream, passing a variety of ruins of the gunpowder works; cross a style and then a stream gully. Access to the next locality involves a steep climb up a grassy slope, just to the north of the gully, and should be attempted only if you are physically fit and the ground is dry. The first outcrop, which occurs on the northern flank of the gully (SN 9193 0857), reveals a 2.5 m coarsening- and thickening-upwards unit (Sub-Cumbriense Sandstone) capped by a mixed ganister and fireclay palaeosol. The palaeosol is succeeded by about 1.25 m of dark-grey shales, which are truncated by an erosional surface marking the base of the Cumbriense Quartzite channel complex. At this locality the base of the channel has eroded out the *Cumbriense* Marine Band. The sandbody, exposed in a vertical face about 15 m high, consists of coarse to medium-grain, predominantly massive arenites with well defined erosional bedding and abundant clay flakes and clasts (facies EC 1), most of which have been weathered out. Isolated sets of low-angle trough cross-stratification, and thin lenses of flaggy facies with hummocky and swaley sets are occasionally developed. (No attempt should be made to climb higher than this point or to extend the traverse along the strike.) The overlying EC 2 channel member is a finer grain facies containing more obvious sets of normal and swaley cross-stratification. Thin mud laminae, usually less than 1 cm thick, often drape the main bedding planes. Thus, at this locality, the Cumbriense Quartzite displays the geometry and architecture of a multi-storey and multi-lateral channel complex (Fig. 5.12). In the axial zone of the complex the sandbody reaches a maximum thickness of 20 m and its erosive base cuts out the *Cumbriense* Marine Band, and almost amalgamates with the underlying Sub-Cumbriense Sandstone (an estimated erosional relief 14 m). Both the EC 1 and EC 2 channels display offset stacking, and the trend of the axial zone of the complex appears to have been controlled by northwest-trending syn-sedimentary faults. A regional study of the Cumbriense Quartzite has shown that it was deposited in an estuarine channel and back-barrier complex, which retreated to the northwest as a result of sea-level rise (Fig. 5.13).

In terms of sequence stratigraphy the transitionally based coarsening- and thickening-upwards Sub-Cumbriense Sandstone is interpreted as a normal (Waltherian) regressive shoreface, which prograded seawards during a highstand of sea level. The Cumbriense Quartzite is interpreted as a multi-storey and multi-lateral incised estuarine channel fill (Fig. 5.13). Laterally, both to the east and west, the erosional base of the channel complex (sequence boundary) rises stratigraphically and passes into mature palaeosol surfaces interpreted as interfluve sequence boundaries. The western interfluve palaeosol is seen below the Henrhyd waterfall at locality *3e*, whereas its eastern counterpart has been identified in the Gelli Isaf borehole (Appendix 6).

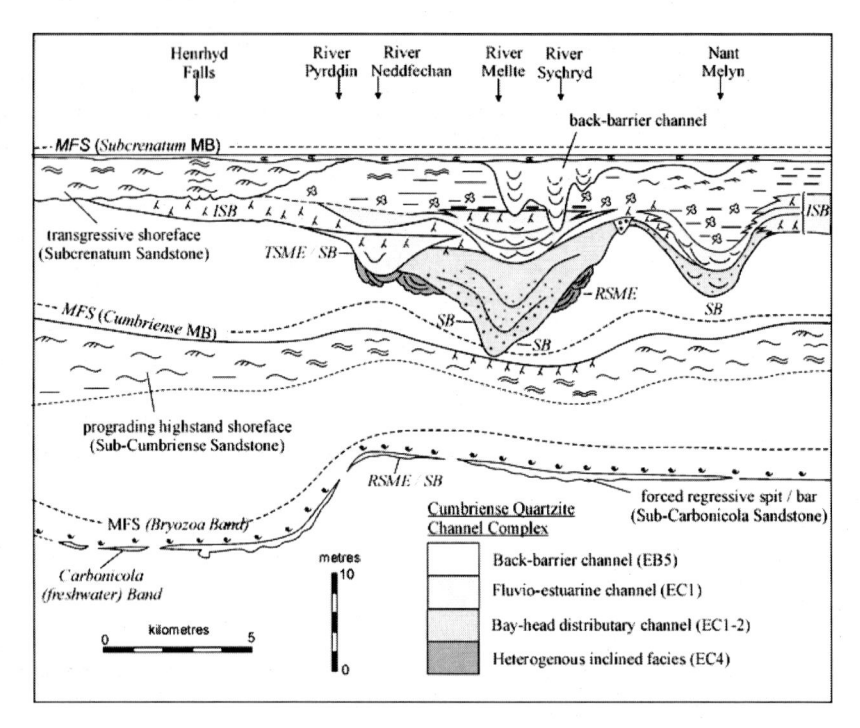

Fig. 5.12 Cross section illustrating the depositional and sequence stratigraphy model for the late Namurian strata (including the Cumbriense Quartzite and Subcrenatum Sandstones), developed along the North Crop.

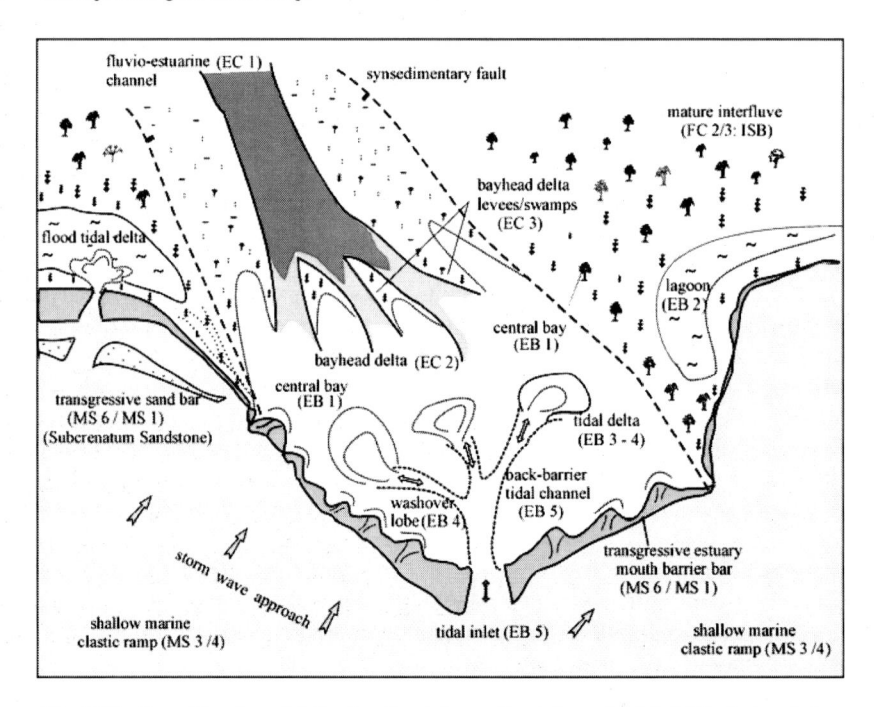

Fig. 5.13 Depositional model for the Cumbriense Quartzite and related lithofacies during rising sea level (transgressive systems tract).

5.4.3 Penwyllt and Ynyswen

Take the A4067 through Ynyswen and about 1 km north of Pen-y-Cae turn right onto the minor road signposted to Penwyllt (Fig.5.3). At the top of the hill, turn sharp right along the track adjacent to the limestone quarry and park in the National Nature Reserve car park (SN 856 156). The karstic landscape is dotted with caves and swallow holes (referred to as shake holes on the OS map), and large areas of Namurian strata, which have collapsed into the limestone as a result of solution. The extensive caves in this area are known as Ogof Fynnon Ddu, while farther to the northwest, on the opposite side of the Tawe Valley, are the popular show-caves of Dan-yr-Ogof. The prominent scree-strewn scarp of Carreg Lwyd, seen to the southeast, is composed of hard quartz arenites belonging to the Twrch Sandstone Formation (Kinderscoutian to Marsdenian), which define the southerly-inclined dip-slope of the hill. The main quarry at Penwyllt is periodically worked for limestone, and permission is required for visits. However, this is not the case with the long-abandoned quarries in the Carboniferous Limestone and the Twrch Sandstone that are discussed in this itinerary. Hard hats must be worn and on no account should you enter the limestone caves or climb the quarry faces.

3a Abandoned limestone quarries. The abandoned quarry, located adjacent to the car park, exposes thickly bedded southwards-dipping (15°) limestones belonging to the middle part of Oxwich Head Limestone Formation. These predominantly dark-grey limestones contain a shelly fauna of micritized gastropods and brachiopods and also a thin (30 cm) mottled and reddened palaeosol horizon. The eastern face of the quarry is cut by a fault, which dips at 72° west; it displays vertical slickensides and has a breccia containing large calcite crystals. About 350 m farther to the south, past the terraced houses belonging to the South Wales Caving Club, is another abandoned quarry (SN 856 153), where the strata exposed occur near the top of the Oxwich Head Limestone. Here the lower part of the succession contains a 40 cm carbonaceous shale, developed above a mottled limestone with *in situ* rootlets. This palaeosol unit is truncated by an interesting coarsening-upwards limestone facies (found as large blocks on the floor of the quarry), which contain about 30 per cent coarse sand grains and displays sets of hummocky and swaley cross-stratification. A cleft (4–5 m wide), seen at the top of the quarry, marks the position of the mostly concealed Oystermouth Formation, which is overlain unconfomably by the Twrch Sandstone–fallen blocks of which contain quartz pebbles up to 8 cm in diameter.

3b Western crags. Return to the caving-club houses; take the path across the garden and continue over the disused railway (passing old limekilns) to the semi-detached houses. Note the large swallow hole, with collapsed blocks of grit, to the north of the houses. Continue for 250 m southwest, over the brow of the hill, to a small but perfectly formed limestone pavement with clint-and-gryke weathering. From this karstic surface there are good views across the Tawe Valley to the limestone hill known as Cribarth, an anticlinal structure that runs parallel to the line of the Swansea Valley Disturbance. Due west, along the banks of the River Tawe, is Craig-y-nos Castle and country park, which was the former home of the opera diva Adelina Patti.

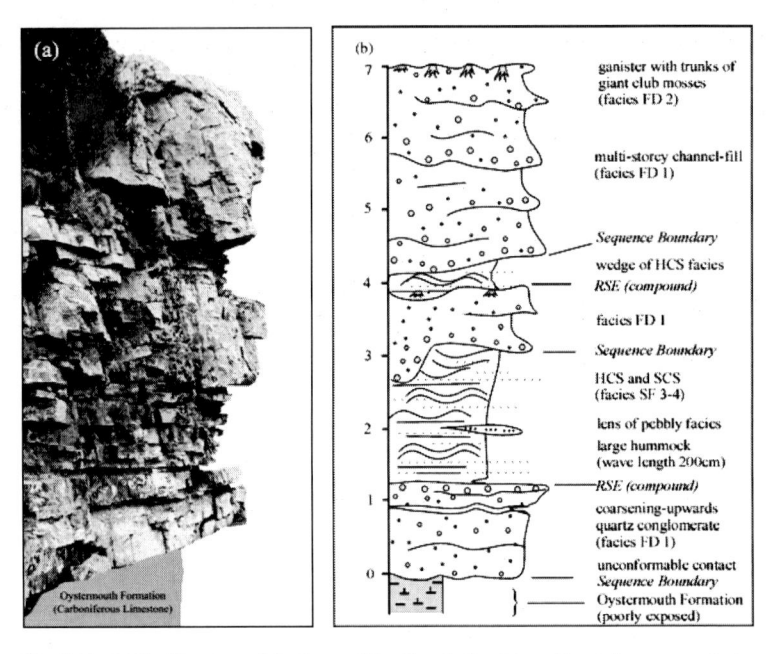

Fig. 5.14 (a) Profile view of the base of the Twrch Sandstone Formation exposed at Penwyllt. (b) Graphic log and interpretation of the same succession.

To the south (SN 853 154), the lowermost 10 m of the Twrch Sandstone is exposed along an impressive southwards-trending crag (Fig. 5.14a). Hollows and shallow caves, at the base of the crag, mark the position of the unconformity between the argillaceous Oystermouth Formation and the overlying grits. The basal 1.25 m of the grit is a conglomerate, composed of a quartz pebbles, which is succeeded by more thinly bedded medium- to coarse-grain quartz arenites with well developed sets of HCS and SCS (Fig. 5.14b). Large hummocks with amplitudes of 30 cm and widths of 200 cm are common, and thin mud drapes are often preserved between the sets. The top 4 m of the succession is composed mainly of very coarse pebbly facies with basal and internal erosional planes marked by pebble lags. Isolated sets of trough cross strata provide southeasterly-directed palaeocurrent data. A lens of hummocky facies, preserved towards the base of the channel fill, can be traced for about 50 m along the outcrop. Stumps of *Stigmaria* and *in situ* roots occur on the uppermost bedding plane of the channel fill. In terms of sequence stratigraphy the succession is interpreted as a lowstand sequence set, composed of fluvial channels incised into forced regressive shoreface sandbodies. Thus in this succession all of the previously deposited transgressive facies, including marine bands, have been removed by regressive surfaces of marine erosion.

3c Abandoned gritstone quarries. Return to the caving-club houses and continue south on the track (passing locality *3a*) to another abandoned quarry cut into the Twrch Sandstone (SN 855 150). The strata exposed here are stratigraphically higher than those at the previous locality, and they are composed of 10 m of hummocky and swaley facies, organized into a coarsening- and thickening-upwards cycle (Fig. 5.15). These hummocky bedforms are the largest examples known to the author. Some giant forms, with wavelengths of 10–30 m and amplitudes of up to 1 m, are compound and contain smaller internal sets (Fig. 5.16). A prominent erosional surface, seen at the top of the cycle, can be followed into an adjacent quarry, where it defines the base of a channel complex with a multi-storey and multi-lateral fill (Fig. 5.17). Basal and internal erosional surfaces are marked by pebble lags (filling flute casts), and they often contain rafts of carbonaceous mudstone and shale. Palaeocurrent readings, taken from the elongate flute casts and trough cross sets, give consistent southerly vectors. Casts of *Stigmaria* occur on the uppermost bedding plane of the sandbody.

In this sequence there is a gradational upwards passage from mudstones of offshore shelf origin to coarsening- and thickening-upwards siltstones and mudstones, interpreted as transitional shelf facies. It is quite possible that a marine band exists in the unexposed beds developed below the quarry floor. The graded pebble beds and giant hummocky sets in the overlying facies were deposited during periods of violent storms or even tsunamis. Higher in the sequence this facies contains normal cross-bedded sets that were deposited during fairweather conditions. These latter facies are incised by a fluvio-deltaic channel, which is assigned to the lowstand systems tract.

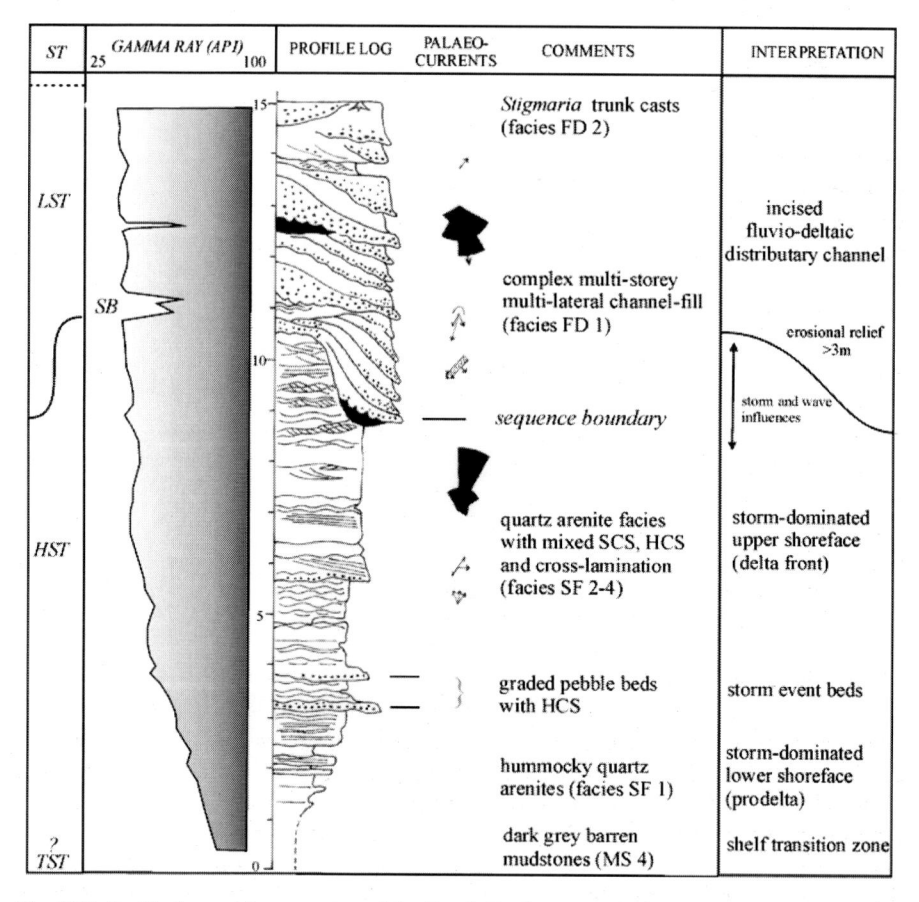

Fig. 5.15 Profile log and interpretation of the Twrch Sandstone succession exposed in the abandoned quarry at Penwyllt (modified from George 2000)

Fig. 5.16 Two giant hummocky sets exposed in the old quarries at Penwyllt (monopod scale is 70 cm).

Fig. 5.17 Complex multi-storey incised fluvio-deltaic channel, illustrated in Figure 5.15 (10–15 m).

3d River Tawe, Ynyswen. From Penwyllt, travel back to Ynyswen on the A4067, where there is parking for a few cars behind an isolated building (opposite the church and cemetery) on the eastern outskirts of the village (SN 838 133). Walk 100 m south through the woods to the River Tawe, and then 500 m up stream to a small waterfall that marks the top of the Twrch Sandstone Formation (Fig. 5.5b). The lowermost quartz arenite contains abundant *Stigmaria* stumps, horsetail rootlets and thin fireclay interbeds. This palaeosol is succeeded by 5–15 cm of bioturbated siltstone, and dark grey–black shales containing a rich marine fauna of *Lingula,* thin shell marine bivalves and goniatites, which represent the *Bilinguis* Marine Band. These shales are incised by a further quartz arenite that has a channelized base and internal hummocky and swaley sets (facies SF 3-4). The top of the unit is marked by a very prominent erosional surface with a pebble lag and 10–15 cm of highly bioturbated sandy siltstone that preserves large horizontal feeding traces of *Zoophycos* and smaller cylindrical vertical tubes. This bed represents a transgressive surface of erosion, which initiated the marine flooding that culminated with the deposition of the *Superbilinguis* Marine Band. The sequence stratigraphy and interpretation of the succession are summarized in Figure 5.7b.

Down stream, the lower part of the Bishopston Mudstone Formation is well exposed in the vertical cliff (SN 842 132 to 840 132), forming the left bank of the river. About 15 m of shales, with a couple of prominent calcareous siltstones, ironstone bands and the *Superbilinguis* and *Sigma* marine bands are exposed but not accessible. The Twelve Foot Sandstone, seen at the top of the cliff face, is best examined with binoculars or in the fallen blocks (Fig. 5.18). Here the sandbody displays its characteristic two-tier architecture, the erosive bases of both tiers displaying gutter casts that are elongated in a north–south direction (Fig. 5.8a). The lower tier (1 m thick) displays good hummocky bedding and a sharp top, and is succeeded by 75 cm of black shales; the upper tier is 2.0–2.5 m thick and displays swales cutting out thin shale units (facies SF 3). The Twelve Foot Sandstone, at this locality, is again interpreted as a forced regressive shoreface sandbody that was deposited during the falling-stage systems tract; the seaward regression of the lower tier being interrupted by a minor sea level rise.

Fig. 5.18 The Twelve Foot Sandstone (4 m thick) composed of two sharp-based tiers with well-developed sets of hummocky and swaley cross strata , River Tawe at Ynyswen.

3e Henrhyd Falls. From Ynyswen continue into Abercraf; turn sharp left onto the A4109 and turn left again after 3 km to reach the village of Coelbren, from where the Henrhyd Falls are signposted. Parking for cars and coaches is available at the National Trust site (SN 845 119). The best point to review the very interesting stratigraphical relationships seen at this locality is on the path, above the left bank of the river, overlooking the falls (Figs 5.19, 5.20). Barren marine shales (MS 4), exposed above the plunge pool, are followed by a bleached fireclay facies (EC 3), which is truncated by a sharp-based sandstone. This sandbody, which can be examined on the path leading down to the falls and on the ledge behind the falls, is the newly defined Subcrenatum Sandstone. At this locality it is 4.25 m thick and is composed of three relatively poorly developed fining-upwards units, each with erosive gutter-cast bases and sets of SCS and HCS. The sandbody is overlain by about 2.5 m of inaccessible black shales containing the *Subcrenatum* Marine Band. These marine shales are truncated at the erosive base of the Farewell Rock, which forms the upper third of the cliff face. Sets of cross bedding, observed in the later channel, indicate southerly palaeoflow directions.

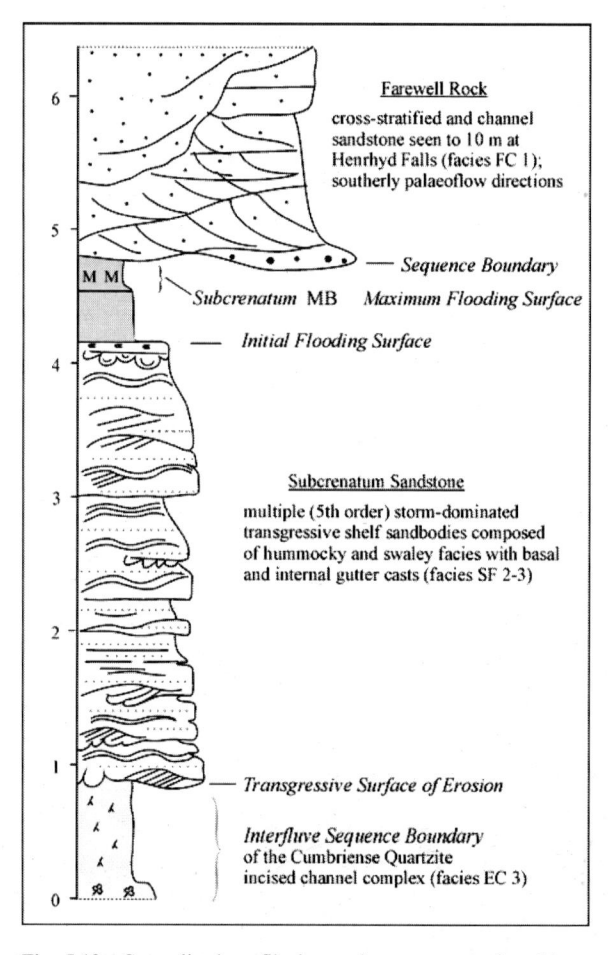

Fig. 5.19 Generalized profile log and sequence stratigraphic interpretation of the Subcrenatum Sandstone and the base of the Farewell Rock.

Fig. 5.20 View of the Henrhyd Falls from the footpath showing the Subcrenatum Sandstone (4 m thick) overlain by *Subcrenatum* Marine Band (<1 m), which is truncated by the incised Farewell Rock. Note that it is possible to walk along the ledge behind the falls (see also Frontipiece A).

Fig. 5.21 Well formed hummocky sets (HCS) and small-scale loading in the Subcrenatum Sandstone at the Henrhyd Falls (Fig. 5.19 at 2.5 m).

This succession is extremely important for the interpretation of the late Yeadonian sequence stratigraphy of the North Crop. Regional correlations indicate that the mature palaeosol, present at the base of the above section, is the interfluve sequence boundary of the Cumbriense Quartzite channel complex located farther to the west (locality *2b*, Fig. 5.12). The sharp erosive base of the Subcrenatum Sandstone is interpreted as a transgressive surface of erosion, with the overlying facies recording deposition on a storm-dominated shelf during a punctuated eustatic sea-level rise. As the transgression continued, the sedimentary surface moved below storm wave-base (sharp top of sandbody), resulting in the deposition of marine shales (*Subcrenatum* Marine Band). Highstand deposits, associated with this flooding event, were later removed by a significant relative sea-level fall, which resulted in the excavation of the Farewell Rock incised valley. This incised valley was filled during a subsequent rise in sea level in a manner similar to the Telpyn Point Sandstone (late Yeadonian) of southwest Wales (George 2001; see also Ch. 6).

Chapter 6

South Pembrokeshire and Carmarthen Bay

6.1 Introduction

South Pembrokeshire lies to the south of a line running from Newgale to Narberth, a boundary roughly coincidental with the ancient Landsker Line. This boundary was marked by castles (Roche, Haverfordwest, Llawhaden, Narberth and Laugharne), which divided the Welsh speakers of north Pembrokeshire from the more anglicized population to the south. For this reason south Pembrokeshire is often referred to as "Little England Beyond Wales". This term is also applicable to the coastal parts of southwest Carmarthenshire extending eastwards from Amroth to Pendine and on to Laugharne. The landscape of the south is more uniform and less rugged than the north because the bedrock is composed predominantly of sedimentary rocks. These rocks have nurtured the beautiful coastline of the Pembrokeshire National Park, which is characterized by sandy bays protected by headlands, and extensive sea cliffs formed of limestones, sandstones and conglomerates. The large sweeping coastlines of St Brides Bay in the west and Carmarthen Bay in the east have been eroded into relatively soft argillaceous sediments preserved in the Pembrokeshire Coalfield Syncline. The contrasting colours of the rocks (green–red–grey) and their variable attitudes (horizontal–vertical–folded) add further interest to the coastal landscape. The Milford Haven waterway, and the branching arms of the Cleddau Estuary dissect the southern half of the area. To the north of the Haven the Marloes and Dale peninsulas and islands of Skomer and Skokholm dominate the landscape. To the south, breathtaking limestone cliffs extend eastward form Linney Head to St Govan's Head and on to Stackpole Quay. Most parts of the coastline, with the exception of the Defence Training Estate (MOD) at Castlemartin, are accessible via the Pembrokeshire Coast Path, which was officially opened in 1970. The path follows well marked, mainly clifftop routes for 300 km from Amroth in the southeast to St Dogmaels, near Cardigan, in the northeast. The trail provides a unique opportunity to access the largely unspoilt landscape of the Pembrokeshire National Park, which is internationally important for its nature reserves and geology.

South Pembrokeshire is a very popular holiday destination, and many of the towns (particularly Tenby and Saundersfoot) and access roads become very busy during the summer months. The area is covered by comprehensive public bus services including the Coastal Cruiser and Puffin Shuttle. There are also railway services from Whiteland to Pembroke Dock via Saundersfoot and Tenby, and from Haverfordwest to Milford Haven. Timetables for these services are available from the tourist information centres, and some include useful information on the logistics of using the coastal bus services to plan walks and field trips within the national park. Further information regarding travel, tide tables and tourist attractions can be found in the national park's visitor newspaper *Coast to Coast*.

6.2   Geological history

The area is composed mainly of Upper Palaeozoic rocks with areas of older strata, exposed only along the lines of major faults and in the cores of anticlines (Figs 6.1, 6.2). Precambrian strata, belonging to the Johnston Complex (dated about 643 Ma), include metamorphosed plutonic igneous rocks (diorites, granodiorites, granites) and acid volcanics, which occur along the Johnston–Benton Fault Zone. In the extreme west of the area, important outcrops of Ordovician and Silurian strata occur to the south of the Musselwick Thrust on the Marloes Peninsula. These shallow marine facies, containing shelly faunas, were deposited around the margin of the basin, in a belt extending from Llandovery to Haverfordwest (see Fig. 2.7). Within the area, the oldest Silurian strata include an impressive development of thick basalts, rhyolites, tuffs and clastic sediments, which reach their maximum development on Skomer Island (the Skomer Volcanic Group).

Most of the remaining solid geology of south Pembrokeshire is composed of ESE–WNW-trending tracts of Upper Palaeozoic strata. To the north of Milford Haven there is a conformable transition from marine Silurian into continental Old Red Sandstone strata; to the south this contact is unconformable. The Lower Old Red Sandstone comprises thick successions of red marls, sandstones and conglomerates that are exposed on the northern flank of the Pembrokeshire Coalfield Syncline and to the north and south of the Haven. The Skrinkle Sandstone (Upper ORS), which is preserved only to the south of the Ritec Fault, records an upward transition from continental redbeds to grey marginal marine facies. The succeeding

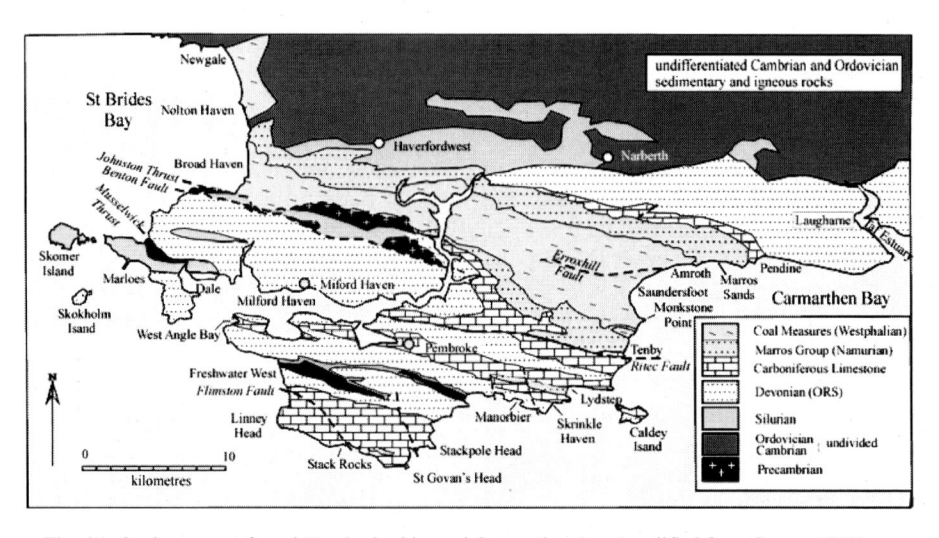

Fig. 6.1 Geology map of south Pembrokeshire and Carmarthen Bay (modified from George 1970).

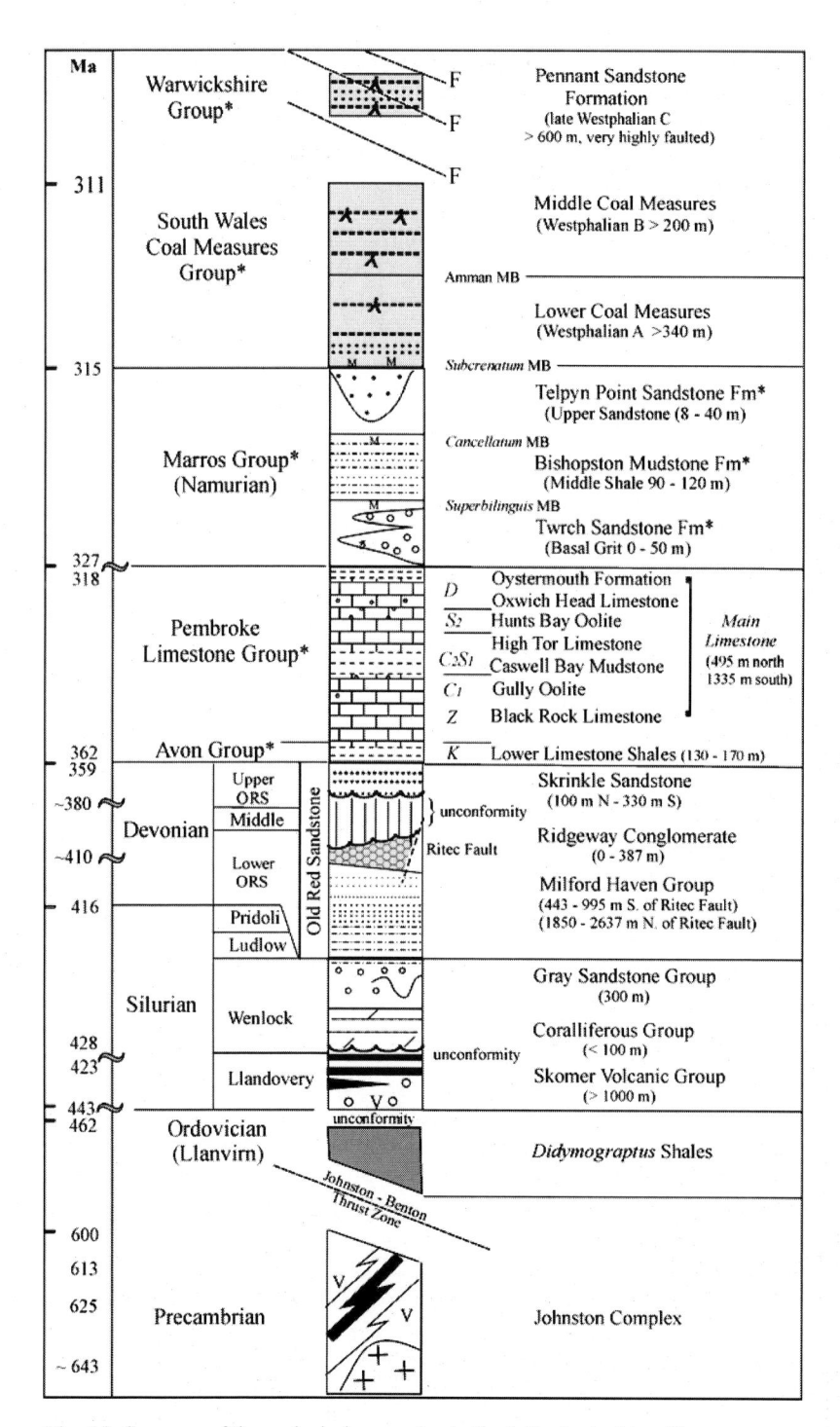

Fig. 6.2 Summary of the geological succession in South Pembrokeshire. The new names for the Carboniferous, proposed by the British Geological Survey (Waters et al. 2007), are marked with an asterisk. In extreme south the Caswell Bay Mudstone/High Tor Limestone is replaced by Pen yr Holt Limestone and the Hunts Bay Oolite is the approximate equivalent of the Stackpole Limestone.

Carboniferous Limestone covers much of south Pembrokeshire, giving rise to impressive cliffs, extending from Stackpole Quay to St Govan's Head and west to Linney Head. Further outcrops of Carboniferous Limestone define the northern and southern margins of the Pembrokeshire Coalfield Syncline. Within this syncline there are excellent sections through Upper Carboniferous (Namurian and Westphalian) strata on the east coast between Amroth and Tenby, and on the west coast between Druidston Haven and Little Haven. Further exposures of Westphalian rocks occur in the small Nolton–Newgale Coalfield. Variscan compression has resulted in these Upper Carboniferous strata being highly faulted and folded, with crustal shortening values of around 30 per cent in the east to 40 per cent in the west, and locally over 60 per cent.

The geology of the area is covered by the 1:25 000 Ordnance Survey maps (Explorer OL36 & 177); the 1:50 000 geological maps for Milford (Sheet 226 & 227), Haverfordwest (Sheet 228), Carmarthen (Sheet 229), Pembroke and Linney Head (Sheet 224 & 245) and their respective memoirs (Cantrill *et al*. 1916; Strahan *et al*. 1914; Strahan *et al*. 1909; Dixon 1921). *Geological excursions in Dyfed and south-west Wales* (edited by M. G. Bassett) is the most important field guide to the area; further citations to important publications are given in the itineraries.

6.3 Itineraries

The itineraries are presented in a broadly stratigraphic order (oldest to youngest), but the localities can be combined to limit the amount of travelling. Hard hats are recommended for all sections located at the bases of cliffs and extra care should be taken after heavy rainfall and storms. It is important to refer to tide tables before planning field visit; details of the optimum tidal conditions are given for each coastal itinerary.

6.3.1 Marloes Sands

From Haverfordwest travel 16 km southwest on the B4327 Dale road to Mullock Bridge and then turn right onto the minor road signposted to Marloes; continue for 1.5 km before turning left onto a narrow road that leads to the National Trust car park near Runwayskiln Farm (SM 780 082) (Fig. 6.3). Coaches should continue through Marloes village on the narrow road to Martin's Haven and turn left after 1.5 km onto an unmade track leading to the rear of the car park. Marloes Sands is a SSSI, designated for its superb cliff sections of Silurian strata. The complete itinerary requires about 5–6 hours field time and a low or falling tide. If the tide is high and rising, Martin's Haven and the Deer Park should be visited first. There are no facilities at the beach, so a packed lunch is advisable.

1a Mathew's Slade. From the entrance to the car park, walk back along the road for 150 m to Sandy Lane that leads down to the beach. At the beach, walk 300 m to the east (left), passing the grassy hollow known as Mathew's Slade, which is composed of a small faulted slice of the Coralliferous Group (Fig. 6.3). A little farther to the southeast, steeply dipping and faulted grey–green polymict conglomerates and feldspathic sandstones are exposed; they belong to the upper part of the Skomer

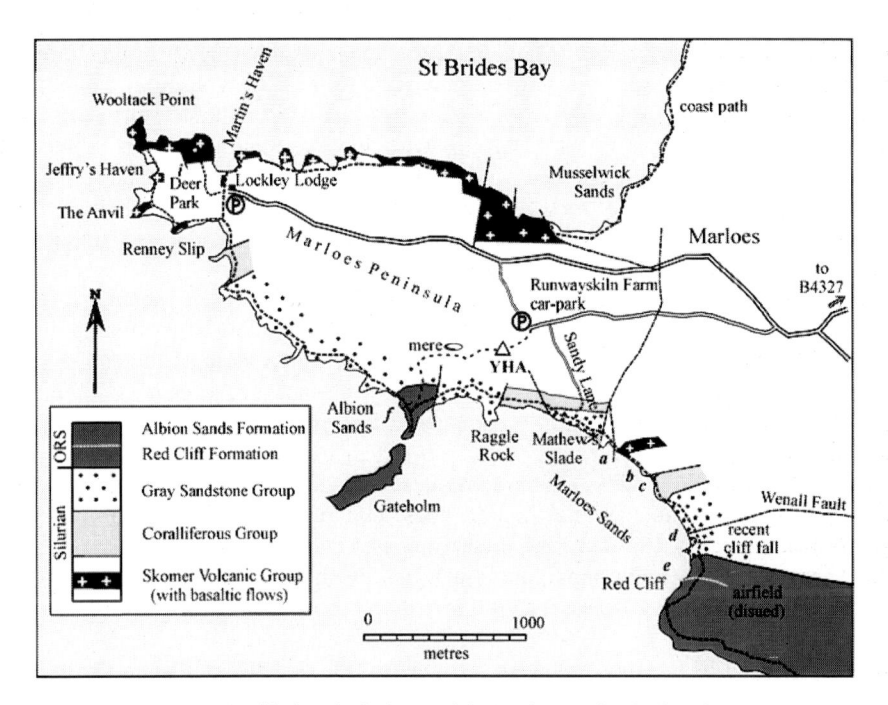

Fig. 6.3 Locality and simplified geological map of the Marloes Peninsula (based on Bassett 1982).

Volcanic Group. These coarse-grain facies have erosional bases and the sandstones appear mottled from intense bioturbation. Thinner-bedded sandstones, siltstones and mudstones, which contain lithic fragments and display hummocky and ripple bedforms, succeed them. The beds young to the southeast and are overlain by two basaltic lava flows (SM 785 074) that are faulted, and permeated by small veins containing epidote. The lowermost flow (~20 m thick), has a vesicular base and a reddened weathered top, and is overlain directly by the vesicular base of the second flow (~19 m thick). These basalts correlate with a much thicker succession of flows on Skomer Island. The reddened tops of the lava flows indicate that they were subjected to sub-aerial weathering before being covered by marine sediments. This important sea-level datum has been used to document the changes in palaeobathymetry and faunal communities in the overlying transgressive facies (see Fig. 2.7). The lava flows are succeeded by a mixed group of grey–green silty mudstones, sandstones, conglomerates and tuffs. One tuff band (6 m thick) is composed of angular to sub-rounded lithic fragments up to 5 cm in diameter.

1b Three Chimneys. Farther to the east, about 12 m above the 6 m tuff (SM 7860 0732), three near-vertical beds of weathered ferruginous sandstones form a prominent feature known as the Three Chimneys. The lowest sandstone bed (120 cm thick) has an erosional base and two further internal erosional surfaces, whereas the bases of the other sandstones (75 cm and 60 cm thick) display low-amplitude load casts. These sandstones are succeeded by about 60 m of silty mudstones, ripple-laminated sandstones, and thin decalcified limestones containing brachiopods (*Leptaena* and rhynchonellids), corals (*Favosites*), bivalves, crinoid debris and the screw-like *Tentaculites* (see Fig. 2.7b).

In terms of early Silurian (Llandovery) palaeogeography, Marloes was located near the northern margin of a large landmass, with basaltic volcanoes centred on Skomer Island (see Fig. 2.7a). In this setting, the coarse-grain sedimentary facies, belonging to the Skomer Volcanic Group, were deposited in high-energy beach and barrier-bar environments characterized by the *Eocoelia* and *Pentamerus* faunal communities. In more protected back-barrier lagoons, finer facies belonging to the *Lingula* community were deposited. Large volumes of volcanic ash frequently choked these marine environments.

1c Marloes cliffs. Continue the traverse to the southeast, noting a broad gully in the cliffs formed by an east–west-trending fault. Above this fault the Skomer Group is overlain unconformably by the Coralliferous Group. The latter comprises conglomerates and sandstones, which pass up into cleaved silty mudstones containing thin lenses of shell debris. The best fossil locality occurs in the roof of a small cave (SM 7872 0724) where the coral *Palaeocyclus* and the brachiopod *Eocoelia* are common, while higher beds contain abundant remains of the brachiopod *Costistricklandia*. The fauna indicates that the facies were deposited in nearshore-shelf (*Eocoelia* community) to offshore-shelf (*Stricklandia* community) environments (see Fig. 2.7; Bassett 1982).

1d Marloes cliffs. About 150 m farther to the southeast, the beds of the Coralliferous Group pass up transitionally into the Gray Sandstone Group, although this contact is obscured by debris. This group comprises interbedded flaggy sandstones, siltstones and mudstones, with rottenstone bands. Some 170 m farther to the southeast, the section is intersected by the eastward-trending Wenall Fault, which cuts out about 200 m of the group (Fig. 6.3). Above this fault there are good wave-polished sections of the upper part of the group on a stack located seawards of a very large landslip. Here the facies consist of erosive quartz arenite sandbodies (3–5 m thick) that fine and thin upwards into ripple-laminated sandstones, siltstones, and bioturbated mudstones containing occasional *Lingula* and crinoid debris (Fig. 6.4). The erosive bases of the sandstones commonly have thin lags containing black mudstone and phosphatic fragments, and a variety of green lithic clasts. The finer facies display climbing ripples, flaser bedding and highly bioturbated (*Skolithos* ichnofacies) units. Some of the beds display mud drapes, which die out laterally, suggesting that they represent lateral accretion surfaces. Hillier (2000) has interpreted the cycles as meandering tidal channels that flowed through northward-opening estuaries, which were cut during periods of falling sea level and were filled during subsequent sea-level rises.

1e Red Cliff. Towards the top of the Gray Sandstone, redbeds begin to appear in the succession. This change in colour marks the incoming of the fluvial facies characteristic of the ORS. An arrow chiselled into the rock face, located 1 m above beach level at the base of the second prominent red bed (SM 7899 0678), marks the position of the conformable contact, selected as a reference point ('golden spike') during the original survey of the area. The overlying Red Cliff Formation (Ludlovian) comprises red mudstones with calcretes and thin fining-upwards sandstones. At low tide it is possible to continue the traverse to the southeast to examine the coarser facies belonging to the Albion Sands Formation. However, if the tides are not suitable, this interesting facies association can be examined at the type locality of Albion Sands or to the west of Raggle Rock.

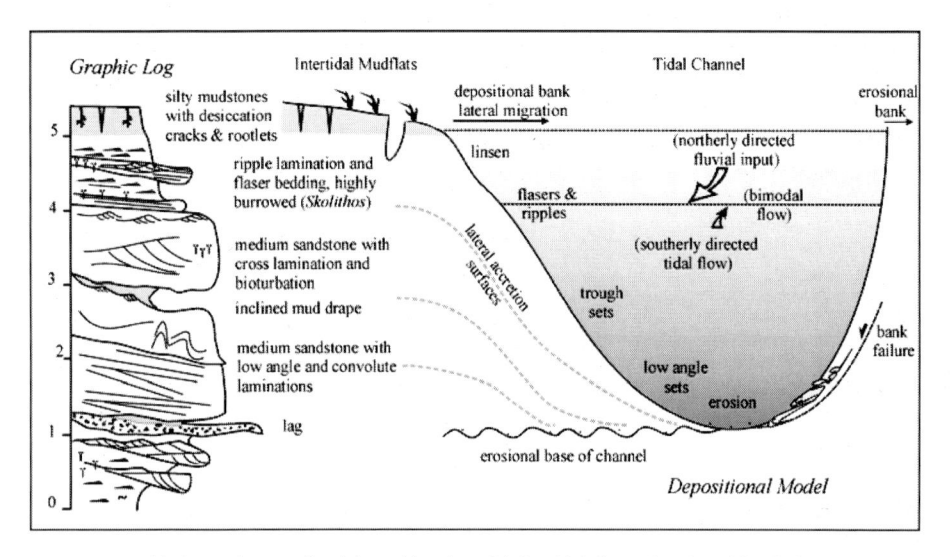

Fig. 6.4 Graphic log and generalized depositional model for tidal channel and mudflat facies recorded in the Gray Sandstone Group (model adapted from Hillier 2000).

1f Albion Sands. Return to Sandy Lane and follow the coast path westwards to a point adjacent to Gateholm Island where a path leads down the promontory (Horse Neck) to the beach (SM 7720 0765). Walk around the point into Albion Sands, named after the paddle steamer *Albion*, which was wrecked here in 1840 en route from Pembroke Dock to Bristol. To the north of the bay the marine Silurian passes transitionally up into the basal ORS (Red Cliff Formation), which comprises a succession (50 m thick) of thin, fining-upwards sandstones and red mudstones with calcretes that have yielded a Ludlovian spore assemblage (Hillier & Williams 2004). These distal, tidally influenced fluvial deposits are succeeded by the Albion Sands Formation (~100 m thick), which is composed of vertically stacked multi-storey sandstone bodies (Fig. 6.5). Intraformational conglomerates, composed of rip-up clasts of red mudstone, are common on the basal and internal erosional surfaces of the sandstones. The trough cross-stratified sandstones are very poorly sorted and contain sub-angular to sub-rounded granules and small pebbles of vein quartz and acid igneous rocks. The interbeds are composed of bright red calcrete-free mudstones with desiccation cracks and cross-laminated sandstone ribs. A prominent 3 m-thick band of mixed argillaceous facies occurs in the centre of the bay. Viewed from the beach the multi-storey sandbodies are seen to have sheet geometries, but individual sandstones are frequently channelized or wedge-out laterally (Fig. 6.6). This type of multi-storey and multi-lateral sandstone architecture is characteristic of low-sinuosity braided river deposits (Fig. 6.7). Deposition occurred in the relatively shallow pools of northeasterly-flowing braid channels that were filled and abandoned rapidly during flood events. The absence of planar cross sets from the sandbodies suggests that any sand-flat facies, developed between the braid channels, were continuously reworked and therefore had little preservation potential. Also, the rapid switching of the channels across the floodplain did not allow enough time for calcrete soils to form. A comparison of the facies, geometry and palaeocurrent modes of the braided river deposits at this locality with the meandering river deposits exposed at Freshwater West (localities *2g–h*), makes an interesting field project.

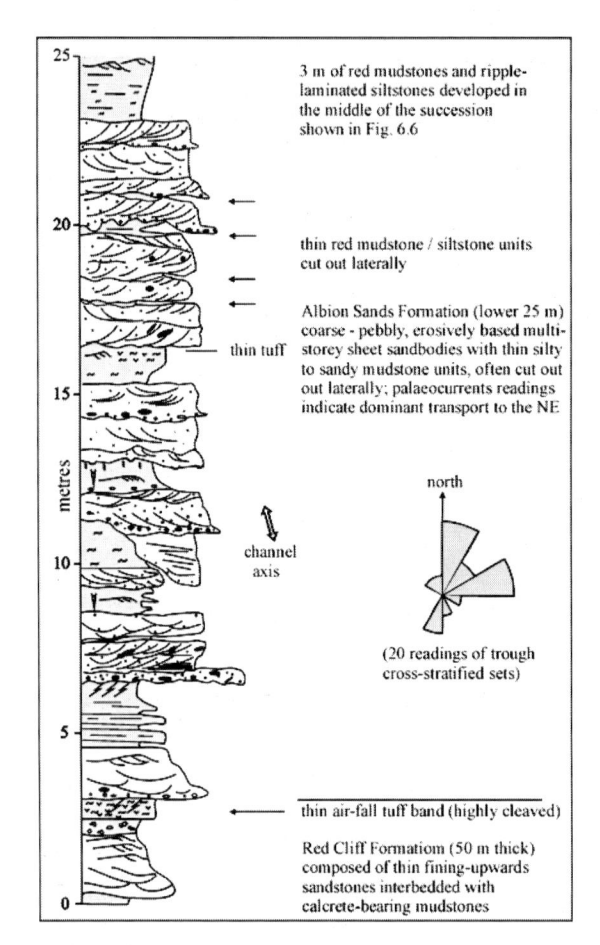

Fig. 6.5 Log and palaeocurrents for the lowermost 22 m of the Albion Sands Formation at the type locality. Note the erosive bases and sharp tops of the multi-storey sandbodies, which are interpreted as braided river channel fills.

Labels on log:
- 3 m of red mudstones and ripple-laminated siltstones developed in the middle of the succession shown in Fig. 6.6
- thin red mudstone / siltstone units cut out laterally
- thin tuff
- Albion Sands Formation (lower 25 m) coarse - pebbly, erosively based multi-storey sheet sandbodies with thin silty to sandy mudstone units, often cut out out laterally; palaeocurrents readings indicate dominant transport to the NE
- channel axis
- north
- (20 readings of trough cross-stratified sets)
- thin air-fall tuff band (highly cleaved)
- Red Cliff Formatiom (50 m thick) composed of thin fining-upwards sandstones interbedded with calcrete-bearing mudstones

Fig. 6.6 View of the Albion Sands Formation showing the multi-storey sheet geometries of the braided river sandbodies.

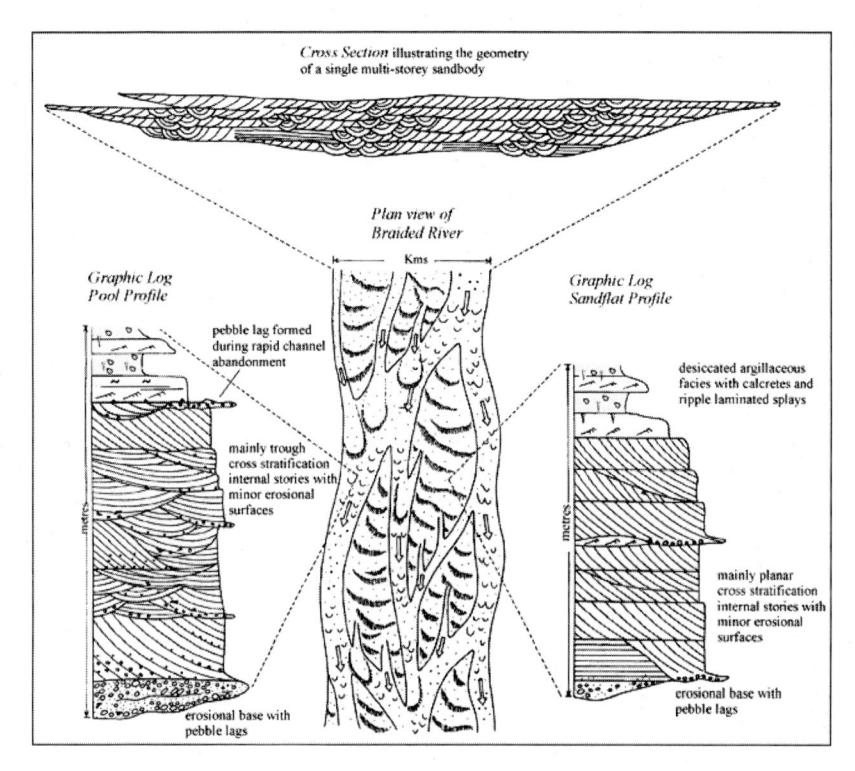

Fig. 6.7 Braided river depositional model and facies sequences (modified from Nemec 1992).

Before returning to the coast path it is well worth examining the sets of quartz veins that cut the sandstones at Albion Sands or those present in the same facies on the eastern side of Gateholm Island (SM 774 078). Many of the sandstones display spectacular conjugate shear zones with *en échelon* quartz-filled tension gashes and pressure-solution cleavages.

1j Martin's Haven and the Deer Park. It is possible to walk 2 km northwestwards on the coast path to Martin's Haven; alternatively you can return to the Runwayskiln car park via the path that crosses the fields to Marloes Mere, and then continues eastwards to the youth hostel. Take the unmade track at the rear exit of the car park and join the minor road that terminates near another National Trust car park at Martin's Haven. There are tourist information centres at Lockley Lodge and the Skomer Marine Nature Reserve office, both of which have interesting information on the wildlife and details of the popular boat trips to and around the islands. Skomer and the neighbouring islands of Skokholm and Grassholm are renown for their colonies of Manx shearwaters, gannets, storm petrels, razorbills and puffins. (N.B the puffins leave the islands towards the end of July). In the autumn, Atlantic grey seals and their pups can be seen in pebbly coves on the islands and at Jeffry's Haven on the mainland.

Since the coves around the tip of the Marloes Peninsula (Deer Park) are mostly inaccessible, some view stops are suggested to summarize the geology and scenery of the area. Enter the Deer Park via the stone gateway; climb the embankments of the Iron Age fort and follow the path to the old coastguard lookout (SM 7588 0922), which stands on a dark basaltic flow extending westwards to Wooltack Point. The basalt is vesicular and has a fine grain texture with small phenocrysts of feldspar. From the lookout there are views to the north across St Brides Bay to the St David's Peninsula and Ramsay Island; from Wooltack Point there are views of the islands of Skokholm (southwest) and Skomer (west). The southern headland of Jeffry's Haven has a natural arch eroded into a basaltic flow. In the small cove farther to the south, the sedimentary rocks of the Skomer Volcanic Group are deformed into a tight faulted syncline. The basaltic flow that forms the prominent headland known as the Anvil, has been correlated with the two flows seen at Marloes Sands. A further basaltic flow forms the northern headland of Renney Slip (see Bassett 1982 for details of the stratigraphy of Renney Slip and the bays to the southeast).

6.3.2 Freshwater West

From Pembroke take the B4319 road to Castlemartin and then travel a further 3.5 km to Freshwater West. The car park adjacent to the beach has toilets and there is often a mobile café open during the summer months. Coaches cannot continue on the B4319 to West Angle because of a narrow bridge just north of the car park (Fig. 6.8). The low rock outcrops on the foreshore to the south of the sandy bay, expose steeply dipping and southward-younging strata ranging in age from Silurian (Wenlock) through the Devonian (ORS) to the Lower Carboniferous (Avon Group). These strata are cut by the Flimston Bay Fault, which has a dextral displacement of 155 m in the centre of the section. To the south of Little Firzenip, two dextral splays of the main fault are developed. Periodically it is possible to see small patches of weathered black shales of Ordovician (Llanvirn) age to the northeast of the main outcrop. In the centre of the bay the oldest of these shales form the core of the ESE-trending Castlemartin Corse Anticline. During very low tides, relics of the postglacial submerged forest are exposed in the bay.

The section is sedimentologically important and has been well documented (see Williams *et al.* 1982). The main aim of the itinerary is to examine the excellent ORS fining-upwards sequences exposed on wave-polished outcrops to the south of Little Furzenip. This part of the section is fully exposed from about mid-tide. Note that the southern half of the section is within the boundaries of the Castlemartin artillery range and permission to visit localities *h–j* should be obtained from the commandant of Merrion Camp. A red flag is hoisted at the lookout on Great Furzenip when firing is in progress. The Pembrokeshire National Park has an office at the camp, where a field ranger can be contacted for information regarding access and guided walks into the firing range (Tel 01646 662213).

2a. Stream. Before examining the geology, a little time should be allocated to examine the sedimentary structures forming in the sandy bed of the stream that crosses the beach. Linguoid ripples can usually be seen forming in the shallow low-energy margins of the stream; these become washed out as current velocities increase and sand is transported as a traction carpet over a flat bed. In the higher-energy central parts of the stream, large amounts of sand can be seen in suspension and the water surface displays standing waves that periodically break up stream. The flow velocities (cm/sec) characteristic of the various bedforms and surface features can be roughly calculated by measuring the time it takes for floating

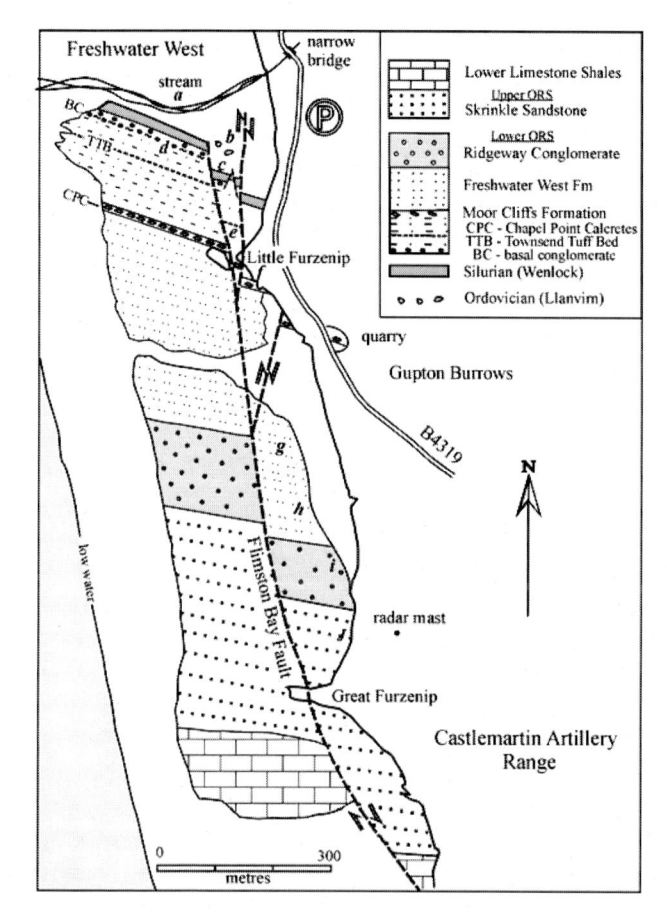

Fig. 6.8 Geological and locality map for Freshwater West (modified from Dixon 1921; Williams *et al.* 1982).

markers to travel a measured interval. The flow conditions are summarised in Fig. 6.9, but note that the water in this stream is too shallow for the formation of dune bedforms.

2b–d Beach. North of the main foreshore outcrop, small patches (about 1 m^2) of dark grey to black shales can be seen just below the surface of the sand (SR 8850 9957). These near-vertical cleaved shales (strike 110–290°) have in the past yielded the graptolite *Didymograptus bifidus*, dating them as Lower Ordovician (Llanvirn Series). Farther south, about 10 m of interbedded conglomerates, hard sandstone and fossiliferous limestones form a more continuous outcrop. This succession begins with a basal polymict conglomerate containing abundant angular to sub-rounded fragments of vein quartz (up to 4 cm in diameter) and sandstone, mixed with elongate to rounded mudstone clasts. This conglomerate is succeeded by greenish-grey calcareous sandstones and siltstones, which have generally been homogenized as a result of bioturbation, but occasionally, display parallel and low-angle laminations. The above facies also contain thin lenses of limestones (frequently decalcified to rottenstone), which have an abundant shelly fauna of

(a)

(b)

Fig. 6.9 (a) Relationship between bedforms and flow velocity in a shallow stream transporting fine to medium sand. (b) Stream showing sanding waves and upstream breaking waves in the high flow velocity central zone, passing laterally into lower energy zones with flat bedding and linguoid ripples (flow bottom to top).

brachiopods (rhynchonellids and chonetids) and debris of gastropods and bivalves. This succession is correlated with the Gray Sandstone Group of late Wenlock age, which is more fully exposed at Marloes Sands (see locality *1d*). The base of the conglomerate therefore marks the position of a major unconformity.

The previously described Silurian strata are truncated by a further conglomerate, which marks the unconformable base of the ORS (a time gap of approximately 6 million years). The conglomerate is about 18 m thick and is composed of framework-supported, sub-rounded to rounded pebbles and cobbles (up to 40 cm in diameter) of sandstone, quartzite, vein quartz and more elongate clasts of green silty mudstone. Four erosionally based beds of conglomerate are present, each being capped by green sandstones from which a sparse fauna of lingulids, plants and ostracoderm fragments have been recorded. This conglomerate is also well exposed on the foreshore outcrop to the west of the Flimston Bay Fault.

2e North of Little Furzenip. Above the basal conglomerate the succeeding Lower ORS is composed mainly of red and green mudstones with calcretes and subordinate siltstones and fine sandstones. Within this succession the Townsend Tuff Bed (SR 8849 9947) forms a distinctive yellow-weathering unit (2.8 m thick) containing an interesting sinusoidal bedform with a wavelength of 30 cm and amplitude of 3 cm. It is suggested that this sedimentary structure, and similar ones observed in the Priest's Nose Tuff Bed at Manorbier, were formed as a result of tsunamis, which inundated the continental plain and were triggered by the Plinian volcanic eruptions.

2f South of Little Furzenip. Farther to the south the promontory of Little Furzenip has been detached from the mainland by splays of the Flimston Bay Fault. Just to the south of the headland a thick development of red mudstone, containing dense pedogenic calcrete nodules and tubules, can be correlated with the Castle Point Calcrete that defines the top of the Moors Cliff Formation. Here the calcretes are deformed into features referred to as pseudo-anticlines, which were formed as a result of pressures that built up during the process of soil formation. It has been estimated that palaeosol features such as these took approximately 10 000 years to form.

2g–h Rock platform. Continue 200 m to the south, where the vertical strata belonging to the Freshwater West Formation are free of seaweed and have been polished by wave action. This part of the beach exposes many classic fining-upwards cycles (Fig. 6.10a). Each cycle begins with a low relief erosional surface, generally marked by a thin intraformational conglomerate composed of red–green mudstone clasts and pale calcrete fragments (facies a). Some of the lags also contain small quantities of blue–black scales, spines and bones of armoured fish (ostracoderms). The erosional surfaces of the cycles are overlain by medium-grain sandstones with cross sets (facies b) and parallel laminations (facies c), which pass up into ripple-laminated silty sandstones (facies d). These sandstone facies are succeeded by siltstones and silty mudstones with impoverished ripples and sandy streaks (facies e), which are gradually replaced by mudstones with calcretes (facies f). It is often possible to see evidence of lateral accretion surfaces by tracing the intraformational lags and sandstone facies along the strike. These fining-upwards cycles were deposited in meandering (high-sinuosity) streams, which flowed southwards across an alluvial plain stretching from Pembrokeshire into the Welsh Borderlands (Figs 2.10a, 6.10b).

The argillaceous facies often contain the large vertical cylindrical burrows of *Beaconites barretti*. These trace fossils are thought to have been made by lungfish, which burrowed into the wet river sediments during periods of drought and emerged during the next flood. They may represent part of the evolutionary link between fish and the newly discovered *Tiktaalik roseae*, which is believed to have been the first animal to make the transition from water to land about 375 million years ago, during the late Devonian.

Higher in the succession, two relatively thick green sandstones (5 m and 8 m thick respectively) occur in the succession. The polished surfaces of these sandstones facies display excellent examples of cross stratification, parallel lamination and ripple-drift cross lamination. A 75 cm intraformational conglomerate that truncates the top of the uppermost green sandstone defines the base of the Ridgeway Conglomerate.

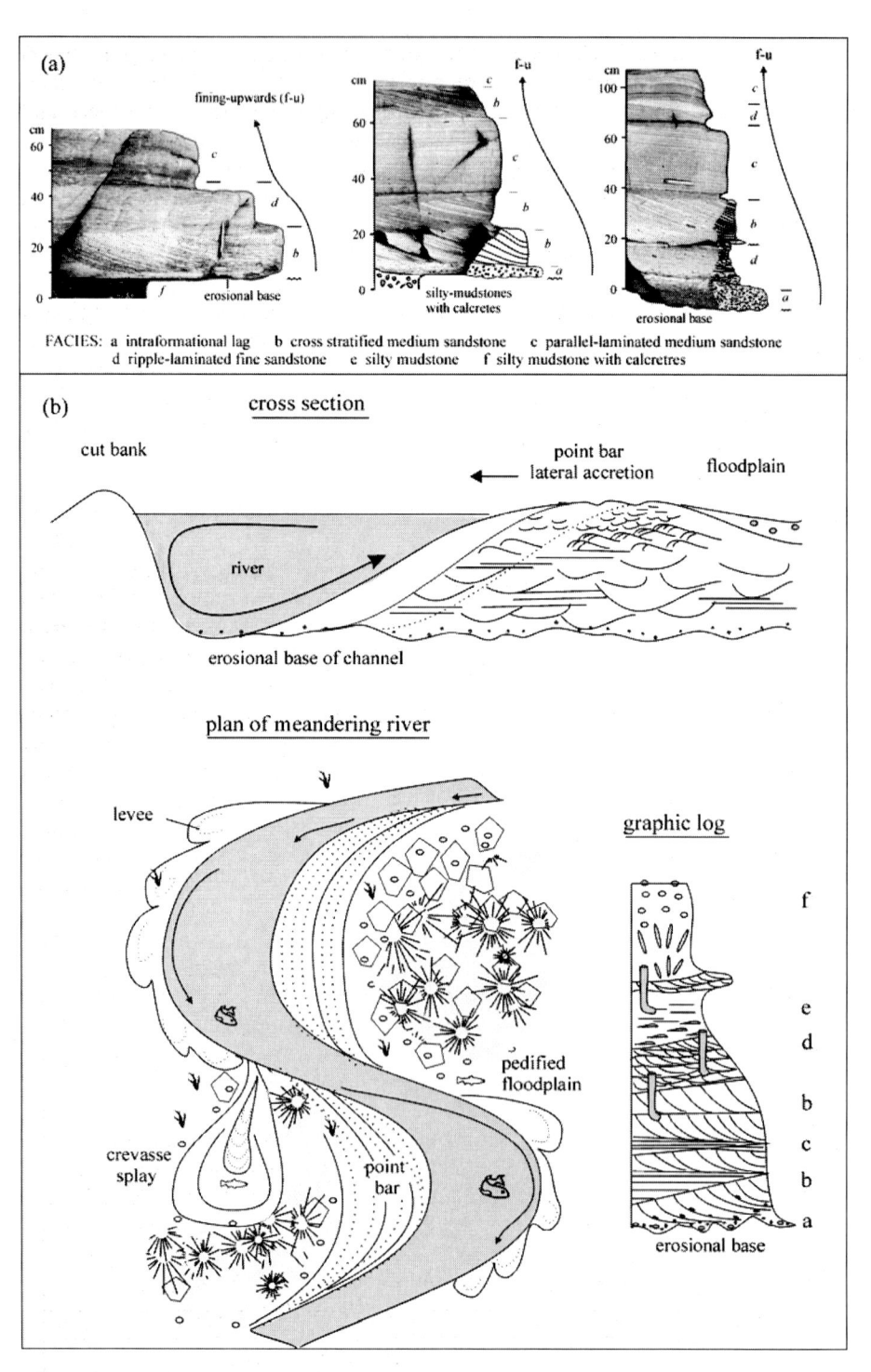

Fig. 6.10 (a) Three examples of fining-upwards cycles from the Freshwater West Formation. (b) Depositional model for the ephemeral meandering river and floodplain facies in the above formation.

FACIES: a intraformational lag b cross stratified medium sandstone c parallel-laminated medium sandstone d ripple-laminated fine sandstone e silty mudstone f silty mudstone with calcretes

2i Rock platform and low cliffs. The Ridgeway Conglomerate (115 m thick) comprises an alternating succession of erosionally based polymict conglomerates and coarse sandstones that are interbedded with bright-red gritty mudstones and calcretes. The conglomerates are composed of sub-rounded to rounded fragments of vein quartz, sandstone, siltstone and elongate clasts of green phyllite. The majority of the conglomerates are clast-supported and they display isolated sets of trough cross bedding and occasional parallel lamination, graded bedding and pebble imbrication. Less commonly the conglomeritic facies are matrix supported and they contain out-sized clasts and irregular gravel lenses. This association of conglomeratic facies is typical of alluvial fans deposited as a result of stream-flood and debris-flow processes. Palaeocurrent measurements indicate that the alluvial fans were derived from a southerly (proximal) source area located in the Bristol Channel (Williams *et al.* 1982; Fig. 2.10a). To the north the conglomerate thickens to over 350 m into the hanging-wall of the Ritec Fault (see Hillier & Williams 2007 for a detailed and up-to-date summary)

2j Cliffs north of Great Furzenip. The disconformable contact between the Ridgeway Conglomerate (Lower ORS) and the Skrinkle Sandstone (Upper ORS) is about 14 m above the top of a 9 m conglomerate bed that outcrops in the cliffs below the boundary fence of the firing range. The basal 55 m of the Skrinkle Sandstone represents a major multi-storey coarsening- and thickening-upwards sequence interpreted as the deposits of a prograding terminal fan (Marshall 2000) (Fig. 6.11). Palaeocurrent measurements indicate that the fan prograded to the southeast. The overlying 20 m of thinly interbedded sandstones, siltstones and mudstones facies of fluvio-lacustrine origin record the abandonment of this fan. The succeeding 40 m of the Skrinkle Sandstone is composed of light grey and pale purple multi-storey sandstone bodies that have been interpreted as a braid plain complex. Unfortunately, the upper 185 m of the Skrinkle Sandstone and its transitional contact with the Lower Limestone Shales, exposed south of Great Furzenip, is in the danger zone of the firing range and it is not possible to visit this area. However, this part of the succession can be seen at the type locality of Skrinkle Haven.

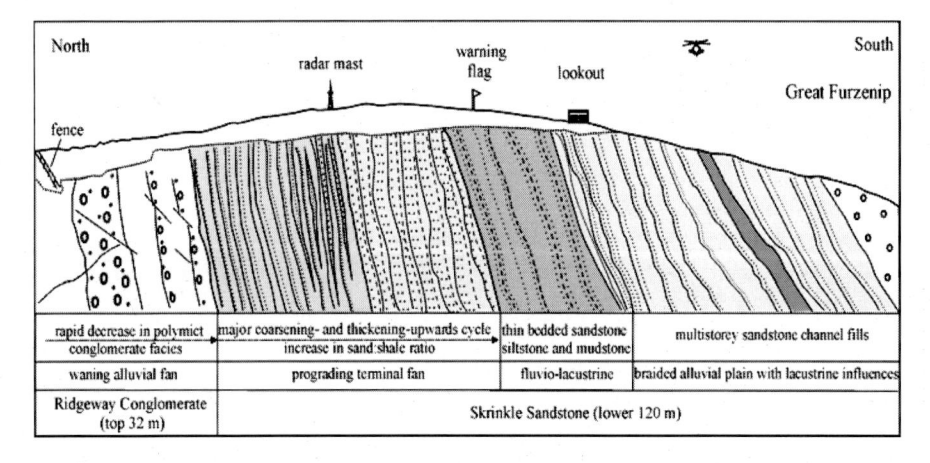

Fig. 6.11 Cliff section at the southern end of Freshwater West showing the facies and interpretation of the Ridgeway Conglomerate and part of the Skrinkle Sandstone (adapted from Marshall 2000).

6.3.3 Manorbier

From Tenby travel on the A4139 through Penally and Lydstep and then take a sharp left onto the B4585 to the small village of Manorbier, which has a large car park located below Manorbier Castle (Fig. 6.12). This well-preserved Norman castle dates from the middle of the 12th century and was the birthplace of the Welsh historian Geraldus Cambrensis, who described the area as "the pleasantest spot in Wales". More recently the castle was used as one of the locations for the film *I Capture the Castle*, based on the novel by Dodie Smith. Opposite the castle is the parish church, with its recently renovated and controversial limewashed tower. The quaint village has a small hotel, public house (Castle Inn), supermarket and café, and there are many campsites located in the surrounding countryside.

This half-day itinerary is important because all of the ORS sites can be studied at high tide. It is most conveniently combined with the itinerary for Skrinkle Haven, which requires medium to low tides and extends the geological traverse from the ORS into the Lower Carboniferous.

3a Manorbier Bay. From the car park, follow the path down the valley to the beach, turn left and continue to the southeast corner of the bay (Fig. 6.12). At this locality (SS 0615 9745) almost vertical ORS strata (Freshwater West Formation) are exposed in low cliffs located at the back of the beach above high-water mark. The section displays six fining-upwards cycles (Fig. 6.13). Each cycle begins with a low-relief erosional surface, marked by a mud-chip lag, and passes up into medium to fine sandstones capped by silty mudstones containing calcretes. These fining-upwards cycles are interpreted as the fills of ephemeral fluvial channels with sinuous profiles. The limited exposure makes it impossible to detect any evidence for the lateral accretion surfaces that characterize meandering channels.

3b Kings Quoit. Join the cliff path that ascends to the headland of Priest's Nose, and walk to an impressive late Neolithic burial chamber known as King's Quoit (SS 059 973). The tomb comprises a massive capstone weighing over 10 tonnes, supported by three uprights, one of which has collapsed. All of the monument stones are composed of red silty mudstone crowded with paler carbonate nodules (calcretes),

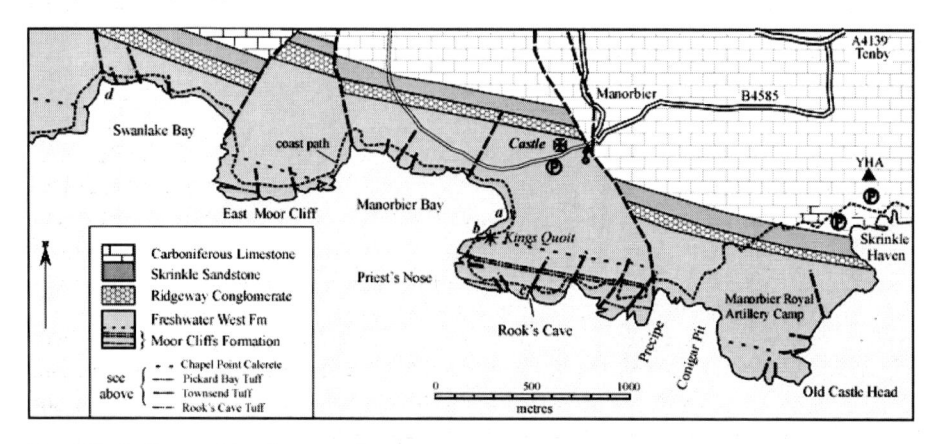

Fig. 6.12 Geology and locality map of the coast between Swanlake Bay and Skrinkle Haven (modified from Williams *et al* 1982).

which were derived from the vertical strata exposed in the adjacent outcrop. It has been suggested that the stones of many burial chambers in southwest Wales were originally placed to replicate aspects of the surrounding landscape. In the case of King's Quoit, the profile of the inland-pointing capstone follows the spur leading down to Manorbier Valley. Also, when viewed from the east, the profile of the capstone would originally (before the collapse of the upright) have followed the skyline of East Moor Cliff.

The thick calcretes exposed at the burial chamber correlate with the Psammosteus Limestone (renamed the Castle Point Calcretes by Williams *et al.* 1982), which marks the top of the Moor Cliffs Formation. From the burial chamber there are good views of Manorbier Castle, the bay and the wave-cut platform (Freshwater West Formation) extending westwards to East Moor Cliff (Moor Cliffs Formation). Here the near-vertical ESE–WNW-striking strata are cut by a number of prominent cross faults. Also seen are gulleys running parallel to the strike, formed as a result of the erosion of volcanic airfall ash (tuff) bands, up to 3.5 m thick. Two of these sheer gullies cross the coast path 30 m and 50 m to the south of King's Quoit. The Pickard Bay Tuff causes the first gully (protected by iron railings), and the second gully marks the position of the Townsend Tuff. A further airfall tuff (Priest's Nose Tuff) is exposed at the next locality. The Chapel Point Calcretes and the tuff bands form important marker beds, which have been used for correlation throughout the area.

3c Priest's Nose. Continue along the cliff path, beyond Priest's Nose, to the cliffs located to the west of Rook's Cave (SS 0618 9703). Here it is possible to leave the main path and cross the grassy slope leading down to a rock platform composed of red silty mudstones with calcretes and thin laterally extensive channel sandstones. A prominent indentation, near the landward side of the platform, marks the position of the Rook's Cave Tuff (1.4–1.9 m thick). The base of the tuff preserves a variety of burrows and faecal pellet trails on the topmost bedding plane of a light green silty sandstone (Fig. 6.14). These delicate trace fossils were preserved as a result of rapid burial by the overlying airfall ash that also preserved the original topography of the contemporaneous floodplain. The tuff displays parallel laminations, and interesting hummocky bedforms occur at two horizons along the outcrop. These bedforms, like those observed in the Townsend Tuff at Freshwater West, were possibly formed as a result of tidal waves (triggered by the volcanic eruptions) inundating the low-lying floodplain. In a detailed study of the Moor Cliffs Formation, Love & Williams (2000) concluded that the sequences were deposited on an extensive low-relief (distal) floodplain supplied by occasional shallow ephemeral rivers and streams. In this setting the mature carbonate palaeosols were preferentially formed in the more poorly drained low-lying areas of the floodplain and also immediately above the tuff horizons. They also deduced that the Devonian climate was sub-tropical monsoonal. From this locality it is possible to continue 1 km eastwards along the coast path to a point (SS 0690 9703) were steps allow access to the small coves of Precipe and Conigar Pit. The sedimentology and geometry of the sandstone facies developed in the Freshwater West Formation between Manorbier and Conigar Pit have been described by Williams *et al.* (1982) and more recently by Hillier and Williams (2004). Farther eastwards the coast path is diverted around Old Castle Head and Manorbier Camp, whence it continues to Skrinkle Haven.

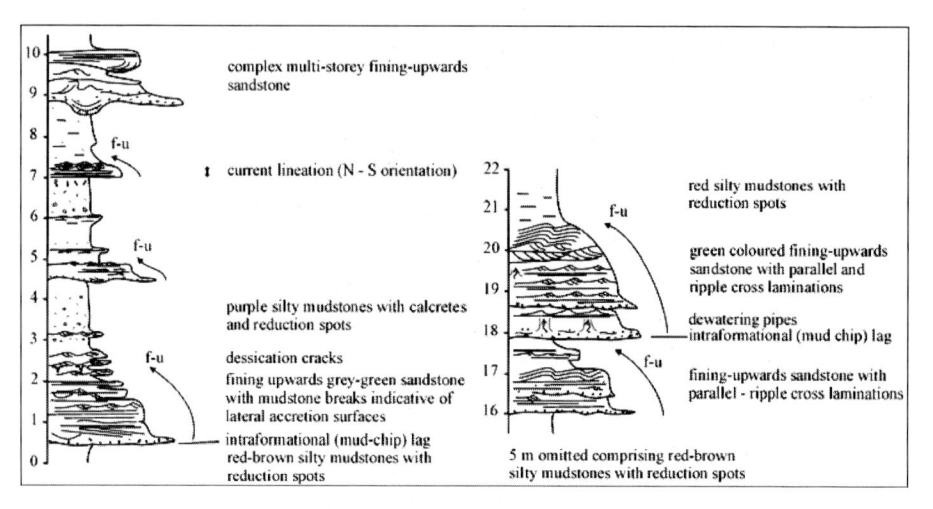

Fig. 6.13 Fining-upwards cycles developed in the Freshwater West Formation at Manorbier beach,

Fig. 6.14 Faecal pellets and burrows preserved on the bedding plane of a green silty sandstone developed below the Rook's Cave Tuff.

3d Swanlake Bay. Swanlake Bay is reached by making the quite strenuous 2 km walk westwards from Manorbier along the coast path, or by driving 1 km west of Jameston and following the footpath through West Moor farm. A large north-northwest-trending fault intersects the ORS strata exposed at the western corner of the bay (SS 0438 9806). This tear fault has a dextral displacement of about 300 m and an 8 m downthrow to the east. To the east of the fault plane there is an interesting wave-polished section through a variety of ORS fluvial facies with excellent examples of intraformational lags composed of reworked calcrete nodules and red mudstone clasts. These lags are succeeded by multi-storey fining-upwards cycles displaying a range of sedimentary structures and mature calcrete palaeosols (Figs 6.15, 6.16).

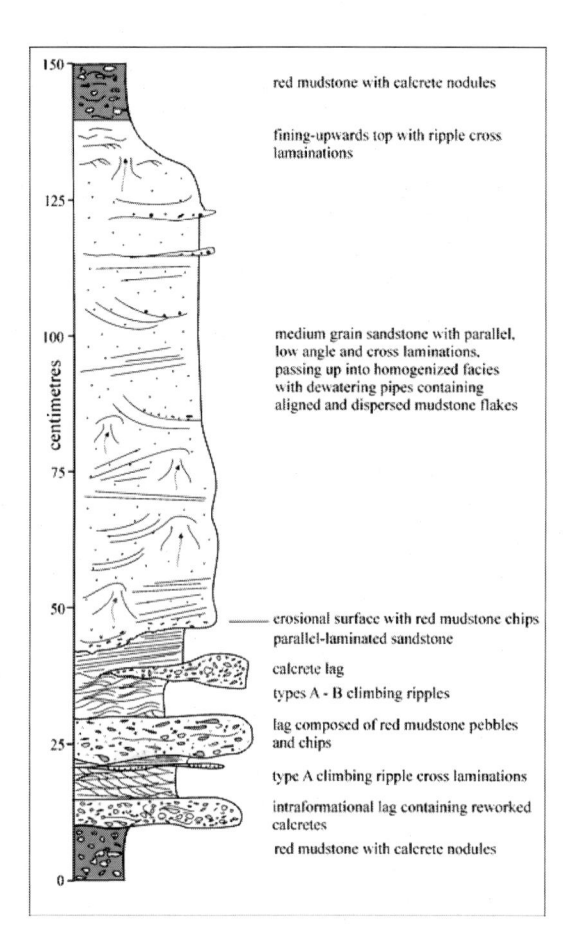

Fig. 6.15 Profile log of a fining-upwards cycle with multiple lags exposed at Swanlake Bay (Freshwater West Formation).

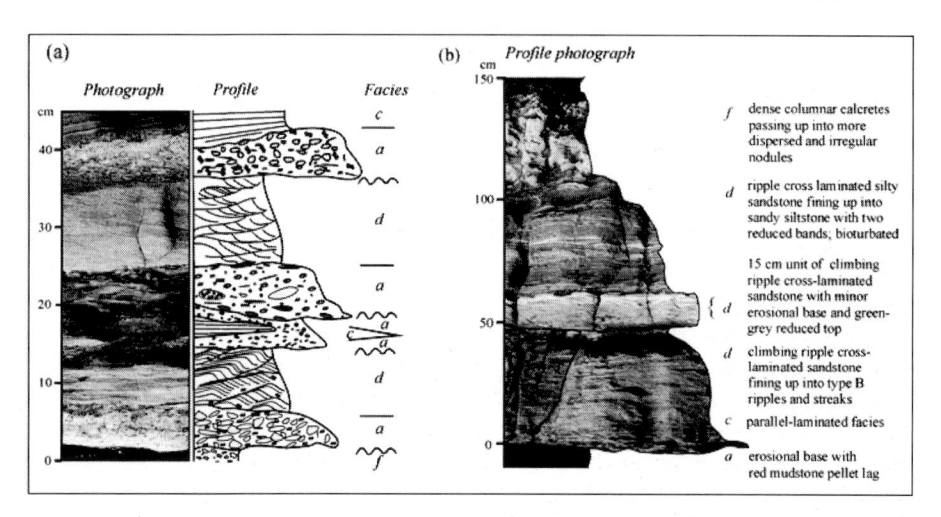

Fig. 6.16 (a) Multiple lag developed at the base of the fining-upwards cycle illustrated in Fig. 6.15. (b) A further fining-upwards cycle seen at Swanlake Bay (see Figure 6.10 for key to facies a–f).

6.3.4 Laugharne

The pretty village of Laugharne is located 7 km south of St Clears, on the A4066 to Pendine. The village is on the tourist route because of its association with the Welsh poet Dylan Thomas (1914–1953) who lived here for the final four years of his short life. Many of the characters and anecdotes that appear in his most popular work, *Under Milk Wood,* were based on his experiences in Laugharne and New Quay, another small seaside town in Ceredigion. His Boat House is open to the public and his favourite hostelry (the bar at Brown's Hotel) still serves a pint of "warm Welsh bitter". The poet is buried in St Martin's churchyard on the outskirts of the village. Laugharne Castle, a fine example of a Norman fortification, is also open to the public (see Fig. 1.3). The writer Richard Hughes, best known for his novel *A high wind in Jamaica*, was a previous owner of the castle.

The aim of this short itinerary is to examine outcrops of typical ORS facies that are exposed on the banks of the Taf estuary. Laugharne provides a convenient lunch stop when travelling from Carmarthenshire into south Pembrokeshire, and *vice versa.*

4a Steps below Dylan's Walk. From the car park, located on the Strand, cross the small bridge over the stream and take the path (signposted to Dylan Thomas Boat House), which runs below the castle and parallel to the low cliffs on the banks of the estuary of the River Taf (see Fig. 1.2). These cliffs expose an interesting and easily accessible section of Lower ORS strata that dip at 10° to the SSW. Follow the path to the second set of steps leading up the cliff to Dylan's Walk and the Boat House (SN 305 110). Here the oldest strata comprise red mudstones and siltstones with immature calcretes, which are exposed from the end of the path to the walls of the Boat House. An erosionally based 1 m-bed of medium to fine sandstone, exposed just to the left of the "dangerous cliffs" sign and lifebuoy, displays small cross sets and good examples of climbing ripples, both of which indicate south-westerly palaeoflow (Fig. 6. 17). This sandstone is succeeded by a very dense calcrete bed (50 cm thick), and a further 13 m of red silty mudstones that display both columnar and fanned calcretes. The latter structures, referred to as pseudo-anticlines, were formed as a result of internal pressures that built up in the soil profile during pedogenesis.

4b Steps below Market Lane. Walk back along the path to the first set of steps, where a more complex succession of ORS facies occurs in a cleft in the cliffline. Exposed at the base of the cliff is a thin bed (15 cm) of intraformational conglomerate, composed predominantly of reworked calcrete nodules. Higher in the succession a very interesting and internally complex sandbody is present (Fig. 6.17, 17.5–21.0 m). This sandbody begins with a basal mélange, which is composed of thin irregular to lensoid sandstone stringers, interbedded with silty–sandy mudstones containing desiccation cracks, dyke injections and isolated intraformational clasts. The erosional surface is succeeded by a fining-upwards multi-storey sandstone composed of parallel-, cross- and ripple-laminated facies that often pass laterally into one another. Vertical burrows of *Beaconites*, up to 10 cm in diameter and with downward-extending meniscate laminae, are very common in this sandstone. Circular cross-sections of the burrows, seen on the erosional base of the lowermost sandstone, display concentric meniscate laminae and small mudstone chips.

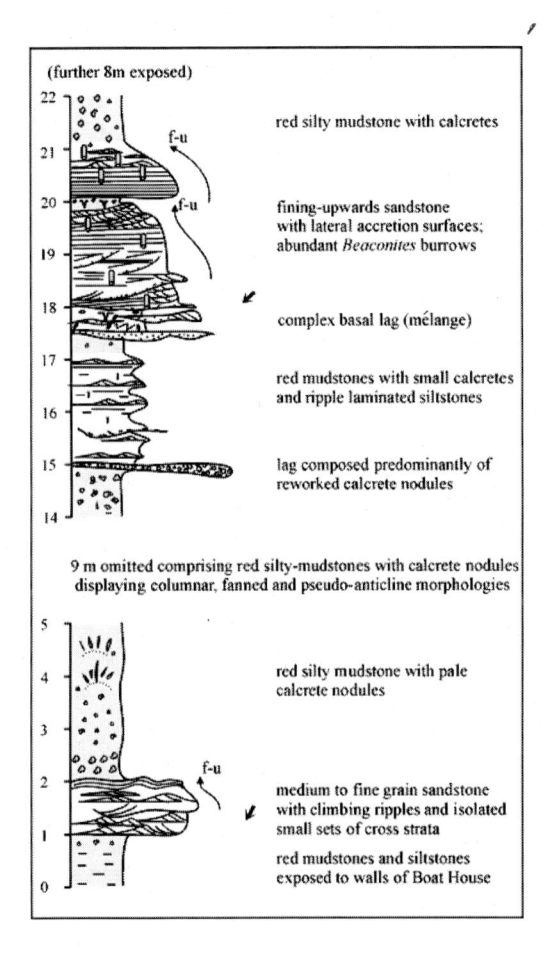

Fig. 6.17 Graphic log of ORS fining-upwards cycles, associated with thick calcretes, exposed along the banks of the Taf Estuary at Laugharne.

The sandstone is overlain by 9 m of red silty mudstones with calcretes, which is truncated by a further thin sandstone with ripple-drift cross laminations (type A ripples passing up into type B ripples).

The minor sandbodies described above are interpreted as the deposits of meandering ephemeral rivers (Fig. 6.10b). Channel migration resulted in the deposition of fining-upwards cycles with lateral accretion surfaces. The ephemeral nature of the rivers is suggested by the isolated lags devoid of channel fills and also the occurrence of desiccation cracks and calcretes in the floodplain facies. The thick mature pedogenic horizons were formed during extended periods of drought when rivers were diverted to other parts of the floodplain. Also, the *Beaconites* possibly record arid periods when riverbeds dried out and the lungfish burrowed into the wet fluvial sediments for refuge

6.3.5 Skrinkle Haven and Lydstep

From the B4585 take the minor road that leads to Manorbier Royal Artillery camp; turn sharp left at the guard post and continue to the modern-looking youth hostel that was converted from an army building. There are many parking areas, but those farthest to the east (located on old gun positions) have the best views (Fig. 6.18).

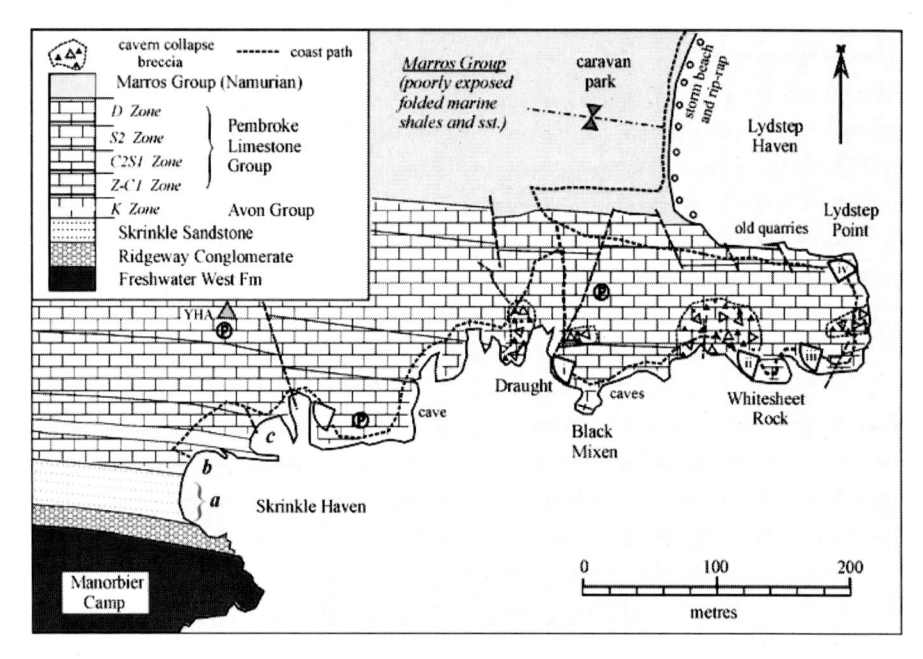

Fig. 6.18 Geology and locality map of the coast between Skrinkle Haven and Lydstep Haven (modified from Dixon 1921).

Skrinkle Haven comprises three coves (west, middle and east) divided by two impressive projecting vertical units of limestone (Fig. 6.19). Access to middle cove is via a sturdy set of metal steps, but unfortunately the path down to the west cove has been closed to the public for some time. The latter point of access should be taken (if and when it has been re-opened) to examine the section in stratigraphical order. During spring low tides a cave in the Church Doors Limestone allows access between the west and middle coves, and an impressive natural arch in the Horsesback connects the middle and east coves (Fig. 6.20).

5a West Cove. Access from the middle cove through the small cave in the Church Doors Limestone is possible only on a falling spring tide and allows a study time of 1–2 hours at low water. In the southern corner of the cove, fallen blocks of rock obscure the contact between the Ridgeway Conglomerate and the overlying Skrinkle Sandstone. The Skrinkle Sandstone Group is about 90 m thick and consists of a variable succession of clastic facies that can be divided into five units. The basal unit (12 m thick) is composed of a coarsening- and thickening-upwards grey sandstone capped by a pinkish-red fining- and thinning-upwards unit. This distinctive cycle is interpreted as a prograding to laterally migrating terminal alluvial fan. The second unit is composed of 20 m of red mudstones, siltstones and thin sandstones with occasional calcrete horizons, which are interpreted as mixed floodplain and fluvial channel deposits. At least five conglomeratic to coarse sandstone units appear in the overlying third unit (about 36 m thick) and these are interpreted as sheetflood and braided channel deposits. The fourth unit comprises 10 m of interbedded red and green sandstones and marls with decalcified calcretes, interpreted as mixed fluvial and coastal plain deposits. The uppermost unit (10 m thick) begins with a 30 cm bed of variegated shale containing *Lingula*, and bivalves,

Fig. 6.19 Skrinkle Haven with the prominent Church Doors (C D Lst) and Horseback limestones. Note the natural arch in the latter and the cave (indicated by a star) through the former, which connects the middle and west coves.

including *Modiola* and a variety of nuculids, which is overlain by a coarsening-upwards sequence. This final cycle is interpreted as a transgressive barrier-bar sequence, which preceded the Dinantian (Lower Carboniferous) marine transgression (for a detailed account see Marshall 2000).

5b West Cove. The transitional contact between the Skrinkle Sandstone and the Lower Limestone Shales (Avon Group) is recorded by a relatively rapid change from brown to grey facies, and by the appearance of limestone and calcareous shales containing marine fossils. These fossils include crinoid ossicles, bivalves, brachiopods (*Chonetes*, *Productus*), bryozoans and abundant phosphatic remains of fish debris and internal casts of *Orthoceras*. About 25 m above the base of the group a conspicuous bed of calcareous sandstone has well preserved traces of *Rhizocorallium*. Shales and impure limestones, with further fish remains and phosphatic debris, overlie this bed.

5c Middle Cove. The Church Doors Limestone (15 m thick) is composed predominantly of fine-grain cross-laminated oolitic limestones. This facies is succeeded by about 40 m of lime mudstones, darker grey to black shales and thin limestones, which form a topographic hollow extending inland (Fig. 6.19). The paler mudstones are very highly bioturbated, with the branching traces of large *Thalassinoides* conspicuous on the bases of some beds. The thin erosively based limestone beds contain abundant shell debris, intraclasts and phosphatic fragments. Higher in the succession the transitional contact between the Lower Limestone Shales and the Black Rock Limestone (*Zaphrentis* Zone) is recorded by an increase in the proportion of limestone and chert beds. To the north of the steps the faulted and calcite-veined limestones contain solitary rugose corals (*Zaphrentis*) and

Fig. 6.20 Impressive natural arch in the Black Rock Limestone forming the Horsesback at Skrinkle Haven.

Fig. 6.21 View, looking due west from locality (ii) Lydstep Peninsula, of the vertical Carboniferous cliffs showing saddle and ridge topography.

brachiopods (*Spirifer*), and many bedding planes are covered with large *Thalassinoides*. In the past, specimens of crinoid calyxes have been found in the fallen blocks at the back of the cove. The limestones forming the Horsesback spur and natural arch (Fig. 6.20) are slightly younger because of the presence of a cross fault.

5d Lydstep Peninsula. The Lydstep Peninsula (National Trust) can be visited when the tides are too high to complete the Skrinkle Haven itinerary. This impressive flat headland, enclosed by sheer cliffs, is crossed by marked paths leading to many scenic viewpoints (Fig. 6.18, localities i–iv):

i. A view of the U-shape dry valley and Lydstep caverns with disorientated blocks of collapse breccia forming the western face of Draught Cove; a large blowhole can be seen farther to the west.

ii. Looking westwards from a grassy hollow (saddle) in the promontory, there are good views of the vertical limestone strata extending to Skrinkle Haven; a large mass of cavern collapse breccia can be seen in the foreground.

iii. An excellent view of the saddle-and-ridge topography caused by differential weathering of thin-bedded gastropod limestones and thicker bedded facies developed in the High Tor Limestone. Also note the cavern collapse breccia deposits (Fig. 6.21) and the raised beach present in the centre of the adjacent cliff.

iv. A view to the northwest looking across the storm beach of Lydstep Haven to the limestone cliffs and caves extending eastwards to Giltar Point.

Lydstep Haven has been eroded in Namurian shales, with minor cherts and sandstones, which are preserved in the core of the Lydstep Syncline. The geology of the southern limb of this syncline continues eastwards to Caldy Island.

6.3.6 *Tenby (South Beach)*

The historic old town of Tenby, is partially enclosed by medieval walls, a well preserved section of which can be seen on either side of the Five Arches; the original south gate to the town. Quay Hill runs down to the pretty harbour and has some of the oldest buildings, including the Tudor Merchant's House (National Trust), which has an interesting collection of antique and modern Welsh oak furniture. The ruins of the castle can be seen on Castle Hill, which is also the site of a small museum with interesting geological displays. The impressive new lifeboat station and public viewing gallery, houses a £2.5 million state-of-the-art Tamar class lifeboat (the Haydn Miller).

Tenby is a very popular and busy holiday resort, and parking can be difficult during the summer months. The most convenient parking for this itinerary is either along the Esplanade or in the Rectory Field car park, which has a pedestrian exit leading down to South Beach (Fig. 6.22). The itinerary requires about 2 hours field time and medium to low tides.

6a South Beach. The cliffs at the southwestern end of the beach expose an east–west strike section of light-grey weathered, fine-grain bioclastic and oolitic limestones, belonging to the Hunts Bay Oolite (~150 m thick), which dip 65–70° southwards. Many of the bedding planes are crowded with the remains of brachiopods (*Composita ficodea, Cyrtina* and *Productus*). In the first cleft of the cliffline, irregular veins and patches of coarsely crystalline calcite, with radiating and columnar structure, infill fissures and cement together detached blocks of limestone and fault breccia. The inside of a small keyhole-shape cave has a draped fill composed of layers of calcite and finer red and buff sediment. These features are interpreted as Triassic pothole and fissure fills, which possibly represent the early stages in the formation of a collapse breccia (gash breccia) deposit. Farther along the beach, directly below the steps leading down to the beach from the Atlantic Hotel, a limestone bed at the top of the cliffs is crowded with the valves of *Gigantoproductus giganteus*.

6b South Beach. Walk 50 m farther to the northeast, passing caves (e.g. Bacon's Hole) in the cliffline, to a gully located just past a set of steps leading down from the Esplanade (SN 1345 0018). Here the Hunts Bay Oolite is composed of many relatively thin (3–5 m) shoaling-upwards cycles, which are best seen from the ledge above the gully. Each cycle has a sharply defined base overlain by light-grey oolitic grainstone facies with parallel and low-angle cross laminations. These grainstones are capped by buff weathering and more thinly bedded dolomitized limestones (<1 m thick). Some of the dolomitized beds have algal laminations, which are most easily seen on the wave-polished outcrops at beach level. The cycles are interpreted as high-energy (foreshore to beach) grainstone shoals passing up into lower-energy intertidal facies with algal stromatolites (shoaling- or shallowing-upwards cycles).

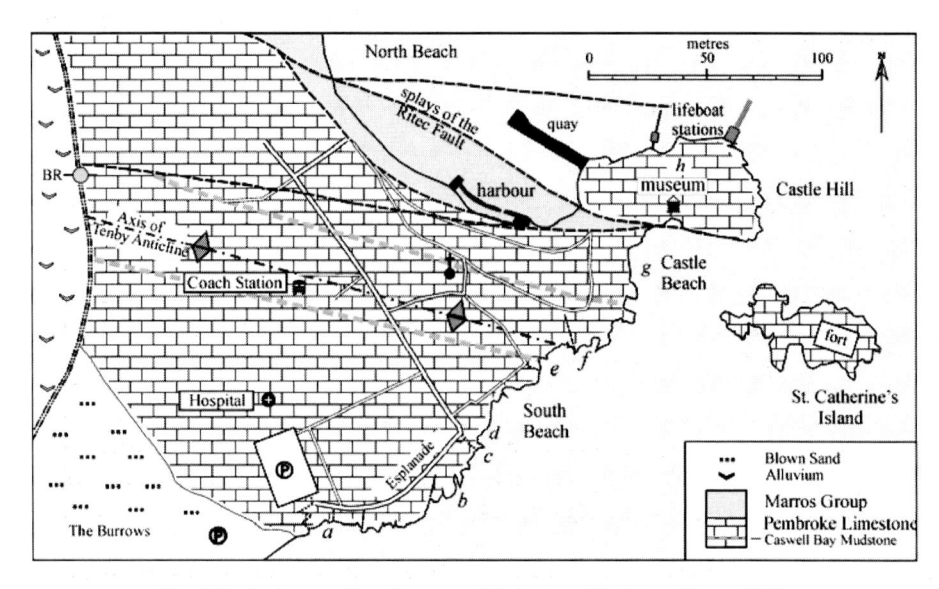

Fig. 6.22 Geology and locality map of Tenby (modified from Dixon 1921).

6c South Beach. Some 25 m farther to the northeast, at a second set of steps leading down from the Esplanade, a prominent vertical bedding plane displays calcite-mineralized slickensides with associated grooves and ridges. The surrounding veined limestones contain abundant productids, *Composita* and *Cyrtina*, the cross sections of which show the distinctive V-shape spondylium and septum extending down from the umbo. These limestones are located very near the base of the Hunts Bay Oolite.

6d South Beach. About 10 m farther north, very close to the base of the Hunts Bay Oolite, large colonies of *Siphonodendron (Lithostrotion) martini* can be seen on a ledge and at beach level. The colony shows evidence of erosion and is overlain by a reddened surface containing eroded blocks of limestone and smaller displaced colonies of *Siphonodendron* and *Syringopora*. This surface was formed during a significant shallowing of the shelf sea, which resulted in wave erosion and probably a short interval of emergence.

6e South Beach. Continue for 75 m northeastwards, across a dip section of the High Tor Limestone, to a cave (SN 1356 0027) where a succession of variegated and thinly bedded limestones and mudstones, belonging to the Caswell Bay Mudstone, are exposed (Fig. 6.23). At this locality the strata towards the top of the cliff-face have been stabilized with rock bolts and steel netting. This distinctive facies (the *Modiola*-phase or lagoonal-phase of Dixon 1921) is about 20 m thick and is composed of thin lensoid, wedge and nodular beds of fine grain limestones and dolomites interbedded with lime mudstones and shales. Some of the nodular horizons are considered to represent altered chicken-wire anhydrite, typical of present-day sabkha environments. These lagoonal and supra-tidal facies are devoid of macrofossils, but show evidence of bioturbation. A palaeokarst surface, defining the contact of the facies with the underlying Gully Oolite (only about 25 m thick), can be seen in the cave at the northern end of the outcrop. Here, the poorly developed palaeokarst surface displays some reddening and veining, and

Fig. 6.23 Interbedded limestones and shales belonging to the Caswell Bay Mudstone Formation (20 m thick) exposed at South Beach Tenby. The base of these lagoonal facies rests on a palaeokarst surface marking the top of the Gully Oolite.

relatively small dissolution features. The top of the lagoonal facies is truncated by a transgressive erosional surface marking the base of the High Tor Limestone. The depositional interpretation of the succession is similar to that already discussed for the type locality at Caswell Bay (see Fig. 4.17). However, the relatively thin oolitic facies, poorly developed palaeokarst and thick mudstone facies seen at Tenby, suggest that sea-level changes were less extreme and lagoonal conditions persisted for a longer period on this part of the carbonate ramp.

6f Gunfort Gardens. The ENE-trending axis of the Tenby Anticline intersects the cliffs just to the north of Gunfort Gardens. The core of the fold is composed of well-jointed dolomitized bioclastic limestones belonging to the Black Rock Limestone (formerly the *laminosa* Dolomite), which is cut by a small thrust.

6g Castle Beach. On the northern limb of the Tenby Anticline the succession is repeated, but a protective wall conceals the faulted and more easily eroded Caswell Bay Mudstone. The overlying High Tor Limestone (~100 m thick) contains a coral fauna, including *Siphonodendron, Siphonophyllia, Michelinia, Syringopora* and *Zaphrentis*, as well as gastropods and large crinoid stems. A small buckle fold and thrust forms a conspicuous feature in the western cliff of Castle Beach.

6h Castle Hill. The Carboniferous Limestone strata forming the Castle Hill were originally mapped as the *laminosa* Dolomite, which occur in the lower part of the *Caninia* Zone (Dixon 1921), but have since been reassigned to the younger Oxwich Head Limestone (*Dibunophyllum* Zone). The hill is isolated by splays of the Ritec Fault, evidence for which is recorded by the sheared and brecciated dolomitic limestones forming the southern margin of the headland. From the top of the hill (Prince Albert memorial) there are good views of the Upper Carboniferous strata preserved in the Pembrokeshire Coalfield Syncline (Fig. 6.1).

6.3.7 Stackpole Quay

Stackpole Quay can be reached from Pembroke on the B4319 via St Petrox and Stackpole village and then by taking the narrow road leading down to the quay (Fig. 6.24). Alternatively travel east on the A4319 to Lamphey; turn south on the B4584 to Freshwater East and continue for 2 km on the narrow road through East Trewent to Stackpole Quay. The Coastal Cruiser minibus also runs a daily service to the quay, where there is a large National Trust car park with toilets and a restaurant. The quay was built by Lord Cawdor to export limestone from the adjacent quarries and to import coal for Stackpole Court. A renovated square limekiln can be seen at the side of the road adjacent to the main quarry.

The aim of this itinerary is to examine the lithology, structure and geomorphology of the Carboniferous Limestone forming the spectacular cliffs to the north and south of the quay. Structurally the strata are quite complex, with many east–west-trending asymmetrical, eastward-plunging anticlines and synclines, the axes of which are displaced by a dextral wrench fault (Fig. 6.24). High-angle reverse faults, associated with inverted fold limbs, add further complexity to the geology. A detailed guide to the structural geology of the area is provided by Hancock *et al* (1982). Most of the foreshore localities are accessible during all but

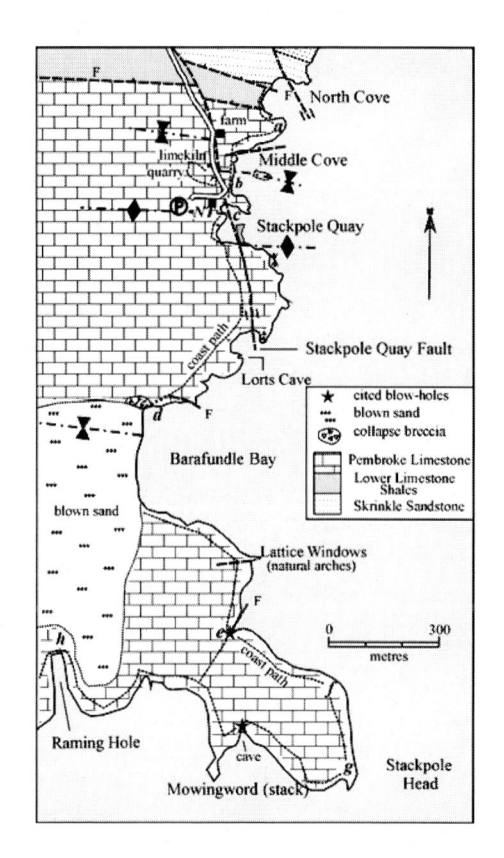

Fig. 6.24 Geology and locality map of the coast around Stackpole Quay (after Dixon 1921, Hancock et al. 1982).

the highest spring tides. The complete itinerary requires about 3 hours field time, but you may wish to picnic and have a swim at picturesque Barafundle Bay.

7a North Cove. Walk 300 m northwards along the coast path to the top of the cliffs overlooking North Cove (SR 9946 9594). Although the cove is inaccessible, the main details of the geology can be seen from the clifftop. The northern cliffs are composed of red conglomerates, sandstones and mudstones belonging to the Skrinkle Sandstone. In the northwest corner of the cove, grey limestones and shales (Lower Limestone Shales) are thrust over the redbeds. The shales dip steeply to the north and are therefore inverted.

7b Middle Cove. Return along the path to the beach in front of Stackpole Quay farm (holiday cottages). A stack in the centre of the bay is composed of bedded limestones (High Tor Limestone) deformed into an asymmetrical syncline with a WNW-trending and eastwards plunging axis (Fig. 6.25). The steeper northern limb of the fold dips at 50–80° and the southern limb dips at 36–45°; the bedding is cut by a pressure-solution fan cleavage. In the southern corner of the cove, the thick-bedded limestones contain an abundant fauna of large caninids (probably *Siphonophyllia gigantea*) up to 6 cm in diameter, *Zaphrentis*, *Michelinia* and *Syringopora*, large dolomitized gastropods, productids (some with geopetal infills) and abundant shell and crinoid debris. Many of the thin interbeds of lime mudstone preserve large *Thalassinoides* and other trace fossils.

Fig. 6.25 Asymmetrical syncline with well developed fan cleavage in the High Tor Limestone exposed in a stack at Middle Cove, Stackpole Quay.

7c Quay Cove. Marked by a 2 m-wide crush zone of breccia and gouge, the Stackpole Quay Fault intersects the northern corner of the cove. The reddened fault plane dips steeply to the east and has multiple sub-horizontal slickensides with calcite mineralization. Regional mapping has demonstrated that the wrench fault has a dextral displacement of 75–100 m (Hancock *et al.* 1982). To the east of the fault the limestones dip at 50° northeast and are cut by subsidiary faults and fractures. On the western side of the fault, cleaved limestones and calcareous mudstones display buckle folds, which define the core of the Stackpole Anticline. Here the repetition of a 9 m-thick band of calcareous mudstone has proved to be a very useful marker in analysing the structure of the area. The mudstone contains a good shelly fauna with large caninids present on the vertical bedding planes exposed in the cliffs (an old quarry face) directly opposite the quay. South of the quay the succession and structural features are repeated on the east side of the wrench fault.

7d Barafundle Bay. Rejoin the coast path, behind the National Trust restaurant, and make the ascent over the hill to Barafundle Bay. During the heyday of the Stackpole estate, the walled steps down to this beautiful bay were lined with trees and ornamental columns. The northern cliffs of the bay comprise steeply dipping (80° south) light-grey limestones (Oxwich Head Limestone) containing abundant recrystallized brachiopods (*Cyrtina, Productus*) and crinoid debris. Near the base of the steps, the solid limestones are replaced by a 5 m-wide pocket of collapse breccia composed of disorientated blocks of limestone set in a reddish clay matrix. Windblown sands that occupy much of the bay conceal the axis of a major east–west-trending syncline.

7e Stackpole Head (SR 992 945). Continue across the sandy bay and rejoin the coast path leading to Stackpole Head. At this locality a fenced-off blowhole occurs near to the coast path and continues seawards as a rectilinear feature with partially collapsed walls (see Fig. 6.29d). The blowhole cave is still being excavated along a northeast–southwest-trending fault.

7f Stackpole Head (SR 995 944). From this point there is a good view of three natural arches, known as the Lattice Windows, in the cliffs to the northwest. This thin promontory of limestone is located just north of a small reverse fault. Across Barafundle Bay the cliffs contain many caves cut into the steeply dipping (75–80° south) Stackpole Limestone Formation. Farther to the north, it is also possible to see the point of exit of the Stackpole Quay Fault, which is indicated by red staining and calcite mineralization.

7g Stackpole Head (SR 9955 9414). From the southern tip of the headland there are fine views westwards to Broad Haven, Church Rock and St Govan's Head. Looking down into the adjacent cove a pot-holed palaeokarst surface can be seen on a ledge (covered with guano) located about halfway up the cliff face (Fig. 6.26). Viewed through binoculars, the pot-holes are seen to be filled with orange–buff–grey clays. To the west of the nose the palaeokarst surface intersects the coast path near another blowhole; the adjacent headland terminates at the Mowingword stack.

Fig. 6.26 Bedding plane exposure of a pot-holed palaeokarst surface in horizontally bedded Carboniferous Limestone at Stackpole Head.

7h Raming Hole (SR 9873 9448). This feature is an impressive NNW-trending 20 to 30 m-gash in gently eastward-dipping limestone strata. A further 500 m to the west the coast path skirts around the very large blowhole (about 50 m wide) named Sandy Hole (SR 9840 9406).

6.3.8 Stack Rocks to St Govan's Head

This part of the Pembrokeshire Coast Path is on the Castlemartin artillery range and access is restricted during firing exercises. Details of any access restrictions can be obtained on a 24-hour telephone line (01646 662367), and information is also published in the local press and on notice boards located near the car parks. A National Park ranger, employed to monitor the wildlife and advise the local community and visitors, can also be contacted by telephone (07866 771188). The car park on the cliffs above Broad Haven is not on the firing range and this attractive sandy bay is always open to the public. The glorious walks around the lily ponds and surrounding woodland (administered by the National Trust) can be accessed from this car park or from the village of Bosherston. These freshwater ponds (supporting a very varied bird population) were created in the 18th century by the Earl of Cawdor to enhance the grounds of his Stackpole estate. Stackpole Court was demolished in 1963 and the stable block was converted into flats. In its heyday the house was supplied with large quantities of coal landed at Stackpole Quay and transported to the house over the eight-arch stone bridge.

The limestone cliffs along the Castlemartin Peninsula display some of the finest limestone geomorphological features in the UK, and they also provide the habitats for colonies of seabirds (fulmars, razorbills, guillemots) and nesting pairs of choughs, peregrines and ravens. Great care should be taken when approaching the unfenced sheer cliffs, which are up to 40 m high and often unstable, especially near faults and the many blowholes. Also, the minor paths along the cliff edge have many protruding metal pegs, which are used as abseil anchor points and for marking rock climbs.

The most convenient starting point for the itinerary is the car park at Stack Rocks, although the localities can be reversed if you wish to start from St Govan's chapel (Fig. 6.27). Timetables for the Coastal Cruiser service, which calls at the above car parks, and also those at Broad Haven, Bosherston and Stackpole Quay, can be obtained from the tourist information centres. Flimston chapel can be visited on route to the Stack Rocks. Within the chapel enclosure two large erratic boulders mark the grave of Lt.-Col. F. W. Lampton and his wife, the former being a well-respected local geologist, who presented his extensive collection of Carboniferous Limestone fossils to the Geological Survey. About 500 m to the southeast, on the opposite side of the road, are the completely overgrown Flimston pipeclay pits, which produced good-quality earthenware pottery and closed in the 1870s. These predominantly white clays are tentatively correlated with the Paleogene strata of the Bovey Basin, both localities being located along the Sticklepath–Lustleigh Fault.

The main aim of the itinerary is to study the relationship between the geomorphology and geology of the limestone cliffs seen from the coast path. It requires 4–5 hours field time and is not tide dependent.

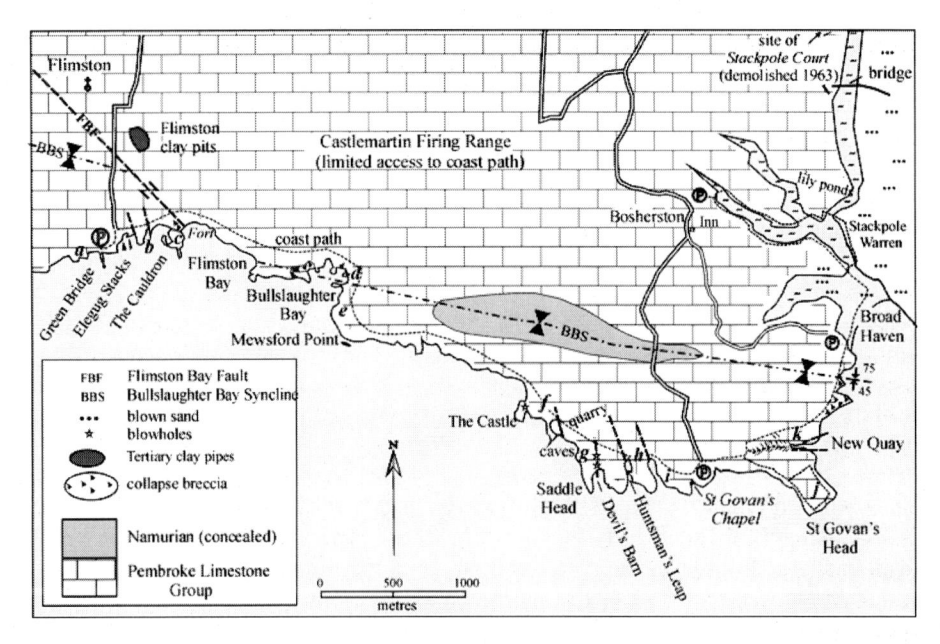

Fig. 6.27 Geology and locality map of the coast between Stack Rocks and Broad Haven (modified from Dixon 1921).

8a The Green Bridge of Wales. From the car park, walk 200 m to the southwest to the viewing platform due west of the Green Bridge. This textbook example of a natural arch is one of the most photographed coastal features in Pembrokeshire (Fig. 6.28a). The main arch, which has a span of about 25 m, is eroded into gently dipping, well-bedded limestones with thin interbeds of mudstone (zaphrentid-phase of Dixon 1921) belonging to the top of the Pen-y-Holt Limestone (equivalent to the High Tor Limestone Formation). These pass up into massively bedded limestone belonging to the Stackpole Limestone (equivalent to the Hunts Bay Oolite Formation), which forms the clifftop above the arch. More resistant and thicker bedded limestone facies also form the wider pedestal of the structure and a small isolated stack that represents the collapsed remains of a former arch. Eventually the main arch will also collapse to leave two stacks. Looking farther east the flat top of the 60 m platform, extending to Saddle Point, can be seen.

8b Stack Rocks. Rejoin the coast path and walk 275 m to the east to a small promontory, which gives the best view of the two impressive limestone pinnacles known either as Stack Rocks or Elegug Stacks (Fig. 6.28b). Elegug was the local Welsh name for the guillemot, which nests on the guano-covered stacks. The smaller Elegug Spire has a prominent palaeokarst surface along its centre. What appears to be the same surface is just visible at the top of the larger Elegug Tower, indicating that a fault, with a downthrow to the east, separates the two stacks. Both stacks would have originally been impressive natural arches similar to the Green Bridge (compare Figs 6.28a and b). The sheer vertical walls of these geomorphological features were formed as a result of differential marine erosion along vertical joints and small faults.

8c The Cauldron. Some 300 m farther to the east is another promontory, the neck of which is protected by the ramparts of an Iron Age fort. A huge blowhole (~50 m in diameter and 50 m deep) occupies the western half of the promontory. Extreme care should be taken if you follow the narrow path that leads around the vertical faces of the hole and ends at a narrow chasm. This is not a locality for a windy day. The northwest-trending Flimston Bay Fault, which has a dextral displacement of about 750 m, intersects the coast in the western corner of the bay.

8d Bullslaughter Bay. Rejoin the coast path and walk 1.5 km to the eastern side of Bullslaughter Bay (SR 943 943), where the complex geology of the area is clearly displayed (Fig. 6.30). The axis of the Bullslaughter Bay Syncline intersects the western point of the cove, which is flanked by a thick unit of collapse breccia (Fig. 6.28c). Small natural arches occur in the cove and farther west, on the headland known as Moody Nose.

8e Mewsford Point. From this locality further details of the geology of Bullslaughter Bay and Moody Nose can be seen. The most prominent structural feature is a fault dipping steeply to the north, with the limb of an eastward-plunging anticline exposed on its southern (upthrow) side. Farther seawards, the axis of the Bullslaughter Bay Syncline is clearly discernable (Fig. 6.30).

8f The Castle. The cliffs between Mewsford Point and the Castle have numerous rock climbs, including the triple-overhang known as Ripper Cliff. A series of caves, located well above high-water mark, have yielded evidence of a Stone Age settlement, which is also recorded by the fort defending the neck of the Castle. A blowhole occurs below the fort on the eastern side of the headland.

8g Saddle Head. Take the track that leads from the radar installation to the coast path; cross the cattle grid and then bear right onto a track leading to another installation on Saddle Head. The old quarry between the two points is cut by a small fault infilled with red and green clay, fine sandstone and stalagmitic deposits of Triassic age. The eastern margin of Saddle Head (SR 956 930) is pierced by a spectacular series of blowholes with an associated natural arch, which are located along the line of a small fault. Rock climbers refer to these features as the Devils Barn.

8h Huntsman's Leap. Rejoin the coast path and walk 200 m to a point where two narrow chasms, with vertical limestone faces, almost intersect the path. The first chasm, known as Huntsman's Leap (SR 9616 9306), is 20 m wide on its landward side and tapers to less than 2 m towards the sea, where the vertical sides (40 m high) are connected by only a thin slither of debris (Fig. 6.29c). The second gully (Stennis Ford), located 200 m to the east, is 30 m wide and tapers to 10 m seawards. Both of these features were eroded along NW–SE-trending faults and probably evolved as a result of the roof collapse of former blowholes.

Fig. 6.28 (a) The Green Bridge of Wales, a magnificent natural arch with a span of approximately 25 m; note the flat top of the 60 m platform extending southeastwards to Saddle Point. (b) Stack Rocks (Elegug Stacks), which represent remnants of former natural arches; note the lighter coloured palaeokarst surface developed along the centre of the smaller stack. (c) Bullslaughter Bay intersected by the axis of the Bullslaughter Bay Syncline; note the thick collapse breccia deposits in the adjacent cliffs.

Fig. 6.29 (**a**) St Govan's Chapel nestling in a cleft in the Carboniferous Limestone. (**b**) View looking eastwards from the chapel towards St Govan's Head. (**c**) Huntsman's Leap, a magnificent 40 m-high vertical chasm eroded along a NE–SW trending fault. (**d**) Partially collapsed blowhole located on a NE–SW-trending fault on Stackpole Head.

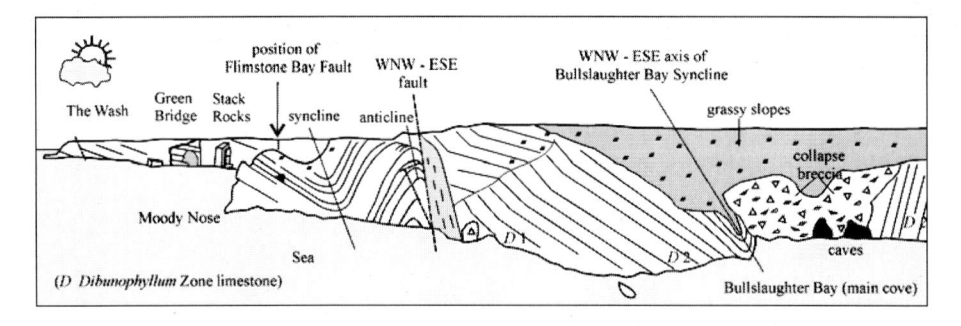

Fig. 6.30 Field sketch illustrating the geology of Bullslaughter Bay looking westwards from Mewsford Point.

8i St Govan's Chapel. The rock platform, located at the top of the cliff to the west of the chapel (SR 965 929), is a convenient and safe area to examine a representative section of late Dinantian limestone forming the cliffs along the coastline. A prominent bedding plane, with rounded to elliptical depressions, stained by iron oxide and draped by a thin ferruginous clay, marks a palaeokarst surface. This surface is developed within a clean grainstone limestone facies with well-developed large-scale cross sets, which are best seen in the adjacent headland. The palaeokarst is overlain by a 15 cm nodular palaeosol with a rubbly top, which is truncated by a massively bedded coarse limestone facies. The palaeosol may be at the same stratigraphical horizon as the one observed at Stack Rocks (Fig. 6.28b). In terms of sequence stratigraphy, the palaeokarst and palaeosol record a significant fall in sea level (lowstand systems tract), while the overlying erosional surface represents a ravinement surface formed during a high-energy transgression (transgressive systems tract).

Everyone should take the opportunity to visit the charming 13th century St Govan's Chapel, which is wedged into the foot of the cliffs and is reached by 'countless' steps (Fig. 6.29a). Alongside the diminutive stone-vault chapel is a cleft in the cliff, which marks the position of the refuge cell purported to have saved St Govan from attack. This cleft contains red stalagmitic and clay deposits of Triassic age believed to cure eye infections. Looking east there is a natural arch eroded through gently seaward-dipping limestones, and a recent large rockfall can be seen in the cliffs near St Govan's Head (Fig. 6.29b).

8j St Govan's Head. The walk around St Govan's Head offers fine views of the limestone scenery extending back along the coast to Saddle Head. To the northeast the rugged cliffs, beyond the small inlet of New Quay, are composed of a mass of jumbled blocks of collapse breccia. Note that the next headland, composed of limestones dipping 75° southwards, is located on the northern limb of the Bullslaughter Bay Syncline. Farther along the coast is Broad Haven, with the very distinctive Church Rock lying offshore.

8k New Quay. Return to the coast path on Trevallen Downs and take the track that leads down a delightful dry valley to the deep inlet of New Quay, used in the past for mooring small boats. This gully with its crystal-clear water has been eroded along two small east–west-trending faults. If the tide is out, the mass of collapse breccia forming the northern promontory can be examined.

6.3.9 West Angle Bay

The village of Angle can be reached by travelling westwards on the B4320 road from Pembroke or via the B4319 from Freshwater West. At the village, turn left at the T-junction and continue on the minor road to the car park at West Angle Bay. A café and toilets are located adjacent to the car park and there is a general store and public house in the village.

West Angle Bay has been formed by the erosion of early Dinantian limestones, preserved within the Angle Syncline (Fig. 6.31). The purpose of this itinerary is to examine the facies and tectonic structures, which occur within these strata. The bay is also an excellent locality for learning field-mapping techniques. The most interesting geology is seen in small coves located along the northern margin and western termination of the bay. These coves, which are old quarry workings, have been given informal names. Much of the section can be examined at high tide, but lower tide levels allow many of the structures to be traced along the strike.

9a Limekiln cove. At the western end of this cove (SM 8527 0331) a thrust, dipping 52° to the south-southwest, separates the Avon Group (Lower Limestone Shales) in the hanging wall from the cleaner Black Rock Limestone in the footwall. The former facies comprises interbedded shales and nodular sandy limestones containing a fauna of corals (*Cleistopora*, *Zaphrentis*) and crinoid debris. Below the limekiln the facies are composed of light grey limestones up to 1 m thick, with thinner alternations of nodular limestones, shales and cherts. The limestones contain abundant debris of solitary corals, crinoids, and brachiopod shells and spines. Some of the beds have erosional bases and sharp hummocky tops and contain low angle sets of SCS; these beds were deposited during storm events. Finer-grain facies, which have highly bioturbated (*Thalassinoides*) bases and rippled tops, are interpreted as fairweather deposits. The trace-fossil assemblage also includes *Teichichnus* and *Rhizocorallium*. Just to the west of the limekiln a gap in the cliffline marks the position of a minor wrench fault, to the east of which the strata have been deformed into a small syncline and anticline.

9b First north cove. Many WNW-trending periclinal folds are exposed in this small cove. The limestones also display sets of *en échelon* calcite tension gashes, fracture cleavage and crenulations, and the more massive beds exhibit abundant stylolites. During the original geological survey, Dixon (1921) recorded a large fossil colony of echinoderms and crinoids with their arms attached. Today the only fossils seen are *Zaphrentis*, crinoid ossicles and brachiopod shell debris. Swaley and hummocky bedforms occur in the coarser limestones.

9c Second north cove. This cove forms the western termination of West Angle Bay and it faces the entrance to Milford Haven. At this locality the massively bedded limestones display many shot holes with radiating fractures, formed as a result of blasting during quarrying. The polished limestone beds display excellent examples of stylolitic seams running parallel and oblique to the bedding. The position of a southward-dipping (80°) reverse fault is marked by slickensides, grooves and minor brecciation. Farther to the west, two pairs of upright periclines are exposed on the foreshore. The northern wall of the cove forms the faulted contact between cleaved

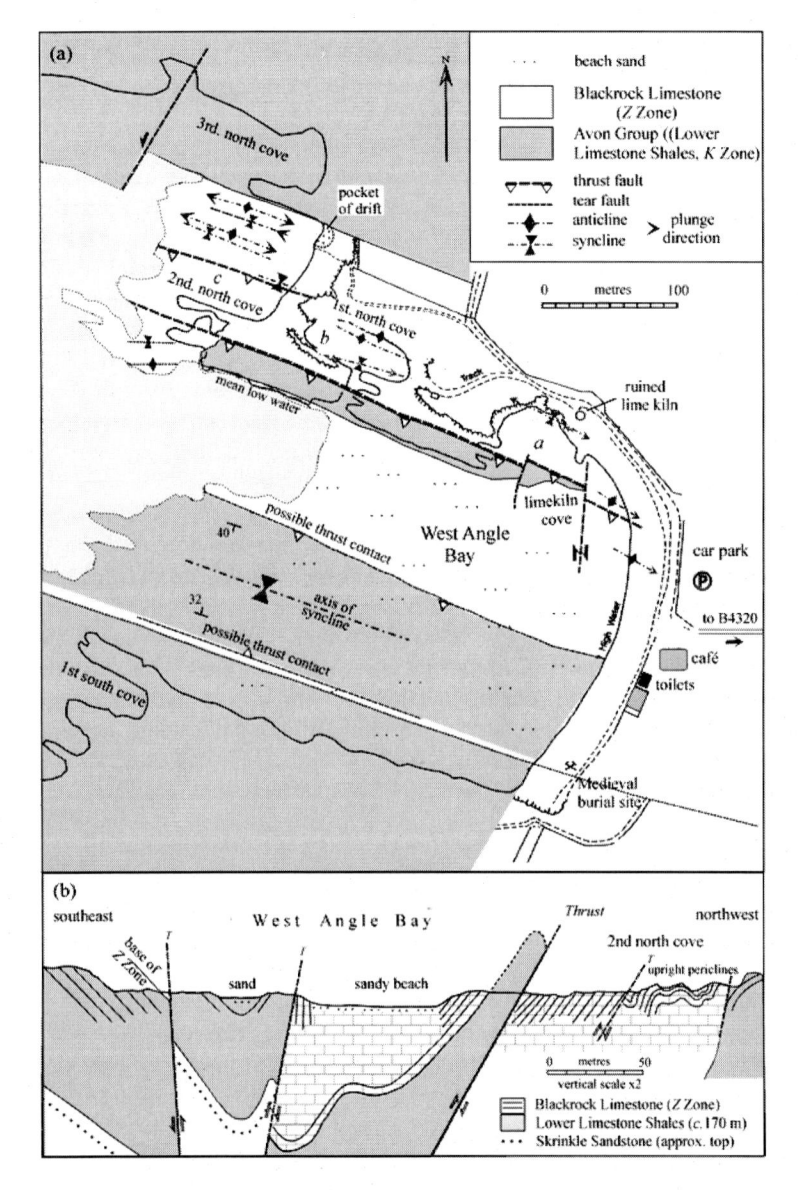

Fig. 6.31 (a) Generalized geology and structural map of West Angle Bay. (b) Simplified northwest–southeast cross section across West Angle Bay (both modified from Dixon 1921, Hancock et al. 1982).

and buckled Lower Limestone Shales (Avon Group) and the more competent Black Rock Limestone. In the northern corner of the cove the limestone facies are very highly weathered and decomposed, and are overlain by a pocket of pebbly drift with cryoturbation structures.

At the southern end of the cove, further upright periclines are exposed to the south of the thrust previously seen at locality *9a*. The steeply inclined bedding planes, seen along the line of the thrust, display very large crinoid stems and occasional crinoid cups with attached arms.

6.3.10 Tenby (North Beach) to Waterwynch Bay.

For this itinerary, the most convenient place to park is at the large North Cliff car park. From here, climb the steps to the Croft where there are good views of North Beach, the prominent stack of Gosker Rock and the harbour (Fig. 6.32). The aims of this itinerary are to examine the quite complex Armorican folding and faulting seen in the Namurian strata exposed in the low outcrops on the beach and in the cliff sections, and to review the sedimentology and sequence stratigraphy of selected facies sequences. Most of the section along North Beach can be examined at medium tide, but note that the section between First Point and Waterwynch Bay should only be attempted during very low spring tides, which allows just enough time (1–2 hours) to complete the walk along the beach to Waterwynch Bay. All of the cliffs along this part of the coast are very unstable; they should be viewed from a safe distance, and great care should be taken when crossing the very slippery, boulder-strewn bays.

10a Tenby Harbour. The poorly exposed early Namurian strata of Tenby harbour are cut by three southward-dipping splays of the Ritec Fault, the most northerly pair of which thrust a wedge of highly sheared and brecciated Carboniferous Limestone into the section (Fig. 6.32). The intermittent Namurian exposures on the beach reveal a succession of black shales with goniatite-bearing nodular limestones, and thin erosively-based quartz arenites displaying north–south-orientated gutter casts and sets of HCS and SCS. The latter are interpreted as forced regressive sandbodies, which were pushed southwards and detached from the main Twrch Sandstone (Basal Grit) facies belt to the north, as a result of significant falls in sea level. The sandstones are important because they display a variety of sedimentary structures, which can be used to indicate the way up of these highly deformed strata. Their usefulness can be demonstrated to the south of Gosker Rock, where the strata have been folded into an inverted syncline (Fig. 6.32). Similar way-up structures indicate that the strata to the south of Barrel-post Rock dip and young consistently to the south.

10b Gosker Rock. Gosker Rock is the prominent southwards-dipping stack in the centre of North Sands. It is composed of 25 m of siltstones that coarsen upwards into ripple-laminated sandstones. Load casts and ripple marks indicate that the sandstone is the correct way up (it dips and youngs to the south). This sandstone also crops out at the southern end of North Cliff, as a result of a displacement of about 125 m along a dextral wrench fault.

10c North Cliff. The highly deformed argillaceous succession, previously exposed along North Cliff, is now mainly concealed and inaccessible as a consequence of the construction of sea walls and the placement of large blocks of rock (rip-rap) along the base of the cliff. However, when viewed from the beach, many asymmetrical and inclined chevron folds with fan cleavage can still be seen (Fig. 6.32). In the past the shales have yielded fossil assemblages including *Anthracoceras, Dunbarella* and *Sanguinolites* (Dixon 1921, George & Kelling 1982).

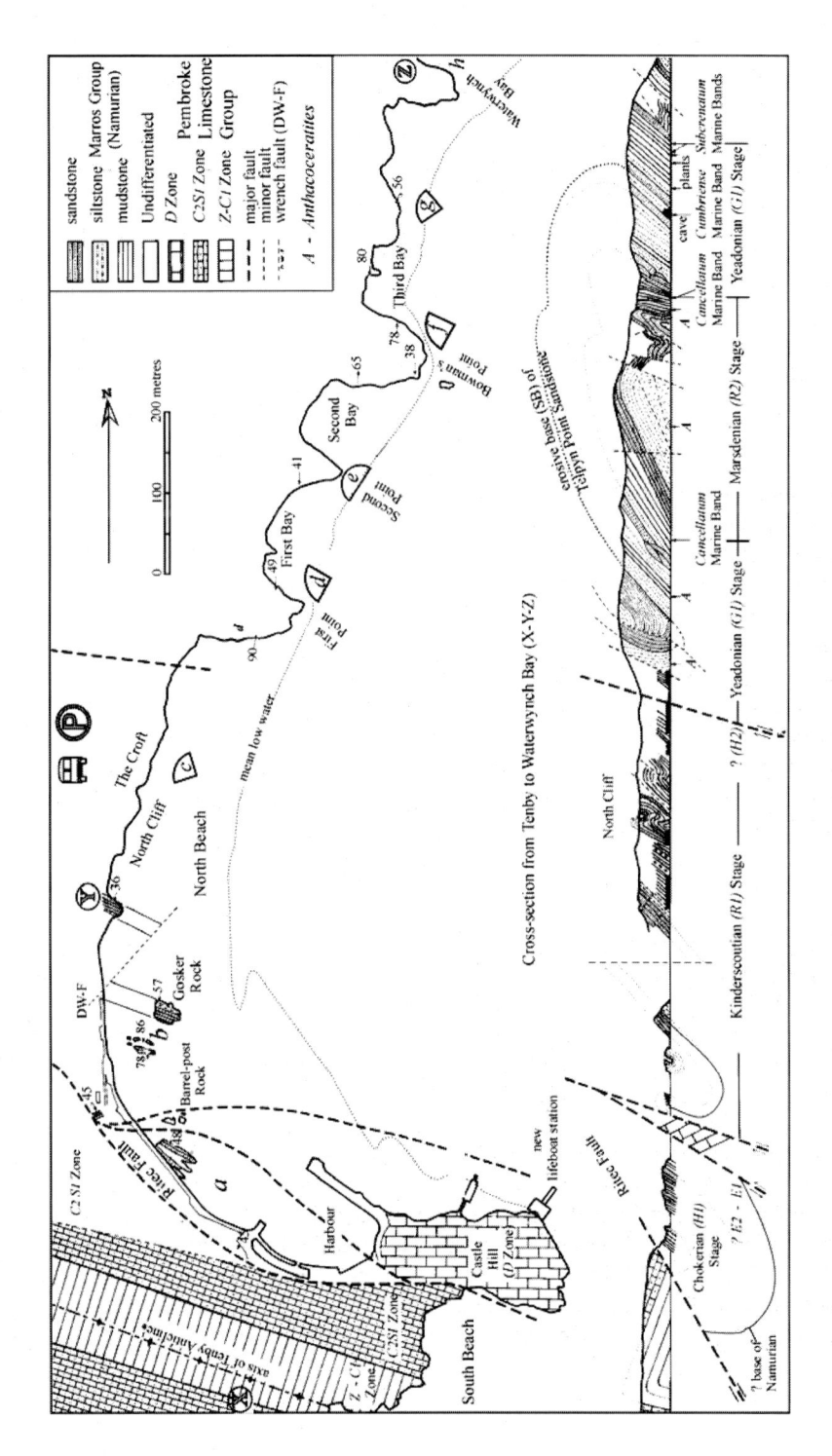

Fig. 6.32 Geology map and cross section illustrating the structure and succession of the Namurian strata exposed at North Beach (from George & Kelling 1982).

10d First Point. At First Point the Telpyn Point Sandstone (late Yeadonian) is over 22 m thick and has been folded into an asymmetrical syncline with a vertical to slightly inverted southern limb. On the southern flank of the point, the prominent erosive base of the sandstone body is marked by a pebble lag, composed predominantly of reworked ironstone nodules. Just below this surface, preserved between two irregular lenses of lag, is the thin plant-bearing shale referred to by Dixon (1921) as the First Point Plant Bed. On the northern flank of the syncline, the main part of the channel fill is composed of massive de-watered sandstones, which pass up into planar cross-stratified sandstones that provide southeasterly palaeoflow directions (Fig. 6.33). The sandstone body is interpreted as an incised valley fill and consequently its erosional base represents a sequence boundary. In terms of this model, the plant bed is considered to represent the remains of a terrace deposit formed during the erosion of the incised valley.

10e First Bay to Second Bay. The oldest strata exposed in the cliffs are mudstones containing *Anthracoceratites*, which occur in the core of an anticline in Second Bay. This marine band is succeeded by a coarsening-upwards sequence truncated by a sharp-based sandstone, interpreted as an incised shoreface deposit. The shoreface sandbody is cut by a fining-upwards tidal channel fill, which records the beginning of the transgression that culminated with the deposition of the *Cancellatum* Marine Band (Fig. 6.34). A further coarsening upwards cycle, developed above the *Cancellatum* Marine Band, is truncated by a multi-storey coarse-grain sandbody interpreted as a lowstand distributary-channel complex (6.35).

Fig. 6.33 Sets of planar cross strata with south-easterly dipping foresets (palaeocurrent direction) developed in the Telpyn Point Sandstone at First Point, Tenby. Note that the 5 m of strata, shown as horizontal in the photograph, actually dip at 45° to the south.

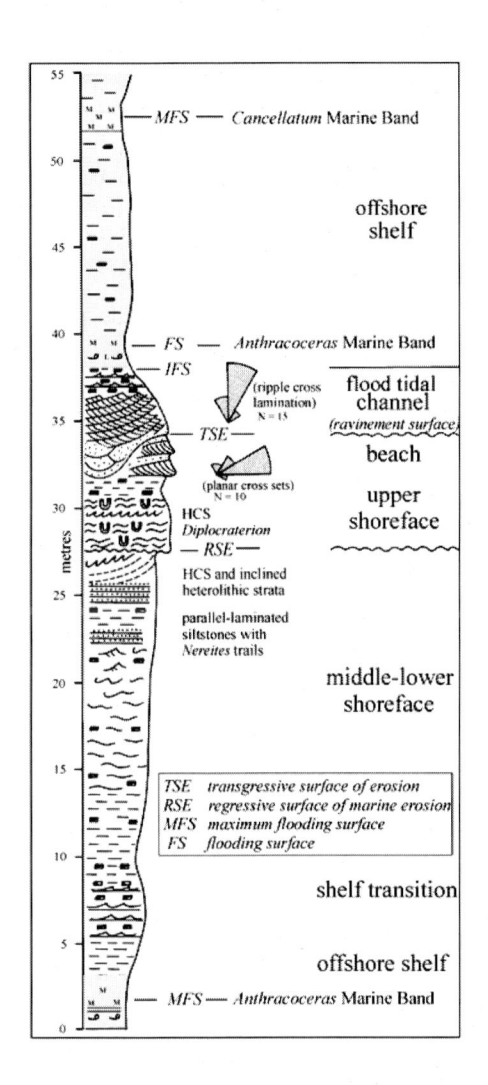

Fig.6.34 Graphic log and sequence stratigraphy of the Marsdenian succession developed between the *Anthracoceras* and *Cancellatum* Marine Bands in First Bay, Tenby (modified from George & Kelling 1982).

10f Bowman's Point. Bowman's Point is composed of the previously described sandstone (Fig. 6.34, 27–37 m), deformed into an impressive anticline/syncline fold pair with chevron geometry. Within the folds the sandstone beds display accommodation effects including minor thrusts and bulbous hinge zones.

10g Third Bay. Between Third Bay and Waterwynch Bay the late Namurian strata are all inverted. In the centre of the bay the *Cumbriense* Marine Band (50 cm thick) occurs above highly bioturbated siltstones exposed in the roof of a cave. The Telpyn Point Sandstone, which is only 8 m thick at this locality, truncates the marine band and must remove it totally at some point to the south. The erosional relief of the sequence boundary and the geometry of the incised valley fill are clearly illustrated in the facies profile of the Yeadonian strata exposed between First point and Waterwynch Bay (Fig. 6.35). Farther to the north the sandstone is succeeded by about 25 m of predominantly argillaceous facies of floodplain and estuarine origin. Two goniatite-bearing shales, which represent the *Subcrenatum* Marine Band, succeed the latter facies.

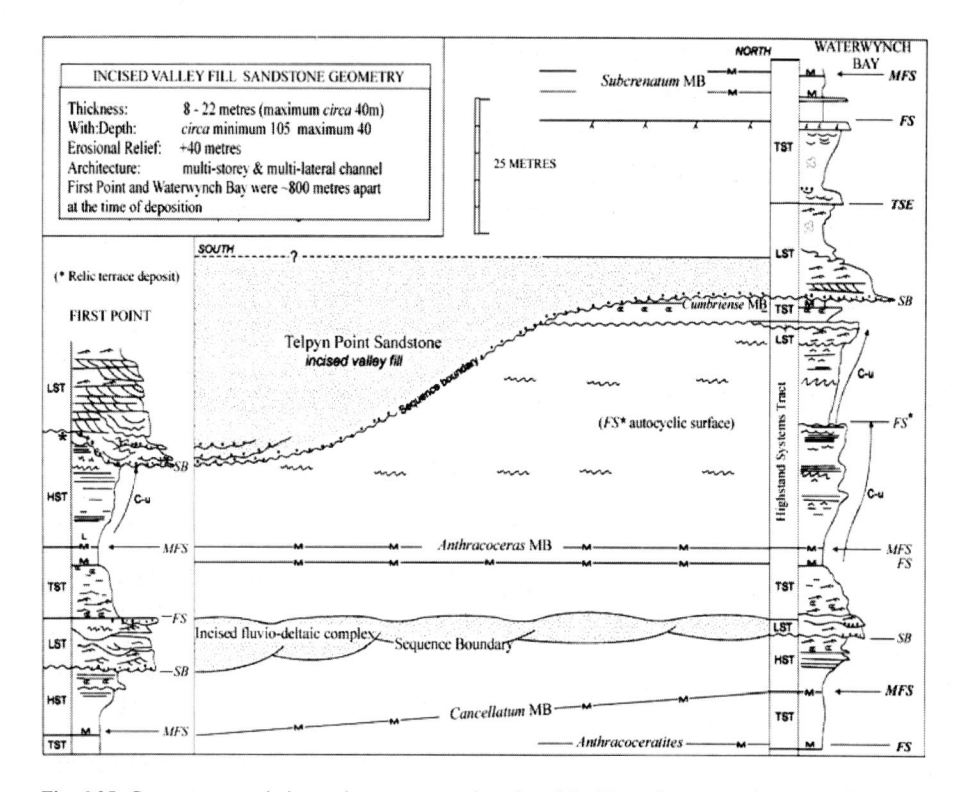

Fig. 6.35 Geometry, correlation and sequence stratigraphy of the Namurian succession exposed between First point and Waterwynch Bay, Tenby (modified from George 2001). Note that at the time of deposition the two localities were less than 1 km apart.

10h Waterwynch Bay. The cliffs to the north of this inlet are composed of alternating mudstones and siltstones of early Westphalian age. They are interesting because they display excellent examples of soft-sediment deformational structures originally interpreted as slumps (Amroth slump sheet). However, detailed studies have shown that very little lateral movement occurred during their formation and consequently they are best described as load casts. They can be correlated with a similar deformed band (2 m thick) present to the east of Amroth (locality *11h*). It is possible that seismic shocking, which accompanied contemporaneous movements on the Ritec Fault, caused the original unconsolidated sediment to become thixotropic and deform into the kidney-shape load casts. North of Waterwynch Bay many thrust faults and folds complicate the Westphalian succession (see Hancock *et al.* 1982).

From Waterwynch Bay it is possible to join the coast path and walk above the cliffs back to Tenby. This route is recommended rather than returning along the beach, where First Point may not be passable. It is also possible to walk along the coast path to Monkstone Point and onwards to Saundersfoot (~4 km). If you take this option you could complete itinerary *13* (Saundersfoot to Monkstone Point), following the localities in reverse order.

6.3.11 Ragwen Point to Telpyn Point and Amroth

The cliffs that extend from Ragwen Point (SN 220 071), across Marros Sands to Telpyn Point (SN 185 072), and end to the east of Amroth (SN 175 073) expose an almost complete Namurian to basal Westphalian succession (Fig. 6.36). It is the type section for the Marros Group (Namurian), previously referred to as the Millstone Grit Series (Waters *et al.*, 2007; Fig. 6.37, Appendix 1). This magnificent section is, in my opinion, the best locality in the United Kingdom for illustrating how the concepts of sequence-stratigraphy can be applied to the interpretation of sedimentary sequences. High-resolution sequence-stratigraphy data, derived from gamma ray profiling using a hand-held spectrometer, have provided a framework for research and for this field excursion (Fig. 6.37). In the following itinerary it has been necessary to simplify some of the quite complex sequence-stratigraphy concepts, but more complete accounts are presented in George (2000, 2001).

Morfa Bychan, a small pebbly beach located between Gilman Point and Ragwen Point, is the most convenient starting point for this itinerary. The bay can be reached in four-wheel drive vehicles via a narrow track leading off the Pendine–Amroth road 100 m west of the Green Man Inn (now unfortunately closed). Alternatively you can park near Pendine church and walk to Morfa Bychan via the lane that starts at the farm 200 m to the southeast. If more than one vehicle is available, one should be left near the New Inn (Amroth) where the traverse ends. The complete itinerary requires at least 6 hours of field time and low spring tides. If possible the itinerary should be timed to begin about 2 hours after an early-morning high tide. If the complete traverse is attempted, it is imperative to round the two promontories of Telpyn Point before the tide reaches the base of the rock outcrops. When the tides are not suitable, the section can be studied in two half-day excursions: Ragwen Point to Marros Sands (localities *11a–e*) and Telpyn Point to Amroth (localities *11f–h*). These two half-day itineraries should also be followed if you only have the use of one vehicle.

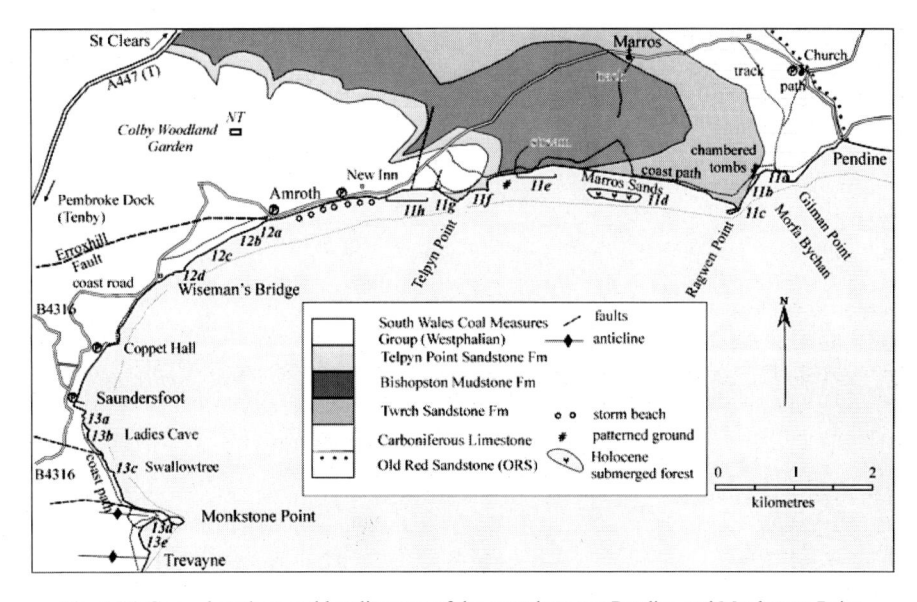

Fig. 6.36 General geology and locality map of the coast between Pendine and Monkstone Point.

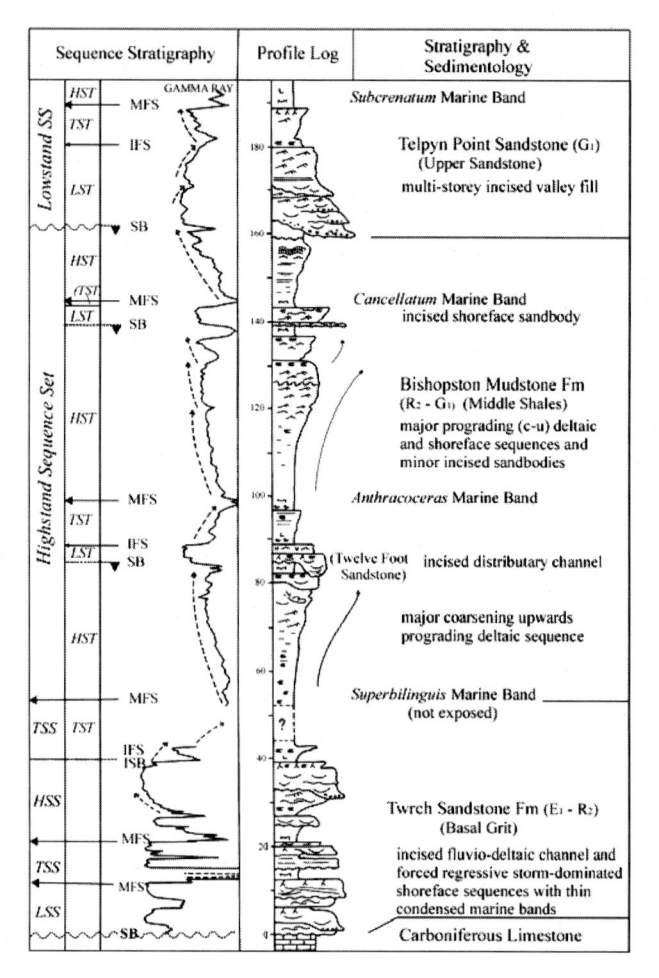

Fig. 6.37 Sedimentology and sequence stratigraphy of the Marros Group (Namurian) exposed at the type locality between Ragwen Point and Telpyn Point to the east of Amroth (modified from George 2000).

11a Morfa Bychan. From Morfa Bychan walk along the impressive storm beach to the western end of the bay; note the concrete wall above the beach was used for target practice by gunboats during the Second World War. From the corner of the bay, continue southwestwards along the boulder-strewn flank of the point, noting the extraordinary amount of fresh water that issues from the concealed contact between the highly jointed quartz arenites of the Twrch Sandstone (Basal Grit) and the underlying impervious Oystermouth Formation (Upper Limestone Shales). Inland the dip slope of the grits forms a large catchment area for this soft water, some of which is being pumped from a well located on the side of the hill.

11b Ragwen Point (SN 2210 0715). The disconformable contact between the Carboniferous Limestone and the overlying Twrch Sandstone can be examined towards the southwestern end of the point. Here the limestones and interbedded calcareous shales of the Oystermouth Formation contain fossil brachiopods and large *Thalassinoides* burrows. In terms of sequence stratigraphy the disconformity represents a sequence boundary. Farther to the northwest towards Haverfordwest the magnitude of the disconformity increases because of overstep and onlap stratigraphical relationships.

11c Ragwen Point. The grit succession at Ragwen Point consists of a variety of quartz arenite sheet and channel sandbodies that erode into marine argillaceous facies and rooted carbonaceous shales. Recent studies of these sequences have shown that the sandbodies were deposited on a storm-dominated marine ramp, during an extended period of time (about 10 million years) when eustatic sea level was continuously fluctuating as a result of repeated Gondwana glaciations and melt-outs. Details of the sedimentology and sequence stratigraphy of the complete Twrch Sandstone succession are summarized in Fig. 6.38 (see Appendix 4 for details of the facies). By making three stops, each beginning on a rock platform, this quite complex succession can be most efficiently summarized.

- Stop 1 is located on the platform that occurs 5 m above the base of the grits on the undulating top of a channel fill facies (FD 1). Above this surface, a wedge of thin-bedded quartz arenites with *in situ* rootlets (FD 2) is capped by a carbonaceous shale (FD 4). This shale is truncated by hummocky and swaley cross-stratified facies (SF 2-3) that are erosively overlain by coarser and cleaner facies (SF 4-5) with mixed swaley and low-angle sets. Thus, at this locality, an emergent palaeosol is truncated by a forced regressive sequence comprising falling stage (SF 2-3) and lowstand (SF 4-5) facies. In this situation, where two regressive erosional surfaces are present, the sequence boundary is positioned at the base of the lowstand facies.

- Stop 2 begins on the uppermost bedding plane of the previously described quartz arenite. This undulating surface is covered with circular and more irregular bleached patches that represent the *in situ* stumps of giant club mosses (Fig. 6.39). Some of the stumps have radiating casts of more distinctive *Stigmaria*. This mature palaeosol surface is directly overlain by 2 m of interbedded shales, bioturbated calcareous siltstones, and two bands of limestone nodules. These limestone nodules are developed within black shales containing pelagic bivalves (*Posidionella*) and conodonts (Fig. 6.40). The latter facies contain phosphatic flakes and nodules that produce extremely high gamma-ray emissions, characteristic of maximum flooding surfaces (Fig. 6.37). Intensely bioturbated siltstones with *Diplocraterion* and *Teichichnus* are considered to represent firmgrounds associated with pauses in sedimentation. Higher in the succession, there are two further forced regressive shoreface sandbodies (SF 2-4), separated by a thin bioturbated shale (Figs 6.39, 6.40). The uppermost of these shoreface facies is truncated and cut out completely by a channel sandbody (FD 1). Thus again, the combined sedimentology and sequence-stratigraphy data indicate that dramatic sea-level changes occurred during deposition of these facies (Fig. 6.40). Actual water depths are difficult to predict, but mean storm wave base was probably around 25 m, and the offshore ramp may well have extended to depths greater than 100 m.

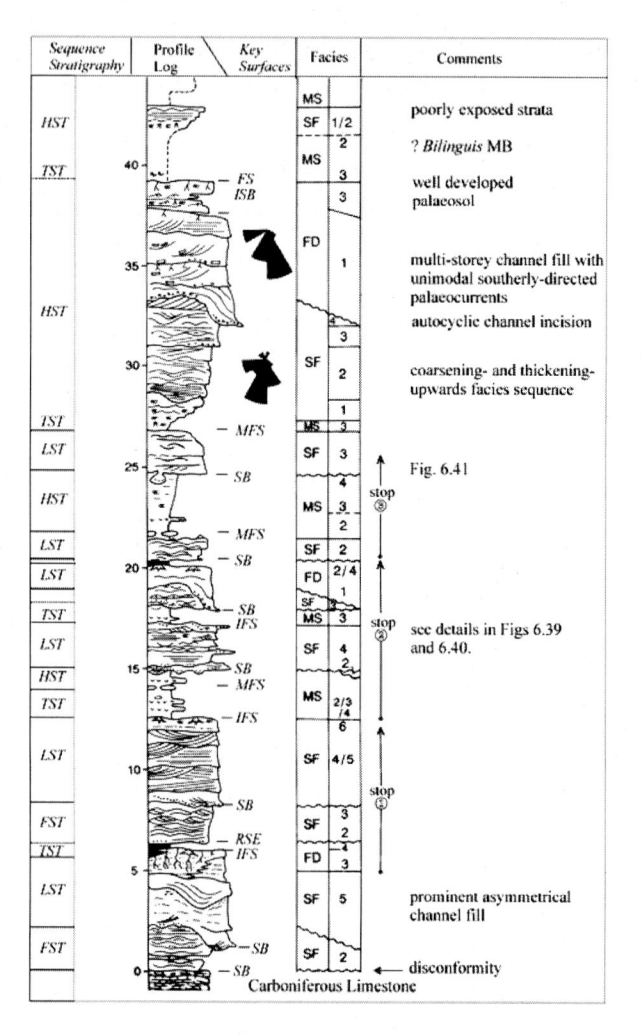

Fig. 6.38 Sedimentology and sequence stratigraphy of the Twrch Sandstone (Basal Grit) succession exposed at Ragwen Point showing the positions of stops 1–3 described in the itinerary (see Appendix 4 for facies) (after George 2000).

- Stop 3 includes the strata from the top of the previously described channel and the succeeding 5 m of strata (Figs 6.38 and 6.40, 20–25 m). The top surface of the channel displays a ganister palaeosol and thin carbonaceous shale, truncated by a 1 m coarsening- and thickening-upwards, forced-regressive shoreface facies. A rapid sea-level rise is recorded by the black mudstones and limestone nodules containing pelagic bivalves (*Dunbarella* and *Posidionella*) that were deposited in an offshore shelf environment. These quickly pass up into shallow-water shales with *Lingula* and barren silty mudstones. Excellent examples of loaded gutter casts occur on the base of the overlying forced-regressive sandstone (Fig. 6.41).

Since the remainder of the Twrch Sandstone (Basal Grit) succession is composed of similar facies, leave Ragwen Point and proceed to Marros Sands.

Fig. 6.39 View of the Twrch Sandstone seen at stop 2 Ragwen Point (Fig. 6.38, 12.5–20.5 m).

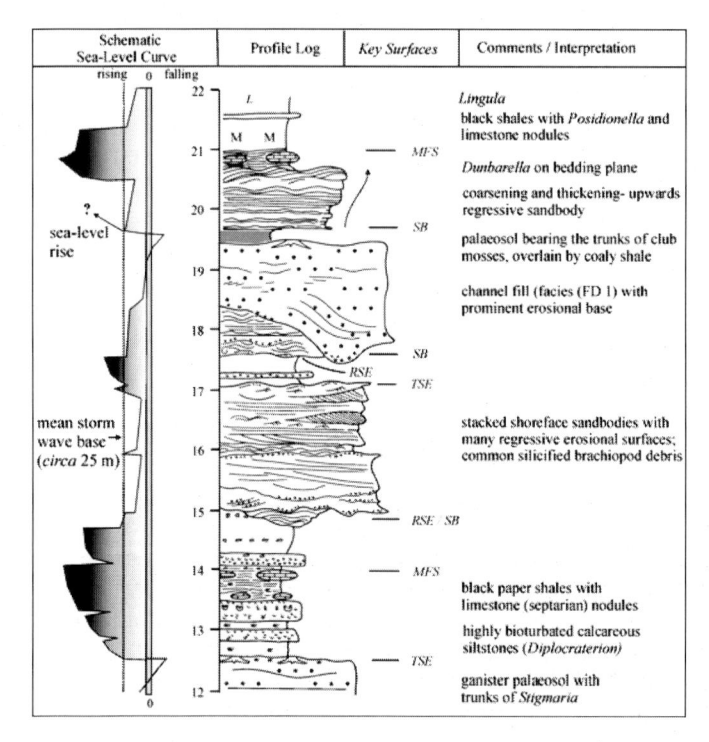

Fig. 6.40 High resolution sequence stratigraphy and schematic sea-level curve for the facies exposed at stops 2–3, Ragwen Point (see also Fig. 6.38).

Fig. 6.41 Loaded gutter cast on the base of an incised shoreface sandbody (facies SF 4) exposed at Ragwen Point (Fig. 6.38 at 25 m). Note the flame structure developed in the underlying mudstones (facies .MS 4).

11d Marros Sands. Low cliffs, composed of clayey and sandy drift, back most of the 2 km stretch of beach at Marros Sands. The top of the drift is penetrated by many ice-wedge pseudomorphs, and from about medium to low tide patches of woody peat, containing well-preserved tree stumps, are exposed on the beach. These deposits represent the best examples of the post-glacial (Holocene) submerged forest to be seen along the South Wales coast. Evidence of Neolithic occupation is documented by the occurrence of flint arrowheads found in the peat, and the presence of chambered tombs on the upper crags of Ragwen Point.

11d Marros cliffs. Towards the western end of the beach, mudstones forming the basal cycle of the Bishopston Mudstone Formation (formerly the Middle Shales) appear in the cliffs (SN 198 076). It is assumed that both the *Bilinguis* and *Superbilinguis* Marine Bands occur in the concealed strata farther to the south. This cycle represents an excellent example of a coarsening-upwards deltaic sequence with a complete record of pro-delta, distal bar, distributary mouth-bar and distributary-channel deposition (Fig. 6.37, 50–82 m; Appendix 5). In the distal bar to mouth-bar facies the upward increase in grain size is associated with a change in sedimentary features from lenticular–wavy laminations to ripple cross laminations, followed by scour and fill structures and minor channels (Fig. 6.42). A variety of penecontemporaneous deformation structures are also common. The distributary channel is filled with heterogeneous fine-grain sandstone and siltstone facies displaying basal and internal channels, hummocks, swales and lenticular bedding. Many of the bedding planes of these structures are draped with silty-mudstones, which are often highly bioturbated and commonly contain streaks and pellets of bright coal and synaeresis cracks. Palaeocurrent measurements are bimodal with

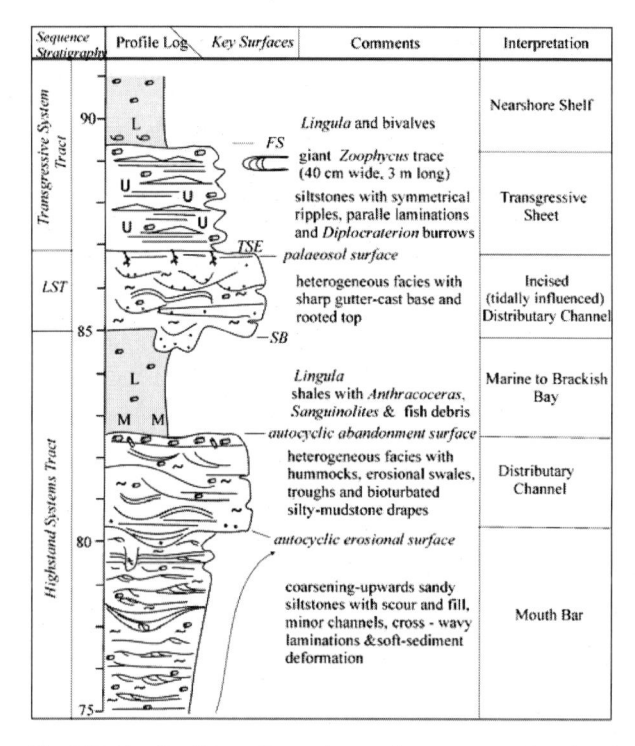

Fig. 6.42 Details of the sedimentology and sequence stratigraphy of the facies developed towards the top of the first coarsening-upwards cycle in the Bishopston Mudstone Formation (Middle Shales) at Marros Sands (see also Fig. 6.37).

Fig. 6.43 Heterogeneous hummocky and swaley facies interpreted as an incised tidally influenced distributary channel fill, overlain by transgressive flat bedded siltstones (Fig. 6.42 85–88 m).

southerly and northwesterly modes. The above features indicate that deposition occurred in the distal portion of a distributary channel influenced by both fluvial and tidal currents. High in the cliffs to the west, what appears to be the same channel displays much greater erosional relief, suggesting that it was located in a more axial position. The abandonment of the distributary channel is recorded by a bioturbated horizon (Fig. 6.42 at 82.5 m), and the succeeding interdistributary bay deposits containing *Lingula, Sanguinolites* and *Anthracoceratites*. The complete coarsening-upwards cycle was deposited as a result of autocyclic delta progradation during a highstand of sea level, which was initiated as a result of the *Superbilinguis* marine transgression.

A second heterogeneous sandbody (Fig. 6.42, 85–87 m; Fig. 6.43) is interpreted as an incised distributary channel; its prominent erosive and gutter-cast base being designated a sequence boundary, which was cut during a significant fall in sea level. The top surface of this lowstand sandbody displays discoloured patches (tree stumps) and rhizocretes, which indicate emergence. It is succeeded by 2.5 m of parallel- and ripple-laminated dark siltstone facies with U-tube burrows, and a very large *Zoophycos* trace that can be followed along the bedding plane for about 12 m. This facies is considered to have formed from the reworking of subaqueous deltaic deposits during a transgressive phase.

At this point, if attempting the complete traverse, you should proceed to the eastern promontory of Telpyn Point (SN 1890 0357) from where you can assess whether it is possible to negotiate the second promontory. If the tide is too high, return to the last locality; join the coast path at the stream and follow the path over the headland to reach the sandy beach to the west of Telpyn Point. Here the upper part of locality *11g,* and locality *11h* can be examined. If you intend to return to Pendine, you should walk back along the beach and join the coast path at Marros Sands.

11e Telpyn Point (east). This promontory is composed of ripple-laminated silty sandstones, interbedded with finer sandy siltstones displaying draped ripples and flaser bedding, which form the upper part of a coarsening-upwards cycle (Figs. 6.37, [100–130 m], 6.44). Northwesterly-directed palaeocurrents suggests that the cycle was deposited during the progradation of a shallow-water shoreface environment. The absence of HCS in the sandy facies indicates that deposition occurred during fair-weather conditions that accompanied a sea-level highstand.

11f Telpyn Point (bay). A very interesting sharp-based sandstone is exposed in the centre of the bay between the two promontories of Telpyn Point (SN 187 073) (Figs 6.45, 6.46). Most of the strata can be examined on the small protrusion, which extends away from the high cliffs. Here the base of this sandbody has a planar erosional surface, defined by a basal lag (6–10 cm thick) containing granules of quartz, mudstone and phosphate. The top surface of the lag, which displays well defined granule ripples and hummocks, is succeeded by a coarsening- and thickening-upwards siltstone–fine sandstone facies, with mudstone-draped sets of HCS and SCS. The highly bioturbated top of the sandstone unit is succeeded by a 2 cm-thick ankeritic siltstone with shell debris, followed by black mudstones containing abundant *Gastrioceras*

Fig. 6.44 Ripple laminated silty sandstone interpreted as fair-weather upper shoreface facies exposed at the eastern end of Telpyn Point, Amroth.

cancellatum. The sequence is a classic example of a forced regressive shoreface sandbody deposited on a storm-dominated shelf during a fall in sea level. In this depositional setting the basal lag probably originated as a result of reworking during a single violent storm surge. The intense bioturbation, seen at the top of the sandbody, occurred during a still-stand, prior to the main transgression that deposited the *Cancellatum* Marine Band. This transgression initiated a phase of deeper water sedimentation throughout South Wales. Parallel-laminated and graded siltstones, with biogenic tracks and trails of *Scolicia*, characterize the sequence above this marine band. In the cliffs, just below the erosive base of the Telpyn Point Sandstone (SN 1860 0724), a 2 m-unit of sinusoidal ripples-in-phase forms a very prominent feature. These ripples have dome-shape crests, although some are elongate or sinuous and some bifurcate laterally. Wavelengths of 40 cm are common and the amplitude of the ripples increases to a maximum of 6 cm.

11g Telpyn Point (west). The gently dipping Telpyn Point Sandstone Formation is exposed from the tip of the point (SN 186 072), and continues for 400 m along the cliffline, extending northwestwards to a small hanging valley (SN 183 073). The erosional base of the sandbody is defined by a mélange composed of pebble lags mixed with, and eroded into, slumped beds with coal streaks and log casts. These chaotic deposits were introduced into the channel as a result of bank failure. The overlying sandstone can be examined on a series of bedding-plane platforms to the west of the point, but be careful because many of these inclined surfaces are very slippery. Working upwards, the sandbody is composed of about eight discrete stories, the upper three of which display fining-upwards

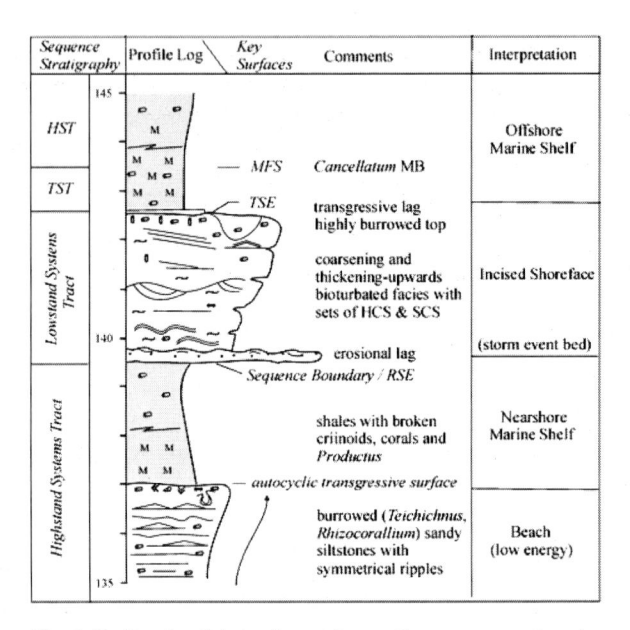

Fig. 6.45 Details of the sedimentology and sequence stratigraphy of the facies below the *Cancellatum* Marine Band at Telpyn Point

Fig. 6.46 Photograph of the Sub-Cancellatum Sandstone depicted in the log shown in Fig. 6.45.

trends (FC 1→ 4). Some of the unbedded sandstone facies (FC 2) show evidence of de-watering, and undulating bedforms interpreted as high-energy antidunes and standing waves have been recorded (Fig. 6.47 at 175 m). The upper 5 m of the sandbody is composed of ripple cross-laminated fine sandstones with evidence of lateral accretion surfaces and minor erosional truncations. Farther to the west, adjacent to the small hanging valley, the ripple-laminated sandstone is again exposed on an upfaulted platform. Here the top surface of the sandstone has scattered rhizocretions and is overlain directly by silty mudstone with small sets of HCS. The remainder of the argillaceous member (7 m thick) is composed of further silty mudstones that coarsen upwards to a palaeosol and thin coal. The sedimentology and sequence stratigraphy of the complete incised valley fill is summarized in Figure 6.47.

11h Amroth (east). The *Subcrenatum* Marine Band, which marks the base of the Westphalian, is exposed in shale outcrops on the beach (SN 1815 0726). These marine shales are succeeded by black mudstones (seen at the bottom of the cliff) containing abundant *Lingula,* many of which are in their life positions. From this point to the storm beach at the eastern end of Amroth (SN 175 073) the cliffs are occupied by another coarsening-upwards deltaic sequence approximately 50 m thick. Two lines of detached slump balls occur in the distal mouth-bar sediments, and the Amroth 'slump' sheet occurs below ripple-laminated fine sandstones interpreted as proximal mouth-bar sediments (Fig. 6.48). Although the distributary-channel sandstone at the top of the sequence is rather inaccessible, it can be observed in the cliffs along the foreshore, where it displays a strongly erosive base and at one point it almost washes-out the Lady Frolic coal seam.

If you have completed the whole itinerary, you have earned a well-deserved drink at the New Inn, which nestles behind the storm beach at the eastern end of Amroth. Until the 1930s there was a row of cottages between the storm beach and the road, but these were destroyed by repeated violent storms.

6.3.12 Amroth to Wiseman's Bridge

Limited parking is available on the sea front at the western end of Amroth adjacent to the public toilets (Fig. 6.36). The cliff and foreshore section to the west expose strongly folded and faulted Lower and Middle Coal Measures (Westphalian A–B). The cliffs are very unstable and falls occur frequently, thus hard hats should be worn and, where possible, the geology observed at a safe distance from the cliff faces. This section can also be approached from Wiseman's Bridge where parking is available on the side of the coast road or, with the permission of the owner, at the Wiseman's Bridge Inn. In this case the localities should be visited in reverse order. The section can only be studied during medium to low tides and requires about 2 hours field time.

12a Amroth (eastern cliffs). The cliff section (Fig. 6.49) begins with a sandstone (dipping 15° southwest), the top of which is discoloured and displays casts of *Stigmaria, Calamites* and many sphaero-siderite rhizocretions. It is overlain by a plant bearing and rooted shale, which is the seatearth of the Kilgetty coal (mined out at this locality). Farther towards the west, strata above the horizon of the

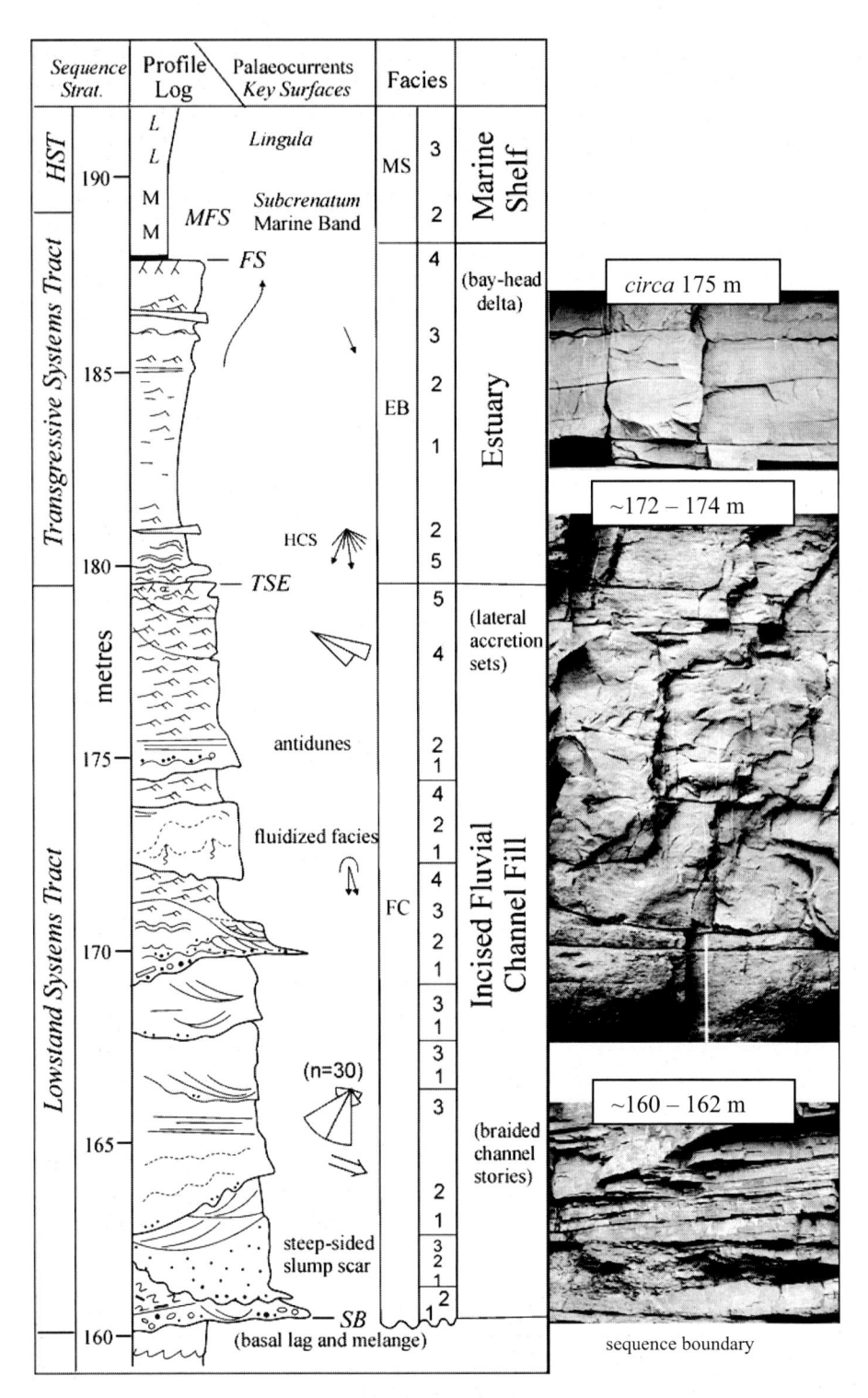

Fig. 6.47 Profile log, sequence stratigraphy and thumbnail photographs of the Telpyn Point Sandstone Formation incised valley fill at the type locality (modified from George 2001).

Kilgetty coal are highly folded and cut by thrusts (Fig. 6.49). Beyond this thrust zone, the coal reappears in a broad anticline, the basal bed of which is defined by a 1 m-thick buff-weathering ankeritic siltstone known as the Amroth Freshwater Limestone (SN 1601 0687). This distinctive lithology contains abundant valves of the non-marine bivalve *Carbonicola bipennis*, previously identified as *Anthracosia regularis*. Some of the mussels are found with both valves attached and in their life positions. The limestone is overlain by mudstones with ironstone nodules and a bed of crushed mussel remains (Fig. 6.50). Above this horizon the facies coarsen upwards into silty sandstones with ripple cross laminations and abundant small-scale

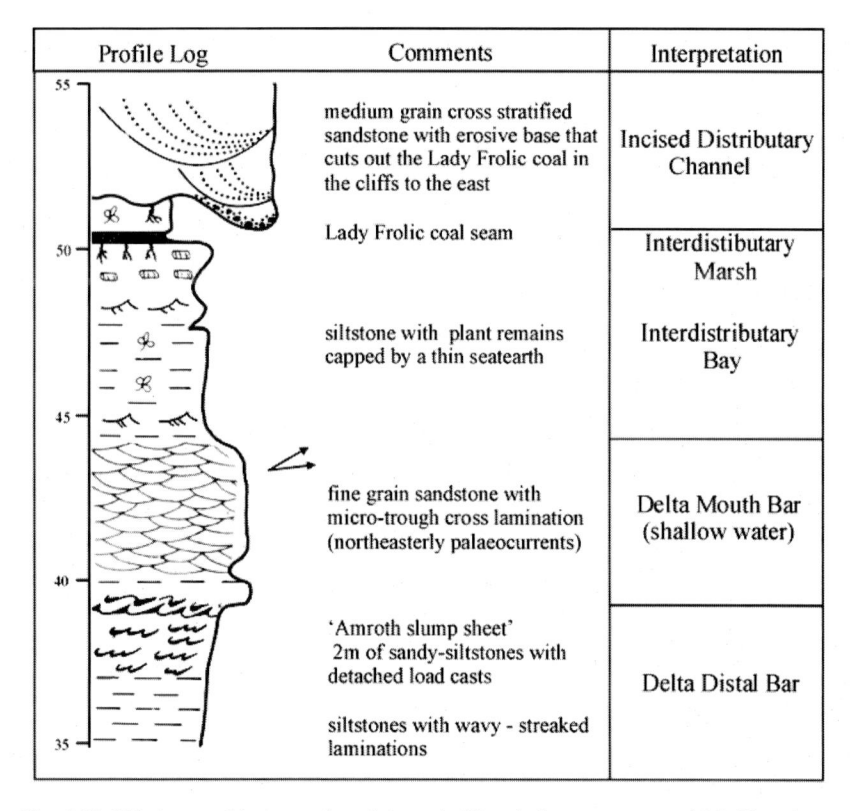

Profile Log	Comments	Interpretation
55	medium grain cross stratified sandstone with erosive base that cuts out the Lady Frolic coal in the cliffs to the east	Incised Distributary Channel
50	Lady Frolic coal seam	Interdistibutary Marsh
45	siltstone with plant remains capped by a thin seatearth	Interdistributary Bay
40	fine grain sandstone with micro-trough cross lamination (northeasterly palaeocurrents)	Delta Mouth Bar (shallow water)
35	'Amroth slump sheet' 2m of sandy-siltstones with detached load casts siltstones with wavy - streaked laminations	Delta Distal Bar

Fig. 6.48 Lithology and interpretation of the early Westphalian strata exposed 35–55 m above the *Subcrenatum* Marine Band at Amroth (modified from George & Kelling 1982).

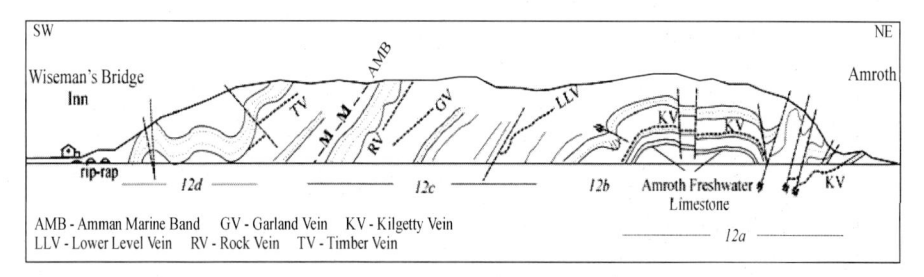

Fig. 6.49 Cliff profile showing the structure and succession of the Westphalian strata between Amroth and Wiseman's Bridge (modified from George & Kelling 1982).

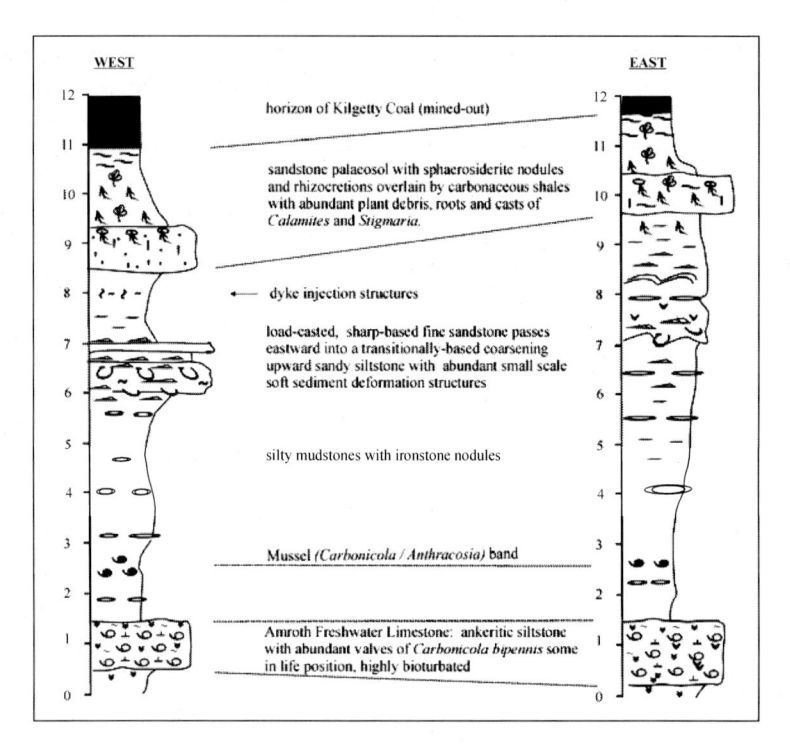

Fig. 6.50 Lateral facies variations in the coarsening-upwards cycle developed between the Amroth Freshwater Limestone and the Kilgetty Coal.

load casts. The cycle is capped by the palaeosol seen previously. At this locality the Kilgetty coal has been mined and an old adit is still visible. Fish scales are common in the roof shale of the coal, and a mussel band with *Carbonicola pseudorobusta* occurs 4.5 m higher in the mudstone succession. About 75 m farther west (SN 1577 0679), on the western flank of the Amroth anticline, the same coarsening-upwards cycle contains a more prominent loaded sandstone facies. This interesting cycle records the progressive infilling of a freshwater lake. The limestone was deposited during a period of relatively slow sedimentation, which encouraged the colonization and growth of mussels. The mussel colony was killed off as a result of increased sedimentation rates, caused by the progradation of a lacustrine delta into the lake. The abundance of soft-sediment deformation structures (load casts and dyke-injection structures) implies that the sand facies were deposited rapidly on very porous clays. In this situation, either an increase in pore pressure or externally generated seismic shocks could have caused the mud to become thixotropic and the sand to deform into load casts. The final filling of the lake is recorded by the rooted palaeosol, plant-bearing facies and the thick peat accumulations, which eventually formed the Kilgetty seam.

12b Amroth foreshore (SN 1570 0670). In the cliffs to the west, a 5 m-thick sandstone, which overlies the Kilgetty coal, is deformed into an asymmetric overfold related to a small sub-horizontal thrust. A bulbous pop-out feature with radiating quartz veins occurs on the overturned flank of the footwall syncline. This fold pair is interesting because it faces south, whereas most of the folds in the coalfield face north (Hancock *et al.* 1982).

12c Amroth foreshore. The succession exposed in the cliff and foreshore for the next 1350 m comprises many cycles with both coarsening- and fining-upwards sandstones, which are capped by seatearth/coal couplets and mussel bands. The main named coals, often referred to as veins, are the Kilgetty, Lower Level, Garland, Rock and Timber seams. Ironstone nodules, present in the interseam strata, were worked for black-band iron ore during the 19th century. Above the Rock seam, a multi-storey channel-fill sandstone displays excellent load casts and flame structures. A fossil band containing *Lingula*, *Dunbarella* and mussels, which occurs 22 m above the Rock seam, has been correlated with the Amman Marine Band. These cycles are interpreted as the deposits of prograding deltas and crevasse splays, which filled freshwater lakes and created extensive peat-forming mires.

12d Wiseman's Bridge. The Timber seam is exposed at the western end of the cliffline near the faulted core of an anticlinal flexure, where it is developed above a 3 m coarsening-upwards cycle (Fig. 6.49). Laterally, the seam is eroded out (washed out) by a sandstone, the base of which has well developed lags composed of ironstone clasts, mudstone chips, coal pellets and some slumped units. The sandbody is composed of up to five crosscutting stories that have fining-upwards heterogeneous fills. Current ripples, best seen on wave-polished fallen blocks, have rounded crests and are encased in mud drapes that thicken into the troughs. All of the ripples record consistent (unimodal) westerly-directed palaeo-flow directions. The sandbody is interpreted as the fill of a westward-flowing fluvial channel that was subject to fluctuating river discharge or weak tidal influences.

At this locality you can either walk back along the beach to Amroth or, if the tide is low enough, continue onwards to the Wiseman's Bridge Inn. From the inn you can return to Amroth on the coast path that follows the road up the hill for 300 m before continuing eastwards across the top of the cliffs.

6.3.13 *Saundersfoot to Monkstone Point*

The busy holiday resort of Saundersfoot was once the main exporting port for anthracite mined from opencast sites and collieries in the area. The high-quality anthracite (93–96% carbon, with as little as 1% ash) was originally transported to Saundersfoot in ox- or horse-drawn carts and loaded onto small sailing ships beached on the sands. The rapid growth of coal mining and related industries (iron ore mining / smelting and brick making) during the first half of the 1800s resulted in the building of Saundersfoot harbour and mineral railway, which were completed in 1834. One branch of the mineral line went from the harbour up the Strand and then through two tunnels to Coppet Hall and along the coast to Wiseman's Bridge. This route now forms part of the Pembrokeshire Coast Path. The Royal Oak Inn has an interesting collection of old photographs of Saundersfoot and also serves good food and real ales.

The objectives of this itinerary are to examine the classic Armorican structural features seen in the Westphalian strata, exposed in the cliffs to the southeast of Saundersfoot harbour, and to briefly review the sedimentology of the Monkstone Point Sandstone (Figs 6.36, 6.51). The itinerary requires medium to low tides and can be completed in about 5 hours; if the tide is high but falling, the localities can be reversed. Long-term parking is available at Saundersfoot harbour or at the sports centre.

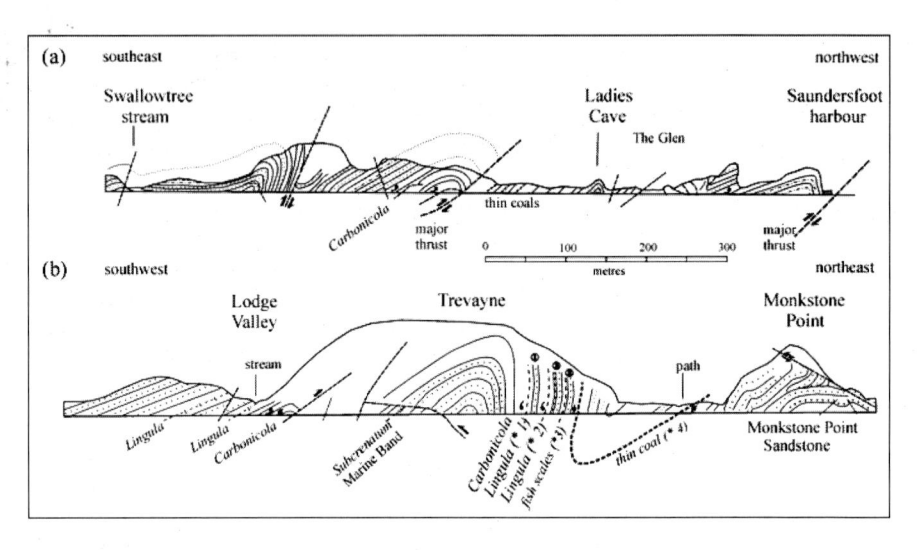

Fig. 6.51 Cliff profiles showing the structure, main lithologies and faunal bands of the Westphalian strata exposed **(a)** between Saundersfoot harbour and Swallowtree Bay, and **(b)** between Monkstone Point and Lodge Valley (modified from Jenkins 1962; Williams 1968).

13a Cliffs south of the harbour. The Lower Coal Measures, exposed south of Saundersfoot harbour, are composed of complexly folded and faulted argillaceous coal-bearing facies with minor sandstones. The cliff above the south side of the harbour wall is composed of siltstones forming the near-vertical limb of an asymmetrical anticline, which has been pushed northwards on a major thrust (Fig. 6.51a). Farther to the south the strata are cut by many relatively small thrusts that usually follow the bedding of the coals and carbonaceous shales and are refracted across the harder sandstone beds. Small normal faults are also common and these have often produced small-scale graben and half-graben structures. In one case, faults have completely isolated the hinge of a tight syncline. Some of the small listric normal faults show evidence that they moved during sedimentation (growth faults).

13b Ladies Cave Anticline. The famous Ladies Cave Anticline can be seen at the small headland (SN 1388 0433), just south of where the coast path leads down to the beach from the Glen. The path, which allows access to both sides of the headland, can be used when the tide is too high to reach the locality along the beach. The anticline is a fine example of a chevron fold with long straight limbs (interlimb angle 65°) and a narrow hinge zone (Fig. 6.52). The fold is asymmetric and its axial surface dips at 75° to the south. Several accommodation features are present, including sub-vertical limb thrusts, and small saddle reefs caused by the flow of mudstones in the hinge zone of the fold. At beach level the mudstones present in the core of the fold display a convergent fan cleavage.

About 250 m to the southeast a major WNW-trending thrust cuts southward-dipping argillaceous coal-bearing facies. Above the thrust, the erosive base of a thick sandstone is exposed in the hanging wall of an anticline. Dark shales just below the erosive base of the sandstone contain *Carbonicola*.

Fig. 6.52 View of the famous Ladies Cave Anticline exposed to the south of Saundersfoot harbour.

13c Swallowtree Bay. If the tide is low, it is possible to walk along the beach to Swallowtree Bay (SN 1416 0379). Alternatively, the bay can be reached via the coast path, which runs through Rhode Wood and descends to the beach just beyond the caravan park. To the south of the bay a late Westphalian A sandstone is deformed into a broad asymmetrical anticline complicated by minor thrusts and accommodation features (Fig. 6.51a). Its geometry is clearly seen in the core of the fold. Just to the north, the overlying strata are displaced by a southward-dipping thrust with a small amount of displacement. The multi-storey channel sandstone (the erosive base of which was noted at the previous locality) displays large-scale sets of cross strata and horizontal beds with current lineation.

13d Monkstone Point. Rejoin the coast path and continue through Trevayne Wood and across the cultivated field to the western end of Monkstone Point, where a signpost warns of cliff falls (SN 1471 0328). There is a good view due east of the erosive base of the Monkstone Point Sandstone exposed in the core of an anticline. Take the right-hand branch of the path that cuts across the headland to the steps leading down into the bay. Near the bottom of the steps a southward-dipping seatearth with rhizocretes, large branching *Stigmaria* and casts of *Lepidodendron* are seen. Carbonaceous shales with abundant plant remains and coal debris overlie this palaeosol. Walk along the storm beach to a cave in the headland where the base of the Monkstone Point Sandstone is exposed in the core of the anticline (Fig. 6.51b). Its erosive base is a good example of a mélange, being composed of multiple lags of quartz and ironstone pebbles mixed with lenses of bright coal and fragments of carbonaceous shale. A prominent pop-out structure appears to be tectonic in origin,

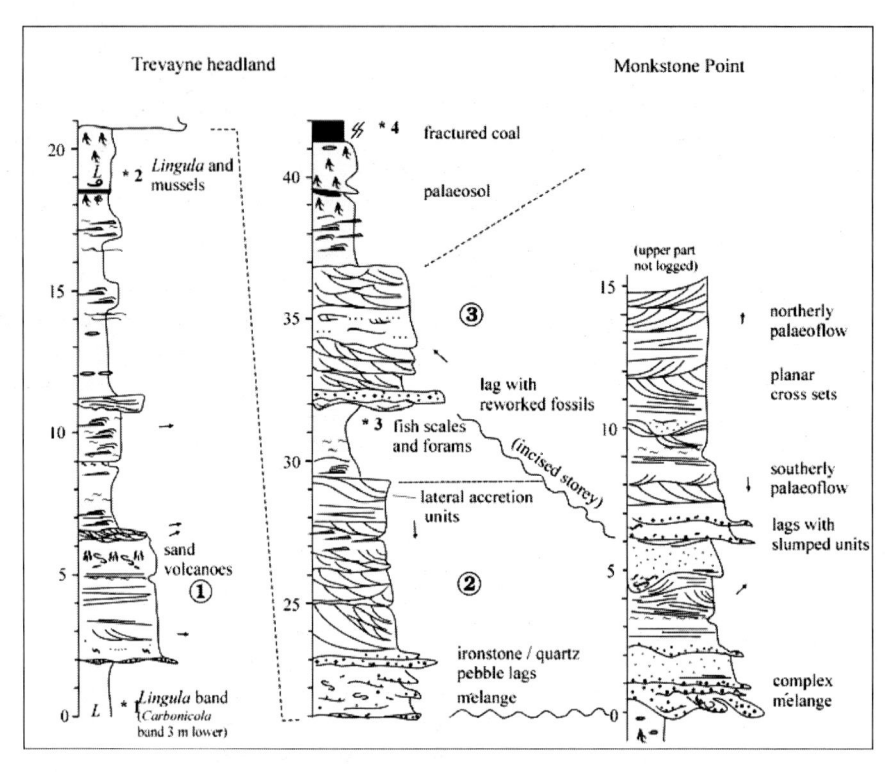

Fig. 6.53 Graphic logs of the Westphalian strata exposed between the northern limb of the Trevayne Anticline and southern limb of the Monkstone Point Anticline (see also Figure 6.51b).

but some of the deformation could be attributable to soft-sediment deformation. The multi-storey geometry of the sandbody is confirmed by the presence of internal erosional lags and mudstone lenses.

13e Trevayne headland. The northern flank of the Trevayne Anticline is composed of three steeply dipping and inverted sandstones interbedded with predominantly argillaceous facies. Details of the facies succession and the positions of fossil bands are shown on the profile logs (Fig. 6.53). All three of these sandbodies have erosive bases and they are interpreted as fluvial-channel fills. Reworked fossil corals have in the past been found in the lag at the base of the third sandstone (Fig. 6.53 at 32 m). This sandstone is capped by a seatearth and a thin coal seam, which defines the inverted southern limb of a syncline that is present below Monkstone beach (Fig. 6.51). The structural interpretation indicates that the second and third sandstones correlate with the Monkstone Point Sandstone. The absence of the mudstone interbed, and the apparent increase in thickness of the sandbody at Monkstone Point, implies that it was deposited in a more axial channel position (Fig. 6.53). The few bimodal palaeocurrent readings taken from the sandbodies suggest that tidal currents may have influenced the fluvial depositional system. In terms of sedimentology the Monkstone Point Sandstone displays many of the features of an incised valley fill. A sequence-stratigraphy appraisal of this sandstone, which is also exposed to the southwest of Lodge Valley, could form the basis of an interesting short field project.

6.3.14 Broad Haven

Enter Broad Haven on the B4341 and park in the large car park at the northern end of the bay, where there is a national park information centre and a modern youth hostel (Fig. 6.54). Alternatively you can travel to Broad Haven from St David's or Milford Haven on the Puffin Shuttle, which also stops at Little Haven and Nolton Haven. (Note that large coaches cannot enter Little Haven because of the narrow road and sharp bends).

The cliffs to the north of the bay between Emmet Rock and the Sleek Stone expose some of the most impressive Armorican folds seen in the Lower Coal Measures of the Pembrokeshire Coalfield. These northward-facing folds are bounded by thrusts that repeat a 25–30 m section of late Westphalian A strata containing three sandstone units. The section is accessible only between medium to low tide; if the tide is receding but too high to round Emmet Rock, the first two localities (*15a* and *b*) from the Little Haven itinerary should be visited first.

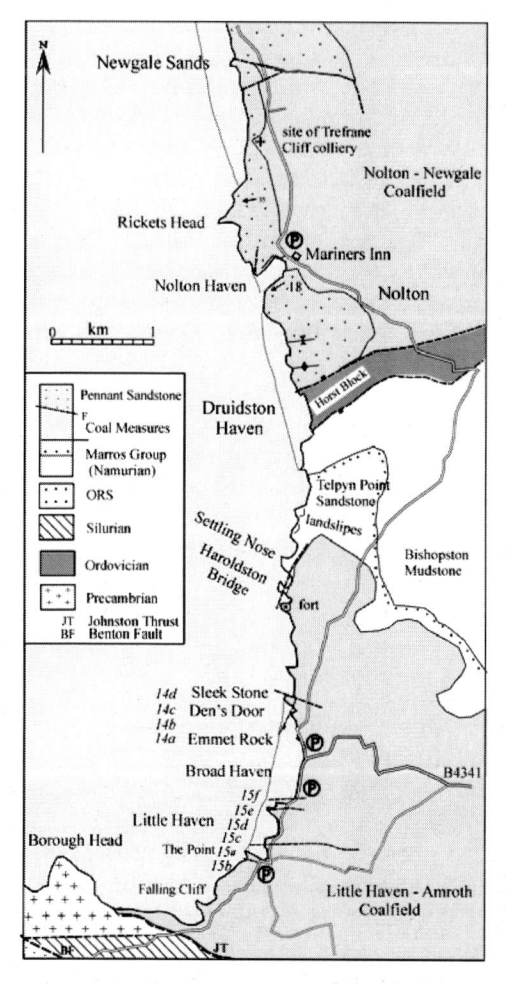

Fig. 6.54 Geology and locality map of the coast between Borough Head and Newgale.

14a Emmet Rock. From the car park walk north along the beach to an old adit in the cliff face. The adit is cut into silty mudstones that contain abundant ironstone nodules, rhizocretes, rootlets, casts of *Calamites* and two thin carbonaceous shales with comminuted plant remains. As there are no records of ironstone mining in the area the adit appears to have been used for drainage and ventilation of subsurface coal workings that once existed, to the northeast, around Belmont.

14b First cove. To the north of Emmets Rock a complex asymmetrical north facing and northwest-plunging anticline is defined by three thin sandstones beds (Fig. 6 55, sandstones 1, 2 and 3) separated by mudstones and shales that thicken into the hinge zone. The northern overturned limb of the fold is cut by a curved low-angle thrust with a small amount of lateral displacement. This is an excellent locality for identifying and analysing way-up structures, which occur in the minor sandstones. Sandstone 1 is a coarsening-upwards ripple-laminated unit with a rooted top that displays pressure-solution cleavage. It reappears to the north of the fold as a result of further thrusting. Sandstone 2 displays well-developed basal and internal load casts, whereas sandstone 3 has an erosional base and rooted top.

14c Den's Door. The next cove is dominated by a large stack with two natural arches (Den's Door), isolated from the cliff face by a northeast–southwest-trending fault. In the adjacent cliff face an impressive north-facing anticline and a thrust fault are exposed (Fig. 6.56). Sandstone 3, which defines the fold, displays good fan cleavage and some pop-out structures occur in the hinge and lower surface of the hanging-wall anticline. These accommodation features are related to a minor splay of the main thrust. In the footwall syncline, the sandstone again displays cleavage and the underlying mudstones exhibit bedding-plane slip and they thicken into the nose of the fold. Sandstone 2 has basal and internal load casts indicating that the limb is the correct way up.

14d Sleek Stone. The Sleek Stone promontory exposes a classic monoclinal fold that trends 110–290° and plunges 10° to the west (seawards). It has a vertical to inverted northern limb, which is cut by a thrust fault, now hidden by cliff-fall debris. The sandstone, which defines the fold, is cut by a set of north–south-tending joints and an east–west sigmoidal pressure-solution cleavage. The sandbody also displays

Sandstone 3: 2.5m thick, basal and internal erosional planes, cross to ripple laminated, rooted top
Sandstone 2: 75cm thick, load-casted base, internal loads, ripple marks and cross laminations
Sandstone 1: 75cm thick, coarsening-upwards, ripple cross laminated, rooted top with plant shale

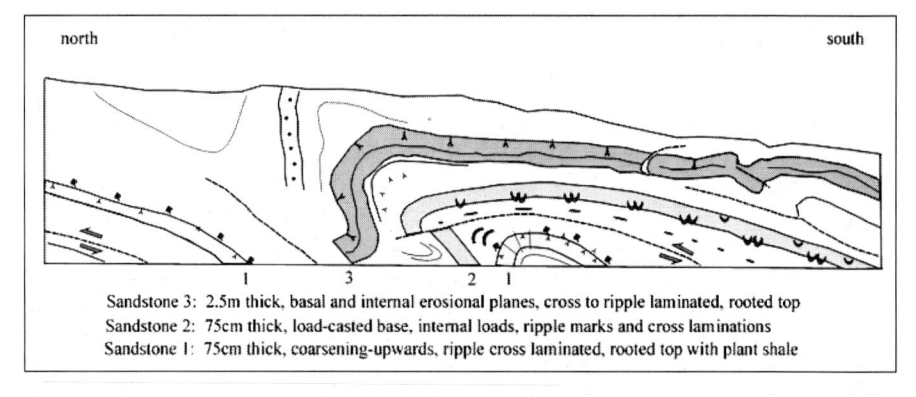

Fig. 6.55 Field sketch of the anticlinal fold exposed in the cliffs to the north of Emmet Rock, Broadhaven. The fold is defined by three sandstone beds (1-3), which contain a variety of way-up structures.

Fig. 6.56 Impressive north-facing anticline and associated thrust fault exposed in the cliffs behind Den's Door, Broad Haven.

micro-trough ripple cross laminations that give easterly palaeoflow directions, and it is capped by a seatearth. North of the thrust the succession is obscured by fallen debris, which includes blocks of a brown-weathering impure limestone with non-marine bivalves that correlates with the Amroth Freshwater Limestone. A coal seam developed in the cliff face is the equivalent of the Kilgetty seam. Farther to the north Westphalian A and Namurian strata are exposed along the rocky shore to the Druidston Haven (Ordovician) horst block (Fig. 6.53). This section is not recommended as part of the itinerary because of the very unstable condition of the cliffs and the lack of access points to the coast path. However, details of this part of the section can be consulted in George (1982, 2001) and George & Kelling (1982).

6.3.15 Little Haven

The picturesque village of Little Haven has a car park with toilets, cafés and a couple of inns / hotels. Unfortunately, the village is accessible only by car or minibus; coaches should be parked at Broad Haven. The cliffs to the north and south of the bay expose Lower Coal Measure strata, which are folded and cut by many strike and thrust faults. In the late 19th century the Little Haven–Broad Haven area was important for the mining of anthracite coal, which was recovered from shallow workings and deeper shafts. In this region, three anthracite seams were worked, each generally less than 1 m thick, but folding and thrusting frequently resulted in rapid thickness changes and the shattering of the coal seams. The anthracite coal was transported from Little Haven on barges loaded on the beach at low tide.

The purpose of this itinerary is to examine the sedimentology and structure of the cliff sections of Westphalian strata (Fig. 6.57). The Point and a small cove to the north can be examined at high tide, although all of the other localities can be studied only at medium to low tide.

15a The Point. From the car park, turn left at the post office and continue past the Swan Inn to the end of the Point, and then take the steps leading onto the headland. The headland is composed of a multi-storey sandstone that displays excellent sets of planar cross stratification and ripple-drift cross lamination, which indicate northeasterly palaeoflow directions.

15b Small cove (SM 855 129). In the small cove to the south, the downfaulted top of the Point sandstone is succeeded by 30 m of argillaceous facies containing four sandstone bodies (Fig. 6.58, A–D). Sandstone A displays a coarsening-upwards trend and has sheet geometry. The overlying sandstones (B, C and D) that occur higher in the cliff face are best viewed from the path leading to the Point. All of these sandstones have erosional bases and channel geometries. The erosional base of sandstone B has a well developed ironstone lag and the overlying channel fill has trough sets indicating a northeasterly palaeoflow. Channel C has a loaded base and is truncated by D to the southwest.

On the walk back to Little Haven, note the northern headland of the bay known as Fox Hole, which is composed of a thick sandstone deformed into an asymmetrical concentric anticline (Fig. 6.57). About 200 m farther inland a southerly dipping reverse fault cuts the cliff section.

13c Little Haven. Strata exposed in the northern corner of Little Haven comprise predominantly argillaceous facies with many palaeosols, carbonaceous shales and thin impure coal seams. The section is cut by a fault that dips at 50° south and has north–south-orientated slickensides and grooves. A drag fold to the south of the fault indicates that reverse slip has occurred.

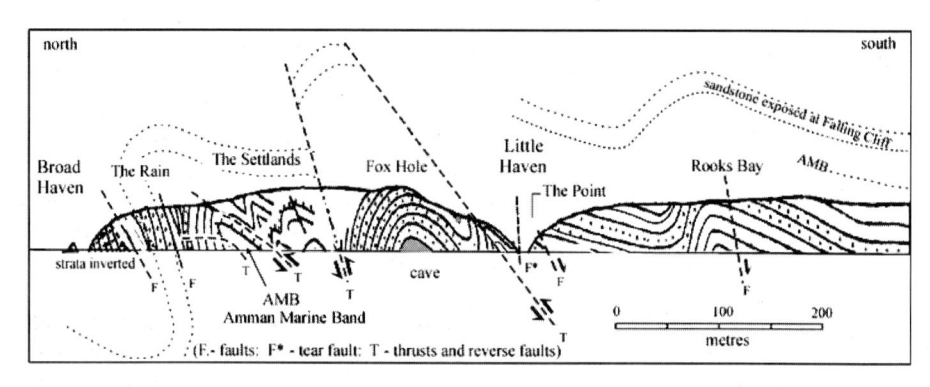

Fig. 6.57 Cliff profile showing the structure and main lithologies of the Westphalian strata exposed between Rocks Bay and Broad Haven (modified from Jenkins 1962; Williams 1968).

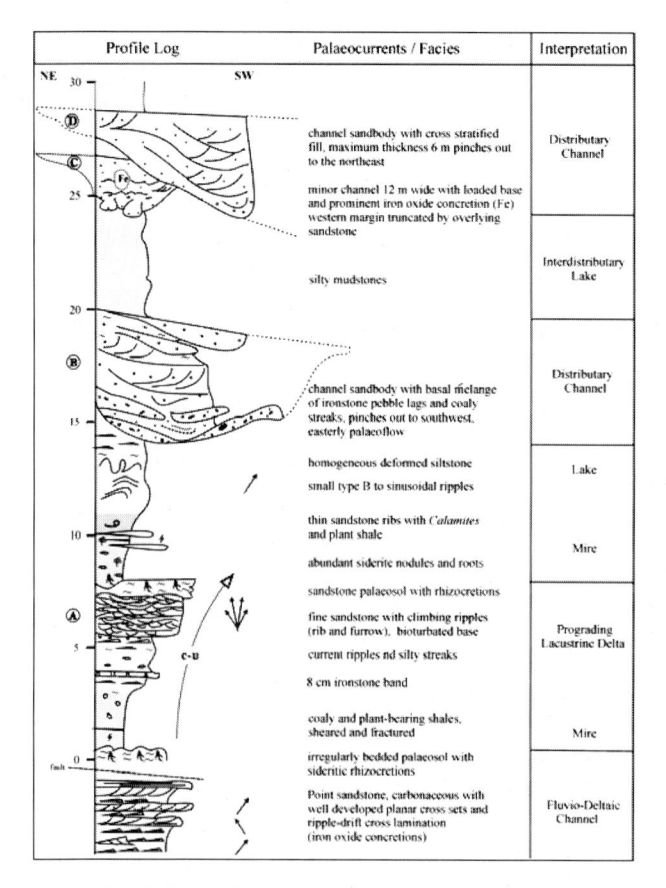

Fig. 6.58 Sedimentology and interpretation of the Westphalian succession exposed in the cove south of The Point, Little Haven (see text for discussions of sandbodies A–D).

13d Fox Hole. Fox Hole headland is composed of a sandbody (about 70 m thick) whose base can be seen in the cave located in the core of the anticline. The erosional base of the sandstone has 1–2 m thick mélange composed of a mixture of lithologies with coaly lenses and ironstone lags eroded from the underlying palaeosol. This major sandstone comprises massive and cross-stratified facies that are overlain by stacked channel fills. Palaeocurrent measurements (corrected for tectonic dip) indicate flow to the northeast. To the north of the headland, the top of the sandstone is exposed on the vertical limb of the WSW-plunging anticline. Here the top 4 m is composed of ripple-laminated facies that fine up into silty sandstones and carbonaceous shales with rootlets, plant remains and a thin mussel band. An adit perched 3 m above beach level appears to have been an old drainage and ventilation shaft for the inland anthracite workings.

15e The Settlands. A major reverse fault cuts the section in the southern corner of Settlands Bay. To the north of the fault the succession continues with shattered shales exposed in an anticline with a core of sandstone. The mostly overgrown strata, exposed in the centre of the bay, are cut by two easily recognized low-angle thrusts. The lower thrust is associated with a north-facing overturned anticline with thickened shales in its hinge zone. Just above the thrust plane, a shale containing non-marine bivalves has been correlated with the Amman (*Vanderbeckei*) Marine Band, which defines the base of Westphalian B. Elongated load casts occur on the bases of two minor sandstones that can be seen in the cliff below the upper thrust.

15f The Rain. The northern headland of the bay is composed of a highly faulted succession of Westphalian B strata that occur above the Amman Marine Band. These inverted strata form the northern limb of a large overturned anticline. A prominent sandstone in this succession can be correlated with a similar sandstone, which occurs at Falling Cliff, to the south of Little Haven. In a small inlet farther to the north, coal-bearing argillaceous facies display many tight, downward-facing minor folds associated with small thrust faults.

6.3.16 Nolton Haven

At Nolton Haven there is a large car park, adjacent to the Mariners Inn, which has toilets, and a notice board with tide-tables and a timetable for the coastal bus service (Puffin Shuttle) (Fig. 6.54). The narrow roads are not suitable for large coaches.

During the late 1800s to early 1900s Nolton Haven was the main exporting centre for anthracite mined from the Nolton–Newgale Coalfield. Many abandoned shafts and the remains of colliery buildings, seen along the coast path from Rickets Head to Newgale Sands, are evidence of the past coal mining industry. At Trefrane Cliff colliery (1850–1905) the seams, at a depth of about 100 m, were worked under the sea, where millions of tonnes of unworked anthracite are still believed to exist. In this area the Pennant Measures (Westphalian C) are composed of thick sandstones, interbedded with argillaceous facies containing many previously mined coal seams (e.g. Folly, Hookes, Cliff, Black Cliff and Rickets Head veins). These strata occur in four fault-defined blocks that young progressively northwards from Nolton Haven to Newgale Sands.

The purpose of this itinerary is to examine and interpret the Pennant Sandstone Formation exposed at Nolton Haven. This type of study is important because it provides field data that can be used to synthesize the palaeogeography of the South Wales basin during the early stages of the Armorican Orogeny.

16a Southern cliffline. The strata exposed along the southwestwardly extending cliffs to the south of the haven, represent the oldest Pennant Measures in the coalfield. In the southern corner of the bay, the succession begins with mudstones containing poorly preserved non-marine bivalves, which are overlain by a seatearth and an unnamed coal. This coal is partially washed out by the erosional base of a multi-storey sandstone (22 m thick), which has six erosional storeys, the upper three of which display fining-upwards trends (Fig. 6.59a). Sets of cross strata indicate southwesterly palaeoflow directions. This sandbody is succeeded by 16 m of argillaceous facies with plant remains, palaeosols and fine-grain sandstones. The minor sandstones have well developed bedforms and lateral accretion surfaces. The base of a second multi-storey sandbody is recorded by a coarse sandstone (3 m thick) with many lag-defined erosional surfaces and deformed units. This sandstone that occupies the cliffs for the remainder of the bay, is estimated to be over 30 m thick.

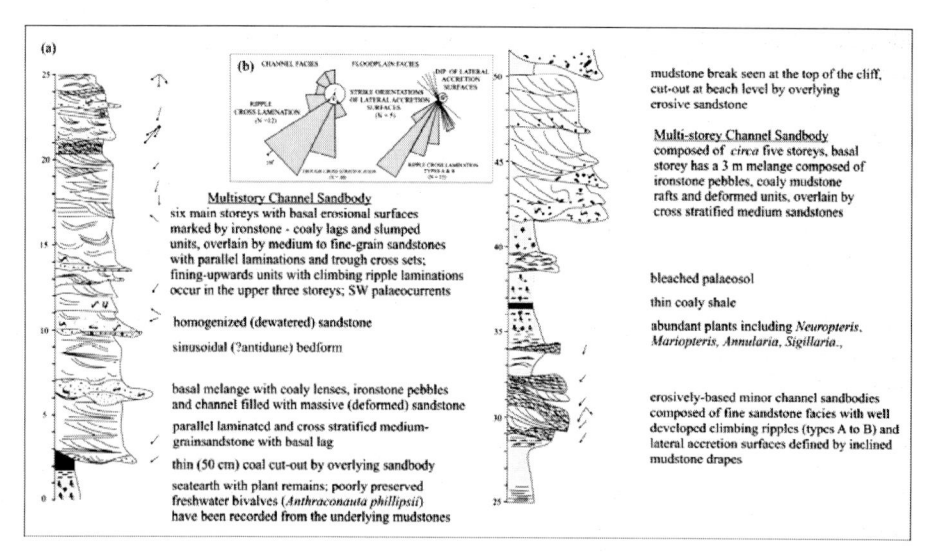

Fig. 6.59 (a) Profile log and palaeocurrents from the Westphalian C strata (Pennant Sandstone Formation) exposed in the cliffs on the southern side of Nolton Haven. (b) Rose diagrams of grouped palaeocurrents readings from the channel and floodplain facies.

The above sequences represent excellent examples of fluvial-channel fills and their related floodplain deposits. The multi-storey sandstone is interpreted as the deposits of southwestwards-flowing braided rivers. Individual braid channels were rapidly filled and abandoned during successive flood events. During avulsion, unstable bank material slumped into the channels to produce the basal lags and mélange deposits. The fining-upwards storeys, recorded towards the top of the sandbody, were deposited on laterally and downstream-migrating mid-channel bars and sand flats. There is no evidence to suggest that either this major channel or the one that succeeds it were incised below their normal base level. Consequently, channel filling and abandonment appear to have occurred as a result of autocyclic processes.

The overlying floodplain facies contains minor fine-grain sandstones with extremely well developed sedimentary structures and lateral accretion surfaces (Fig. 6.59b, 28.0–32.5 m). These inclined mud-draped surfaces dip to the southwest at angles generally 1–10° greater than the tectonic dip, which at this locality is 18° to the southwest. Sedimentary structures, observed on some of the lateral accretion beds, display down-dip transitions from cross to horizontal to ripple laminations, implying that they were deposited during falling river stage. Internal reactivation surfaces, which often cut out the mud drapes, were formed as a result of minor erosion during rising stage. The process of lateral accretion on point bars is generally recognized by the fact that bedforms indicate flow directions roughly parallel to the strike of the inclined surfaces. In the present example, the unusual palaeocurrent relationship (summarized in Fig. 6.59b) appears to be related to the fact that the point bar was migrating downstream. Consequently, the accretion surfaces located on the down stream part of the point bar would have had the greatest preservation potential. These sandstone facies are interpreted as the

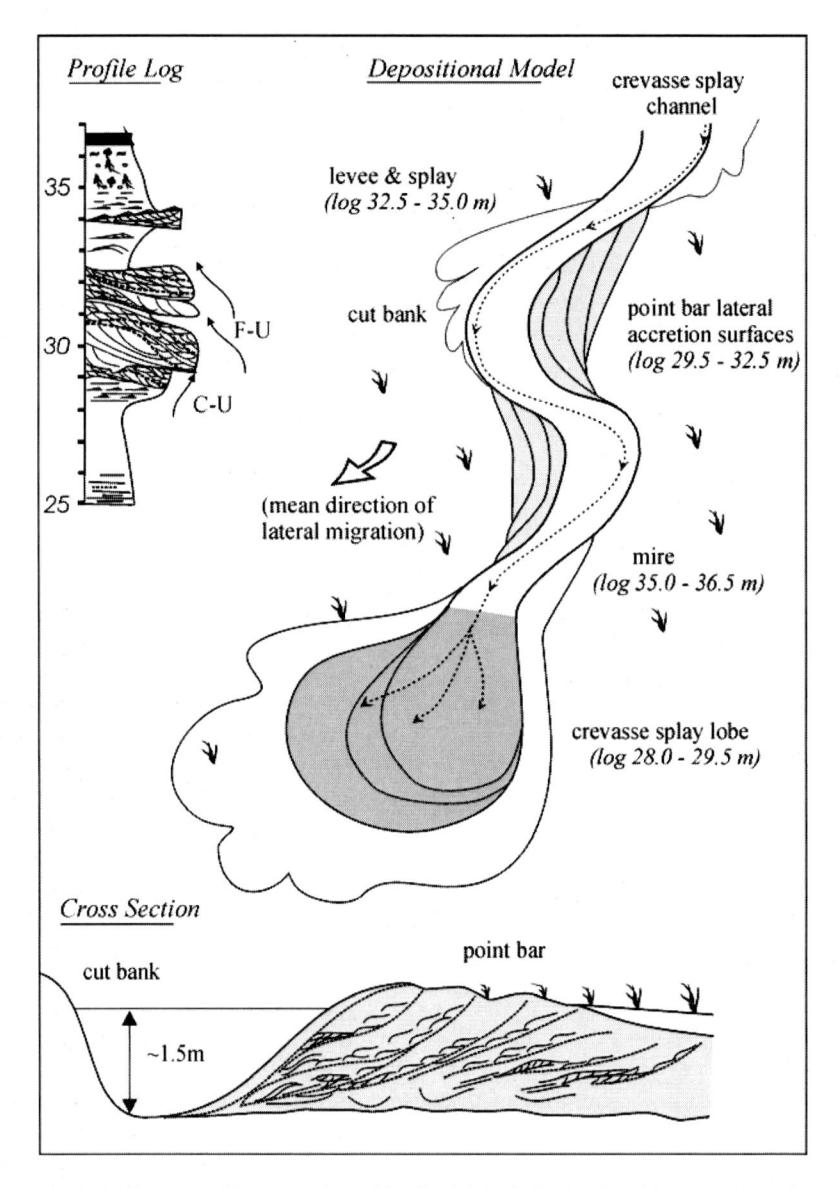

Fig. 6.60 Depositional interpretation of the floodplain facies developed between the major multi-storey channel sandbodies exposed at Nolton Haven (see Fig. 6.59,).

deposits of a sinuous, downstream-migrating crevasse splay system. They are overlain by thin levée deposits and by carbonaceous shales and palaeosols that were formed in topographically lower shallow lakes and swamps (mires).

Chapter 7

North Pembrokeshire and Cardigan Bay

7.1 Introduction

The rugged landscape of north Pembrokeshire has resulted from the differential weathering and erosion of hard igneous rocks that occur within generally softer sedimentary and low-grade metamorphic strata. These intrusive and extrusive igneous rocks form the bare crags and tors of the St David's Peninsula, Strumble Head and Mynydd Preseli, which rise dramatically above the 60 m platform. Differential erosion has resulted in a highly indented coastline, which contrasts with the smooth sweep of Cardigan Bay, formed within a wide outcrop of Lower Palaeozoic sedimentary rocks. Many of the inlets were created towards the end of the Last Glacial Maximum (Late Devensian) as a result of the increased rates of erosion that occurred when melt-waters drained into existing river valleys. Subsequently, during the Flandrian (*circa* 12–6 ka BP) post-glacial rise in sea level, the sinuous valleys were drowned to form rias. Prior to this (around 14 ka BP) west Wales was connected to Ireland, however during the transgression the land connection was progressively flooded by the Irish Sea (see Fig. 2.15). In recent times many of the rias, especially those located on the west coast, have silted-up.

A large part of the coastal zone to the south of the River Teifi, lies within the Pembrokeshire Coast National Park; the guide to the coast path (John 2001) provides a good introduction to the scenery, natural history and culture of this area. To the north of the River Teifi into Ceredigion (formerly part of north Dyfed), the coastal scenery is generally less rugged due to the absence of igneous rocks and the relatively uniform composition of the underlying sedimentary rocks.

Road communications are good with the main towns of Fishguard, Newport and Cardigan linked by the A487 coast road, which continues northeastwards to Aberaeron and on to Aberystwyth. The 'Puffin Shuttle' coastal bus service operates between Newport, Fishguard (Strumble Head) and St David's; timetables being available at tourist information centres. The *Pembrokeshire Coast Path* (John 2001) is an invaluable guide; the area is covered by the Ordnance Survey Explorer (1:25 000) maps for North Pembrokeshire (OL35) and Cardigan & Newquay (198).

7.2 Geological history

North Pembrokeshire is composed of late Precambrian (Neoproterozoic) and Lower Palaeozoic strata with only a thin sporadic cover of drift (Fig. 7.1). The Neoproterozoic rocks of the region are exposed in two inliers located along the axes of the St David's and Hayscastle anticlines. These inliers are composed of thick successions of basic and acid lavas, tuffs and conglomerates intruded by acid plutons (Pebidian Supergroup 587 +25/-14 Ma). All of these Precambrian rocks were uplifted, metamorphosed and deeply eroded prior to the deposition of the basal Cambrian conglomerate around 534 million years ago. Thus, sedimentation began approximately 8 million years later than the Global Standard Section and Point (GSSP) of 542 $\pm$1 Ma for the base of the Cambrian. This transgressive conglomerate unit is succeeded by an almost complete succession of Cambrian shelf clastic sediments, exposed along the classic coastal section of the St David's Peninsula (St Non's Bay, Caerfai Bay, Porth-y-Rhaw and Solva). Deep-water turbidites, and mudstones often containing graptolite faunas, dominate the succeeding Ordovician and Silurian successions present in the Welsh Basin (see Ch. 2; Table 2.1).

During this time period there was extensive igneous activity at a number of volcanic centres. The Trefgarn Volcanic Group (Tremadoc–Arenig) includes the rhyolites forming the picturesque crags of Maiden's Castle and Poll Carn at Treffgarne Gorge, and also those that crop out below Roch Castle. These interesting volcanics have a calc-alkaline geochemistry, and have been interpreted as an island-arc association that was contemporaneous with the Rhobell Volcanics of the Harlech Dome. Strumble Head and Mynydd Preseli are composed predominantly of submarine volcanics and slightly younger, high-level basic igneous intrusions, belonging to the Fishguard Volcanic Group (Llanvirn). St David's Head and the summits of Carn Llidi, Carn Lleithr and Carn Penberry, are composed of an impressive differentiated tholeiitic sill up to 400 m thick.

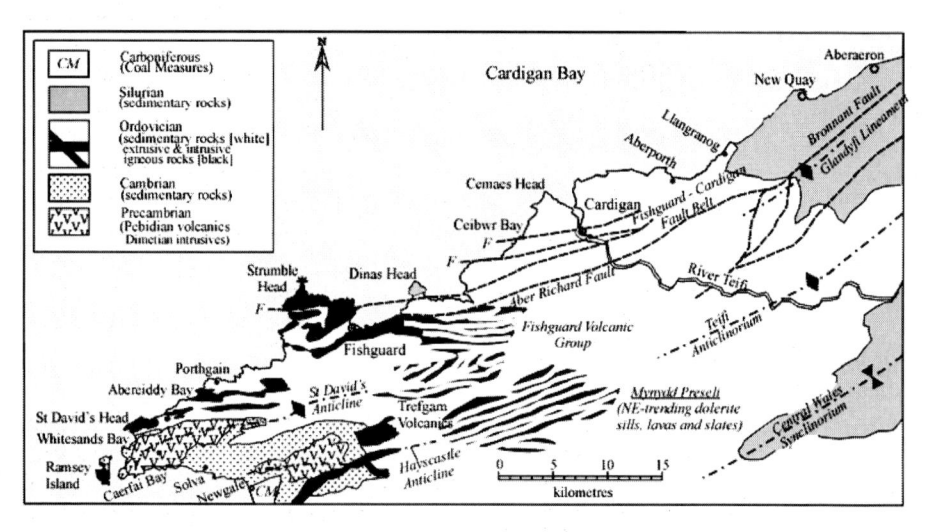

Fig. 7.1 Geological map of North Pembrokeshire (modified from George 1970; Davies *et al*.. 2003)

To the north of the igneous province, the Ordovician and Silurian succession consists entirely of marine sedimentary rocks, most of which are deep-water hemipelagic (graptolitic) shales and turbidites that were deposited within the Welsh Basin, (previously the Welsh Lower Palaeozoic Geosyncline). All of the rocks of the area were deformed during the protracted Caledonian Orogeny, which culminated in Early Devonian times (approximately 400 Ma) and produced tectonic structures with a northeast–southwest trend. This Caledonoid trend is clearly illustrated by the Fishguard–Cardigan and Bronnant fault belts, and the complex folding along the Teifi Anticlinorium and the Central Wales Synclinorium (Fig. 7.1).

The area has been exploited for its geological resources over a long period of time. In Neolithic times the early settlers used large blocks of local stone to construct shelters and burial chambers, such as those at Coetan Arthur, Pentre Ifan and Carreg Samson. The bare and atmospheric Prescelly Hills (Mynydd Preseli) are littered with standing stones, burial chambers and stone circles; they are famous for being the source of the spotted dolerite used in the construction of the bluestone inner circles of Stonehenge around 2500 BC (see Fig. 2.1). In more recent times the Caerbwdy Sandstone (Cambrian) provided the building stone for St David's Cathedral. In the 19th century roofing slate was quarried along the southern flank of the Preseli Hills at Rosebush and along the west coast between Abereiddy and Porthgain, the latter locality also having thriving roadstone and brick-making industries. The absence of limestones and marls in the bedrock has resulted in acid soils, which, in the past, were treated with limestone imported from the south and burnt in kilns located in many of the small harbours. During Victorian times there were around 12 limekilns being fired in Solva. In the past, lead, zinc and copper ores have been worked in small shallow mines scattered across the area.

The geology of a large part of North Pembrokeshire has been published in the provisional (1:50 000) map of the St David's area (Sheet 209). There are also new geological maps and explanatory booklets covering Llanilar and Rhyader (Davies *et al.* 1997; Sheet 178 & 179) and Cardigan and Dinas Island (Davies *et al.* 2003; Sheet 193). Previous itineraries to the area have been published in two out-of-print guides edited by Bassett & Bassett (1971) and Bassett (1982), while the most important geological sites have been documented in the Geological Conservation Review series for the Precambrian (Carney *et al.* 2000), the Cambrian to Ordovician (Ruston *et al.* 1999) and the Silurian (Aldridge *et al.* 2000). The second edition of *The Geology of England and Wales* edited by Brenchley & Rawson (2006) provides an up-to-date synthesis of the stratigraphy and evolution of the rocks of the area, although the book does not incorporate the most recent numerical ages published by Gradstein *et al.* (2004).

7.3 *Itineraries*

The itineraries are presented in stratigraphical order, beginning in the St David's area and progressing northeastwards to Aberaeron. Since many of the localities are beach and cliff sections, only exposed from medium to low tides, it is important to consult tide tables when planning visits. Always wear a safety helmet when working near cliff faces and avoid areas with unstable overhangs. Some of the localities, such as St David's Head, Strumble Head and Ceibwr Bay, can be accessed from the top of the cliffs via the Pembrokeshire Coast Path. During these itineraries take special care when approaching cliff faces and steep slopes, and try to avoid days when the weather is very wet and/or excessively windy.

7.3.1 St Non's, Caerfai and Caerbwdy bays

St Non's, Caerfai and Caerbwdy bays, forming the coastline to the south of St David's, display classic exposures of the Cambrian System (Fig. 7.2). Access to these bays is only possible from medium to low tides; the complete itinerary takes about 4–5 hours and involves a round trip of about 7 km along the clifftop coast path and foreshore.

1a St Non's Bay. From the centre of St David's bear left at the Old Cross into Goat Street and follow the signs for St Non's. There are parking spaces for minibuses and cars on the side of the road terminating at St Non's Retreat (SM 752 244). From this point, on a clear day, there is a good view to the south of Skomer Island with Skokholm just visible between Middle Isle and The Neck. Follow the footpath southwestwards to St Non's Well and Chapel, which marks the site where St David, the patron saint of Wales, was reputed to have been born about 462 *AD*. Note that the ruined walls of St Non's Chapel are built largely of blocks of Cambrian conglomerate and sandstone. Continue across the field to the coast path, and then walk 1 km eastwards to a small gully, eroded by the water that has issued from the sacred well over a long period of time. Here, a steep narrow path leads down through a cleft to the boulder-strewn (very slippery) beach. The cleft, which

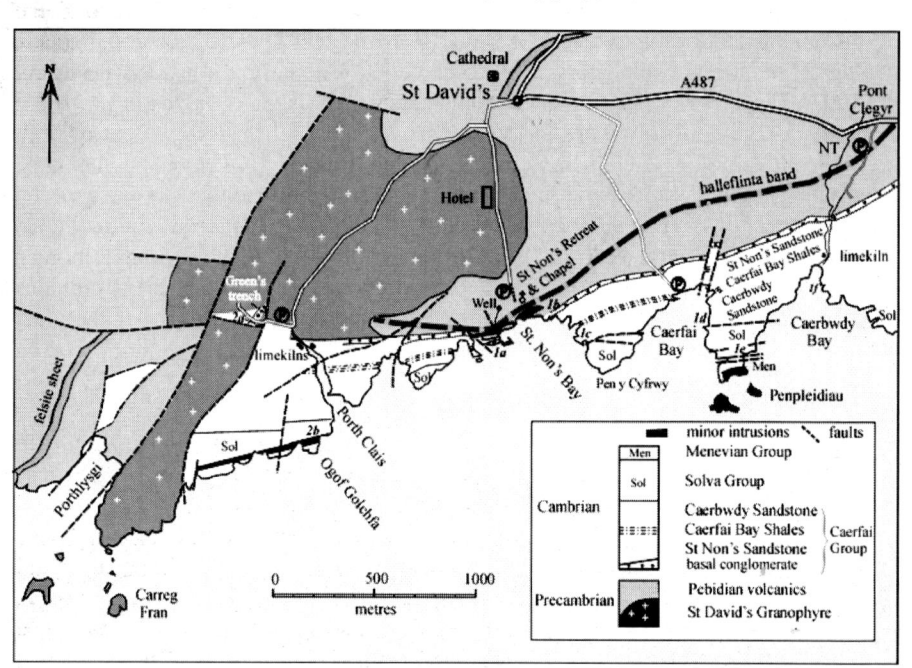

Fig. 7.2 Geology and location map of the coast to the south of St David's (modified from Baker 1982).

is located along a fault intruded by a 1 m-thick basic dyke, intersects steeply inclined silicified acid tuffs, referred to as halleflinta (a Swedish term meaning flint rock). At this locality the blue–green–grey halleflinta is a very fine grain siliceous rock, which breaks into translucent splinters. A 30 m-thick band of this distinctive rock-type has been mapped inland to Pont Clegyr (Fig. 7.2). Seawards, the west wall of the cleft is composed of pods of banded rhyolites and coarser-grain tuffs, which have been intruded by an irregular sheet of highly weathered porphyritic dolerite. The irregular margins of the dolerite enclose lenses of rhyolite, and a tongue of the intrusion extends seawards. These acid feldspathic tuffs and halleflinta belong to the Pebidian Supergroup of late Precambrian (Neoproterozoic) age. Farther seawards, on an east–west-trending stack, purple Precambrian tuffs are overlain unconformably by about 12 m of basal Cambrian conglomerate, which dips steeply (65–80°) to the north, indicating that the strata are inverted (Fig. 7.3). The conglomerate is composed of well rounded to sub-rounded clasts (maximum 25 cm in diameter) of white vein quartz, jasper, pink quartzites and acid tuffs, which decrease in size upwards. Some of the clast-supported units display low-angle cross bedding (southerly palaeoflow) and occasional normal and inverse grading. There is very little evidence of imbrication, although many of the elongated pebbles are aligned parallel to the bedding. Looking eastwards, there are two further fault-displaced stacks composed of conglomerate. The conglomerate appears again above the cave on the far side of the bay, where its base displays a well-defined channel feature; its top is faulted against near-vertical beds belonging to the St Non's Sandstone.

1b Clifftop east of St Non's Bay. The basal Cambrian conglomerate is exposed again, adjacent to the coast path on the headland (SM 7537 2426), from where it continues inland, into the caravan park, as a ridge covered with gorse. Here the well-jointed conglomerate beds are vertical to slightly inverted, and they contain pebbles that are appreciably smaller (2–7 cm in diameter) than those observed at the previous locality.

1c Pen y Cyfrwy. Continue southeastwards on the coast path to a prominent gully on Pen y Cyfrwy headland (SM 7549 2418). This gully, eroded into soft reddish-brown Caerfai Bay Shales, forms a good marker horizon within the Cambrian succession (Fig. 7.2). Looking westwards, the shales produce a prominent gap in the cliffline,

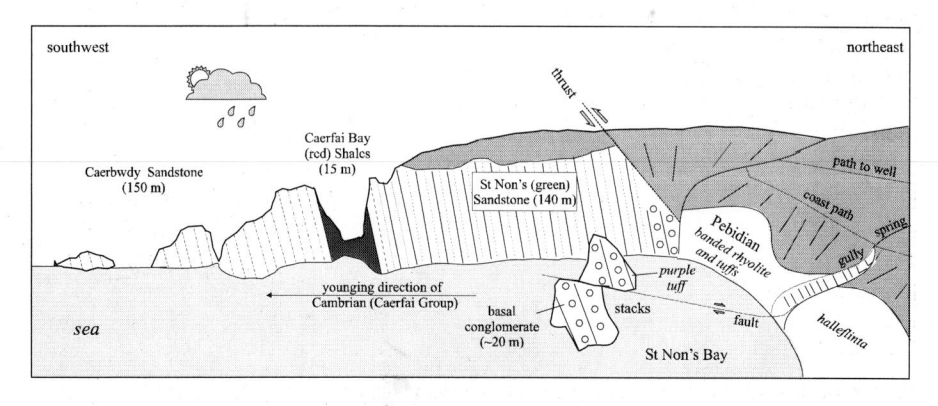

Fig. 7.3 Field sketch showing the geology of St Non's Bay viewed from Pen y Cyfrwy.

the entire headland being composed of slightly inverted strata belonging to the Caerfai Group (Fig. 7.3). Continue for 50 m along the coast path to a small promontory, from where the steeply dipping faulted contact between the Caerbwdy Sandstone and the overlying Lower Solva Beds can be seen to the south. Here, the Solva Beds are composed of thinly bedded sandstones and siltstones, deformed into an open asymmetrical syncline.

1d Caerfai Bay. Continue on the coast path to the car park at Caerfai Bay, seaward of which are overgrown quarries that provided some of the sandstone used to build St David's Cathedral. Take the path down to the beach noting the quartz-veins and shattered sandstone associated with two NNE-trending faults. The cliffs at the northeast corner of the bay are composed of the upper part of the St Non's Sandstone, which owes its green colour to the presence of chlorite and epidote in its clay matrix. On wave-polished surfaces, particularly in the cave, the cleaved green sandstone facies appears mottled as a result of intense bioturbation, although occasional sets of ripple cross-laminae and HCS are still discernable. The trace-fossil assemblage includes abundant *Rhizocorallium* (up to 3 cm in diameter), large *Zoophycos*, *Teichichnus*, possible *Diplocraterion* and a variety of non-specific tubes and escape burrows (Fig. 7.4).

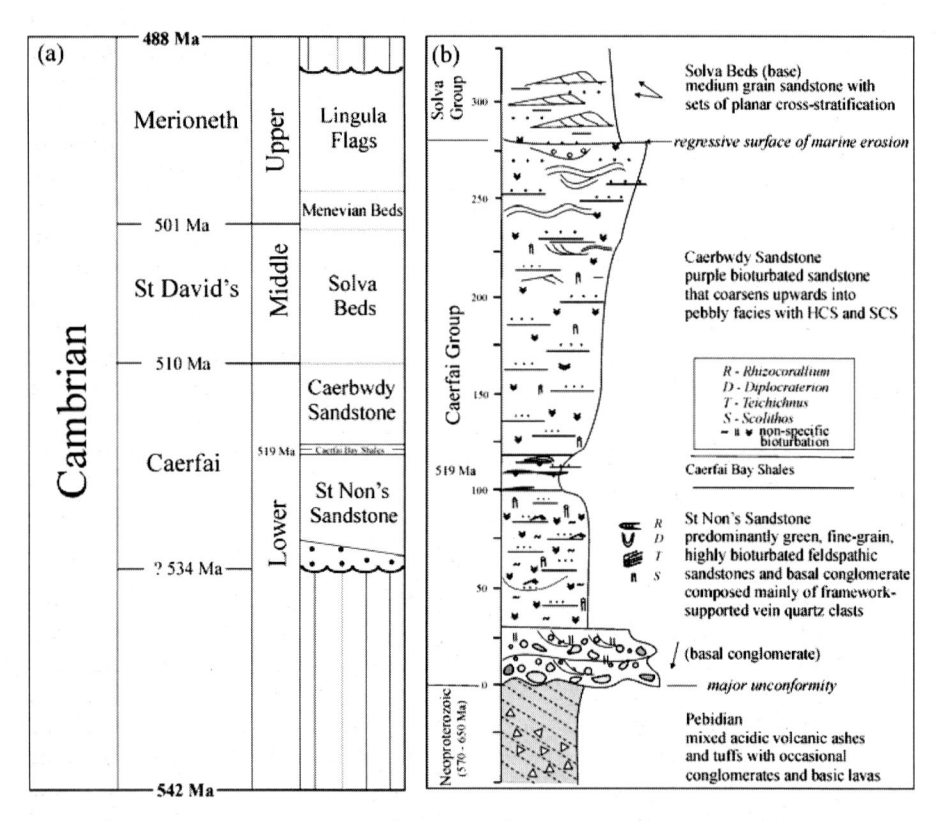

Fig. 7.4 (a) Cambrian stratigraphy and numerical ages. (b) Generalized profile log of the early Cambrian strata exposed between St Non's Bay and Caerfai Bay.

The Caerfai Bay Shales (~15 m thick), well exposed in the centre of a small cove, are composed of brick-red highly cleaved mudstones with thin tuffaceous laminations (1–8 cm thick). Some of the tuffs, which have given a radiometric date of 519 ± 1 Ma, are graded and have erosive bases and hummocky tops. Trace fossils include *Rhizocorallium* and some of the smaller tubes are possibly *Planolites*. Body fossils are absent, although the inarticulate brachiopods *Lingulella* and *Discina*, ostracods and the bradoriid arthropod *Indiana* have been recorded from other localities where they represent the oldest fossils found in the Cambrian of Pembrokeshire. Towards the top of the unit the facies become coarser and pass up transitionally into the purple Caerbwdy Sandstone, which also appears mottled as a result of reworking by deposit feeders such as *Teichichnus*. This sandstone coarsens upwards into a pebblier facies that displays erosional scours with rip-up clasts, and sets of hummocky, swaley and trough cross bedding.

The contact between the Caerbwdy Sandstone and the overlying Solva Group occurs in a small cove (SM 7617 2420), just before the vertical wall of the cliff is reached. Originally, this contact was thought to be a major unconformity, but at this locality it is marked by a curved fault that could be mistaken for a channel feature (Stead & Williams 1971; Fig. 7.4). The overlying Solva Beds contain well-defined sets of planar cross-strata (north-westerly palaeoflow), indicating that the contact was formed as a result of a fall in sea level, and it can therefore be interpreted as a regressive surface of erosion. Unpublished data suggests that this facies was deposited in a shallow-marine delta front environment (see Brenchley *et al.* 2006). Higher beds have in the past yielded a sparse fauna of trilobites (*Condylopyge, Plutonides, Bailiella* and varieties of *Paradoxides*), sponge spicules and *Skolithos*.

The Cambrian succession seen in the area was deposited over a long period of time in a shelf sea enclosed by Precambrian landmasses (see Fig. 2.5a). The initial marine flooding of the Welsh Basin resulted in the deposition of a transgressive conglomerate, which was pushed southeastwards over the Precambrian basement as sea level rose. Continued sea-level rise resulted in the deposition of the St Non's Sandstone in a shallow shelf sea environment. A thriving, mainly soft-bodied fauna destroyed much of the evidence for wave-formed sedimentary structures. The succeeding red mudstones suggest a more rapid rise in sea level to depths below fairweather wave base, whereas the presence of HCS and SCS in the upper part of the Caerbwdy Sandstone indicates a change to a storm-dominated shelf environment. At the beginning of Solva times, a fall in sea level and less stormy conditions marked a return to fair-weather nearshore deltaic deposition. Much farther north in the Welsh Basin (Harlech Dome and Lleyn Peninsula), deep-water turbidite facies were deposited during much of early to middle Cambrian times (see Fig. 2.5a).

1e Penpleidiau. Leave the bay, return to the coast path and walk southwards to the headland enclosed by four impressive embankments of an Iron Age fort. The neck of the headland is composed of southerly-dipping mudstones and siltstones belonging to the Menevian Group, the northern contact of which occurs along a deep gully eroded along a major (slickensided) strike fault. Farther to the south, the argillaceous beds become bleached and spotted as a result of metamorphism at the contact of a columnar jointed quartz diorite sill, which forms the tip of the headland and the small island of Penpleidiau. Looking across to the eastern side of Caerbwdy Bay an impressive monoclinal over-fold and an associated thrust can be seen in the cliffs.

1f Caerbwdy Bay and valley. Continue on the coast path to Caerbwdy Bay, which is the type locality for the Caerbwdy Sandstone. Note the large square limekiln and the remains of old corn mills at the head of the bay. Walking up the valley there are intermittent exposures of the sandstone and the basal Cambrian conglomerate. The band of Precambrian halleflinta, referred to earlier, can be examined in an old quarry near the stream and on the side of the minor road leading off the A487. Higher up the valley, there are also small exposures of the Clegyr Conglomerate (Pebidian), which contains a variety of acid igneous pebbles.

From Caerbwdy Bay you can continue another 2.5 km eastwards along the coast path to Porth-y-Rhaw (SM 7865 2438), which is well-known for being the type locality of the Menevian Group and also for providing the locality from where giant specimens (>40 cm long) of *Paradoxides davidis* were first collected in the 1860s. A comprehensive account of the geology of the site is given by Prigmore and Ruston (1999). (NB parking is very difficult on the A487 at Nine Wells, which is the nearest point of access to the bay).

7.3.2 Porth Clais

From St David's take the minor road that crosses Lower Moor and then descends to the small harbour of Porth Clais, where the National Trust has a car park, café and toilets. The harbour inlet, classified as a ria, was formed as a result of the drowning of the valley of the River Alum during the post-glacial rise in sea level. At the head of the inlet there are four limekilns, the three to the southwest of the river having been restored by the National Trust.

2a National Trust car park. The National Trust car park is located very close to the famous site where Green (1908) dug a trench to prove that the basal Cambrian rests unconformably on the St David's Granophyre. Since this trench has long since disappeared, a new section was cut in 1997 on the side of the track leading up to Porthllisgy farm (SM 7388 2380; Bevins & Horak 2000). This small section shows the basal Cambrian conglomerate dipping at about 75° to the northwest and resting unconformably on the highly weathered granophyre. Please do not hammer or collect at this important Geological Conservation Review (GCR) site. The large intrusion, exposed along the ENE-trending St David's Anticline, forms the moors to the southwest of St David's and extends as a narrow fault block to the headland of Carreg Frân (crow rock) (Fig. 7.2)

2b Eastern side of inlet. The granophyre is also exposed on the side of the path and hillside crags, on the east side of the inlet, just above the picnic tables. It is a medium-grain rock composed largely of quartz and plagioclase feldspar, with alkali feldspar in the groundmass. In thin section the rock displays good granophyric texture, resulting from the intergrowth of quartz and feldspar. The highly fractured and brecciated granophyre is intruded by a very thin basalt dyke (25 cm), close to the foot of the coast path. Walking southwards down to the harbour the sandstones and red shales of the Caerfai Group are exposed.

2c Ogof Golchfa. Cross to the west side of the inlet and follow the steep coast path to the top of the cliffs; continue to the gate at the corner of the field, where the path turns sharply to the west; take the minor path down to the grassy platform to the west of Ogof Golchfa (SM 741 237). Here an olive-green microtonalite sill, with rudimentary columnar jointing, is intruded into vertical Middle Solva beds. The sill (18–20 m thick), which at this locality has a small offshoot, can be traced 750 m to the west (St David's, Sheet 209). At the contact, the country rock has been baked to a hard and flinty, light-grey hornfels. Although the margins of the dolerite are not obviously chilled, the sizes of the feldspar phenocrysts increase towards the centre of the intrusion. The platform is capped by 3.5 m of raised-beach deposit composed of a mixture of grey–purple sandstone and igneous fragments, which decrease in size upwards from boulders to bedded pebbles. Solifluxion head, containing a large amount of angular purple sandstone fragments, overlies the raised-beach deposit.

7.3.3 Solva

Solva can be reached from either St David's or Haverfordwest on the A487. This delightful village becomes very crowded in the summer months and it is often difficult to find space in the free car park at the harbour. At the entrance to the car park there are blocks of Bodmin Granite that represent waste from the construction of the second Smalls Lighthouse (1861) built to replace the original oak tower (1776). In the first part of the 19th century Solva was a busy trading port and the main lime-burning centre for the St David's Peninsula; a group of four restored kilns can be seen to the south of the Ship Inn.

Solva and the surrounding area are composed of Cambrian strata, intruded by Ordovician igneous rocks classified as porphyritic microtonalites by the BGS (Sheet 209). During the Pliocene, the landscape was eroded to form the relatively flat-lying

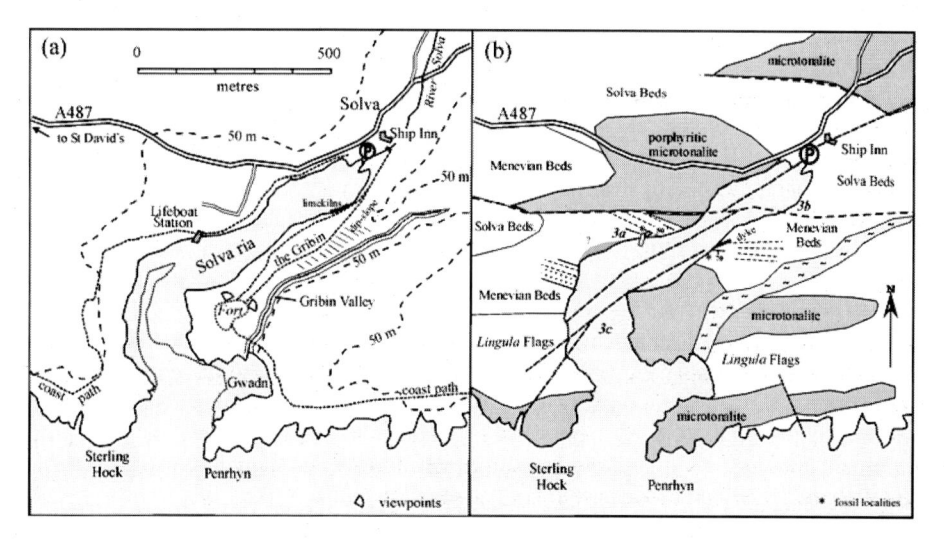

Fig. 7.5 (a) Solva harbour ria and other physical features. (b) Simplified geology map and locality map for Solva (modified from Stead & Williams 1971; BGS Sheet 209).

60 m platform; the harbour inlet was cut when melt-waters, derived from the last Pleistocene (Late Devensian) glaciation, drained into the River Solva and greatly increased its flow and erosive powers. Subsequently the sinuous valley was drowned to form a ria during the post-glacial (Flandrian) rise in sea level (Fig. 7.5a).

3a Lifeboat Station. From the car park walk down the north side of the harbour to the lifeboat station (SM 8014 2410; Fig. 7.5b). Just to the left of the Trinity House granite plinth, the contact between dark grey, well-laminated silty-mudstones belonging to the *Paradoxides davidi*s Zone (mid Menevian), and a weathered porphyritic microtonalite intrusion is seen. The sediments dip 50–70° southwards and are baked along the irregular contact of the intrusion. About 10 m from the intrusion the mudstones contain many white-weathering phosphatic nodules and a fossil fauna including small trilobites (eodiscids and agnostids) and brachiopods. The microtonalite is highly weathered along east-west trending joints and irregular fractures, which are mineralized with haematite and contain knots of vein quartz. On fresh surfaces the porphyritic rock is seen to contain free quartz and phenocrysts of plagioclase defining a flow texture. Farther to the southwest the strata, exposed at the base of the cliff, are largely covered by seaweed and are best viewed from the bottom of the slip. On the walk back to the car park, the path crosses a fresh microtonalite, which also has a porphyritic texture and contains conspicuous plagioclase laths enclosed in a matrix of quartz and dark hornblende.

3b Old limekilns. Cross the wooded bridge adjacent to the Ship Inn and proceed to the low cliffs located to the northeast of the limekilns; here sandstones and siltstones from the middle part of the Solva Group are exposed. These facies at first appear homogenous but on closer examination they display parallel, low-angle and small-scale hummocky (HCS) bedding. Some of the more prominent bedding planes display large-scale undulations associated with low-angle and wedge-bedding, suggesting that they represent giant hummocky and swaley sets. For those who are not convinced, far better exposures of giant hummocky and swaley cross bedding can be seen, at approximately the same stratigraphical horizon, in the cliffs near Pwll March to the northwest of Newgale. An itinerary and geological cross section for Newgale has been published by Williams and Stead (1982).

3c The Gribin. Behind the limekilns you can join the lower of two paths, which ascend the Gribin ridge to reach the Iron Age promontory fort at its southeastern tip (SM 802 239). The headland is composed of a large igneous intrusion previously seen at locality *3a*. Looking to the west, quite complex folding can be seen in the cliff-face formed of faulted *Lingula* Flags. From the summit of the headland there are excellent views of the ria, the V-shaped cross section of the upper Solva Valley and the surrounding 60 m platform. This part of the ria is located along a northeastwards-trending fault zone (Fig. 7.5b). Looking down to the eastern side of the ridge is the Gribin Valley, which is another ria now filled by recent marshy sediments. At low tide it is possible to walk from the bay (Gwadn) along the foreshore back to Solva. Further details of the Cambrian geology at Porth y Rhaw to the west and Newgale to the east are discussed by Stead & Williams (1971).

7.3.4 Whitesands Bay

Travel from St David's on the A487 Fishguard road; turn left onto the B5483 that terminates at the large car park at Whitesands Bay, where there are toilets and a cafe. The long, westwards-facing, sandy beach is very popular with holidaymakers and surfers; there is a hotel and a youth hostel and many campsites in the area. To complete this half-day itinerary in stratigraphical order you require medium to low spring tides, but if the tide is high the St David's Head itinerary can be completed first.

4a Ogof Golchfa. From the rescue station walk 750 m southwards along the beach to a small cove called Ogof Golchfa (SM 7302 2644; Fig. 7.6). On the southern side of the cove the basal Cambrian conglomerate dips 60° to the northwest and rests unconformably on Precambrian (Pebidian) tuffs. The green chloritic tuffs and purple slates have a bedding-plane cleavage and are cut by a band of vein quartz and a massive dark-green dolerite intrusion (about 20 m thick) of probable Ordovician age. The lower chilled contact of the vesicular dolerite is very irregular and it has

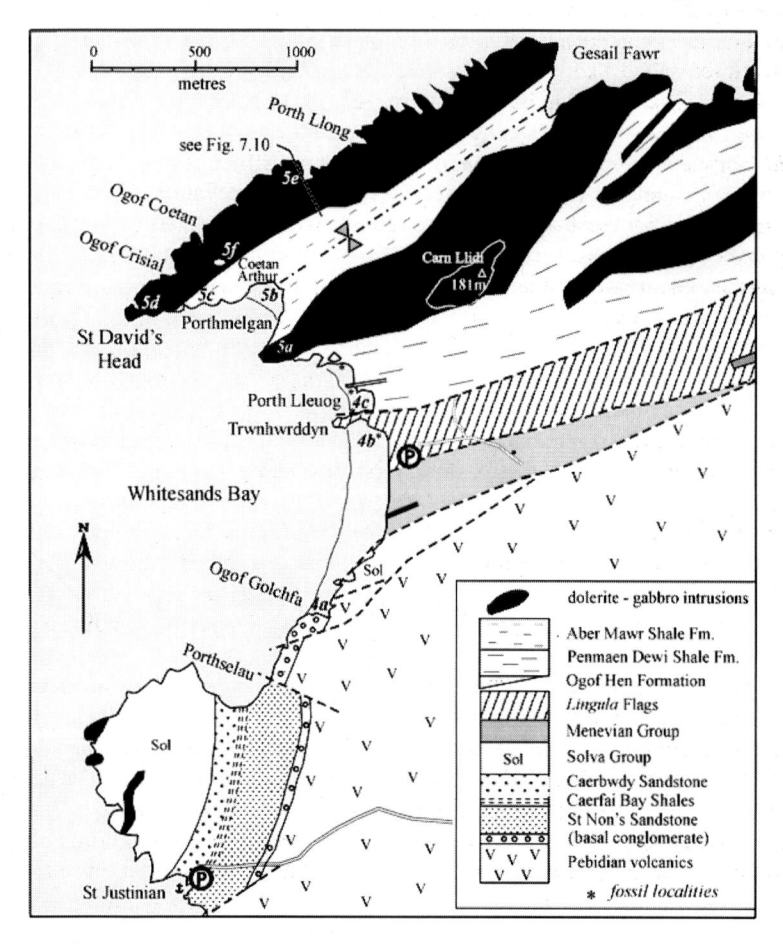

Fig. 7.6 Geology and locality map of Whitesands Bay and St David's Head (modified from BGS Sheet 209).

minor offshoots. At this locality the basal Cambrian conglomerate is over 30 m thick and is composed of two fining-upwards units. Each unit contains rounded pebbles of vein quartz (maximum diameter 25 cm), which show evidence of alignment and imbrication, whereas the finer facies display crude horizontal and cross bedding. Farther north, a thin sliver (1 m thick) of Cambrian conglomerate is preserved on the wall of a narrow fault-defined gully, the top of which contains weathered Pebidian tuffs displaying cryoturbation folds. The fault downthrows green sandstones and siltstones into the section, which have been mapped as the Solva Group on the provisional geological map of the area (St David's, Sheet 209), although they have also been correlated with the St Non's Sandstone (Stead & Williams 1971). These facies display parallel-bedding, current ripples, occasional small hummocks and horizontal burrows. The succession also contains minor basic intrusions, and a thicker dolerite intrusion is seen at SM 7335 2788.

4b Trwnhwrddyn. To the south of the headland of Trwnhwrddyn (ram's nose) the low cliffs and foreshore are composed of an interesting succession of quartz-rich sandstones, micaceous siltstones and flaggy mudstones, belonging to the *Lingula* Flags (late Upper Cambrian). These strata dip 70–75° northwestwards and contain excellent sedimentary structures, including hummocky (HCS) and swaley (SCS) cross stratification, graded bedding, gutter casts, scour and fill structures and load casts (Fig. 7.7a). Here the swaley sets are expressed as low-angle cross laminations, which are usually developed below hummocky bedforms (Fig.7.7b). This is an ideal locality for demonstrating the use of sedimentary structures as way-up criteria. Some of the bedding planes of the finer grain facies are crowded with the remains of *Lingulella,* although these surfaces are often elusive or concealed by holidaymakers. The above association of sedimentary structures indicates that the facies were deposited on a storm-dominated nearshore shelf environment. In other localities to the southeast (Porth y Rhaw and Solva) the *Lingula* Flags are of deeper water aspect (offshore shelf). These regional facies changes reflect the rises in sea level, which culminated in the Late Cambrian highstand.

The northern margin of the headland is defined by a northeastwards-trending reverse fault, which downthrows the Penmaen Dewi Shale Formation (Arenig) against the *Lingula* Flags. In the middle portion of the headland, the contentious unconformable contact between the *Lingula* Flags and the Arenig (Ogof Hen Formation) is recorded by a thin basal gritty transgressive unit and mud-pellet conglomerate (Owen *et al.* 1965; Stead & Williams 1971). At the type locality on Ramsey Island the formation is composed of around 150 m of conglomerates and coarse grain cross-bedded sandstones (Kokelaar *et al.* 1984, Howells 2007).

4c Porth Lleuog. Porth Lleuog is eroded into steeply dipping (70° northwest) soft black mudstones belonging to the Penmaen Dewi Shale Formation, formerly referred to as the *Tetragraptus* Shales. On the south side of the bay the black mudstones contain a sparse fauna of dendroid graptolites, brachiopods (lingulids and orthids) and trilobites. These strongly-cleaved mudstones are intruded by many thin concordant sheets (less than 1 m) of keratophyre – a fine-grain altered igneous rock composed mainly of alkali feldspars accompanied by chlorite, epidote and calcite. Some of the keratophyre sheets pinch-out while others display minor folds and faults. Most have chilled margins, and reaction rims containing altered crystals of iron pyrites. Towards the northern margin of the bay the mudstones are interbedded with a well-defined coarsening- and thickening-upwards turbidite sequence.

Fig. 7.7 Sedimentary structures present in the *Lingula* Flags at Whitesands Bay; **(a)** small load cast and low amplitude hummocks, **(b)** well-defined hummock developed above low angle sets representing swaley cross stratification.

Further north, a prominent seaward extending stack is composed of a 2 m-bed of acid tuff (Fig. 7.8). The mudstones exposed near the contact of the tuff were, in the past, covered with dendroid graptolites (*Callograptus*) and didymograptids (Fig. 7.9), but unfortunately the locality has suffered from indiscriminate collecting . To the north, the succession contains further keratophyres intruded into distal turbidite facies. At the northern end of the bay (SM 7309 2758), the black mudstones contain abundant trilobites including *Bergamia*, *Ampyx* and *Microparia* (Bevins & Roach 1982).

Fig. 7.8 View of Whitesands Bay from the coast path ascending St David's Head.

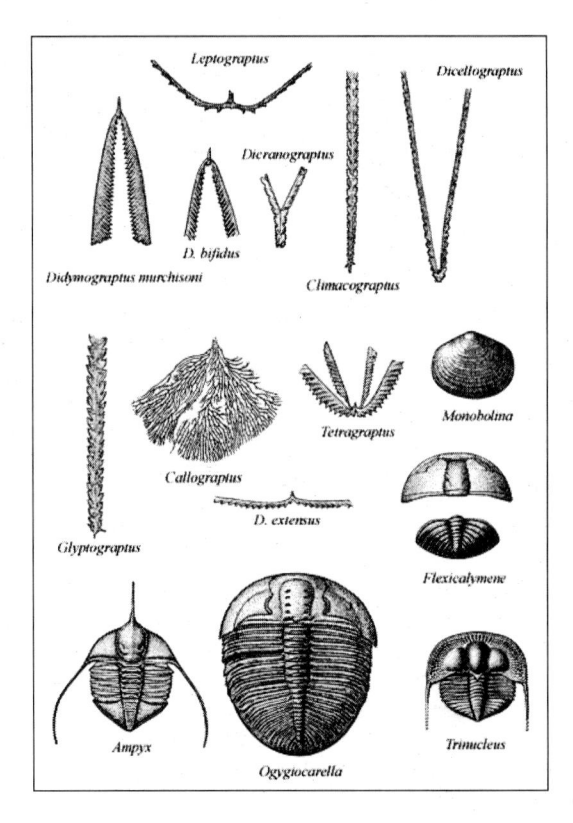

Fig.7.9 Some characteristic Ordovician fossils (reproduced from George 1970; British Museum 1969).

7.3.5 St David's Head

The principle aim of this itinerary is to investigate the nature, rock types and origin of the St David's Head–Carn Llidi gabbro intrusion. Previously this intrusion was interpreted as two discrete igneous bodies, but they are now interpreted as a single sheet, repeated in the limbs of a northeast-trending syncline (Fig. 7.6). Detailed petrographic work on the 400 m-thick intrusion has shown that its composition varies in zones parallel to its margin (Bevins & Roach 1982). The lower and upper margins of the intrusion, composed of quartz dolerite and quartz gabbro, represent undifferentiated magma, whereas the other gabbroic facies were generated as a result of *in situ* differentiation. Magmatic differentiation is the process by which a variety of genetically related rock types are formed from a single parental magma. This process resulted in the concentration of certain minerals in different parts of the magma chamber by crystal fractionation and gravity settling. Consequently, during its cooling and crystallization history, the melt remaining in the chamber became depleted in some components and enriched in others.

From the car park at Whitesands Bay join the coast path and begin the gentle ascent to St David's Head. Stop at the marked viewpoint on the cliffs above Porth Lleuog Bay and look down into the bay to see the prominent stack composed of acid tuff, and the headland of Trwnhwrddyn and Whitesands Bay beyond (Fig. 7.8). Continue to the gate, where three drystone walls converge, and walk another 100 m to the crest of the ridge (SM 7289 2778).

5a Penlledwen. At Penlledwen a quartz gabbro crops out and can be traced along the headland, where the upper contact of the intrusion with the overlying siltstones and mudstones (Aber Mawr Shale Formation) is exposed along the crest of the ridge. Here the sediments dip at 75° to the northeast and are baked to a light-grey hard hornfels in the 2.5 m contact zone (Fig. 7.10). The margin of the intrusion is composed of a dark-grey medium-grain dolerite, containing phenocrysts of plagioclase feldspar and isolated small glassy quartz. This contact of the quartz dolerite sheet can be traced southwestwards, along a rough path, to the tip of the headland where the lower contact can also be seen in the cliffs. From here, the lower margin continues inland and intersects the coast path just south of the gate. The central portion of the intrusion is composed of a very coarse-grain quartz gabbro containing phenocrysts of plagioclase feldspar and pyroxenes, and small quartz crystals. Further craggy outcrops and displaced blocks of the intrusion continue to the summit of Carn Llidi (181 m OD) where the central portion of the body contains a high proportion of olivine.

5b Porthmelgan. Continue northwards on the coast path that descends to Porthmelgan where the cliffs are composed of weathered, faulted and cleaved silty-mudstones. Measurements of bedding and cleavage relationships in the strata on either side of the bay indicate that the inlet is intersected by the axis of a northeastwards-trending syncline. A layer of head, composed of angular and platy fragments, overlies the weathered and cryoturbated top of the sediments. The storm-beach is composed mainly of rounded and wave polished boulders of dolerite, gabbro and pegmatite, derived from the erosion of boulder clay seen at the back of the bay. This is a good locality to preview the various facies of the St David's Head layered intrusion.

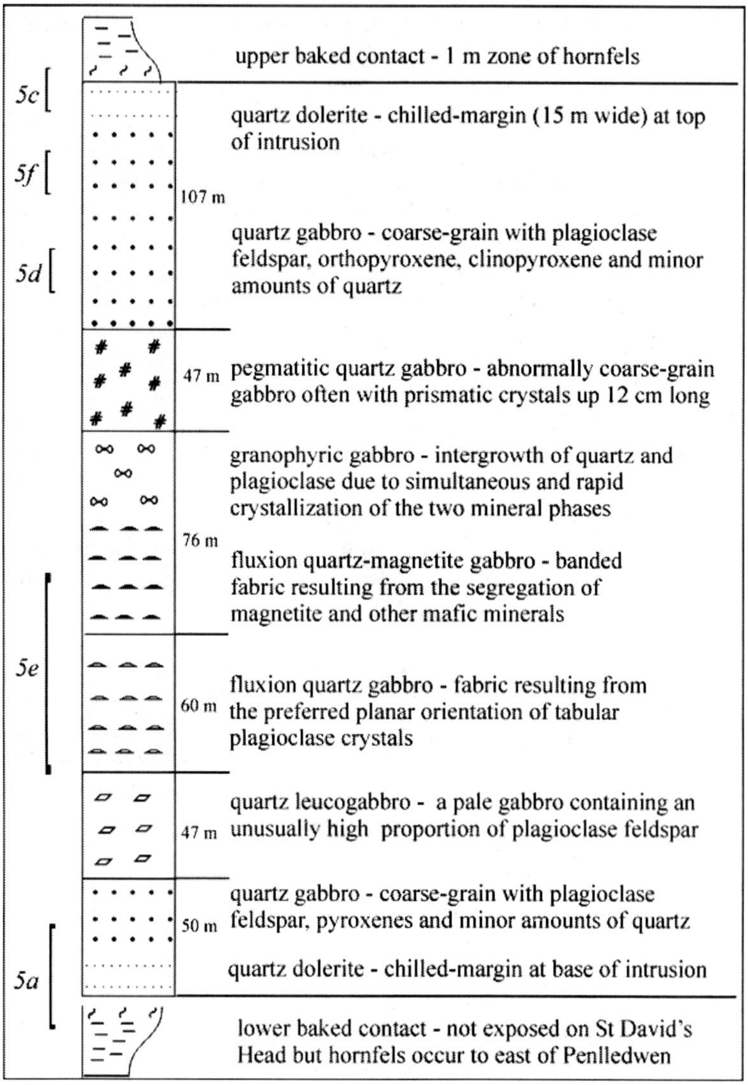

5c [			upper baked contact - 1 m zone of hornfels
5f [			quartz dolerite - chilled-margin (15 m wide) at top of intrusion
		107 m	
5d [			quartz gabbro - coarse-grain with plagioclase feldspar, orthopyroxene, clinopyroxene and minor amounts of quartz
		47 m	pegmatitic quartz gabbro - abnormally coarse-grain gabbro often with prismatic crystals up 12 cm long
5e		76 m	granophyric gabbro - intergrowth of quartz and plagioclase due to simultaneous and rapid crystallization of the two mineral phases
			fluxion quartz-magnetite gabbro - banded fabric resulting from the segregation of magnetite and other mafic minerals
		60 m	fluxion quartz gabbro - fabric resulting from the preferred planar orientation of tabular plagioclase crystals
		47 m	quartz leucogabbro - a pale gabbro containing an unusually high proportion of plagioclase feldspar
5a [		50 m	quartz gabbro - coarse-grain with plagioclase feldspar, pyroxenes and minor amounts of quartz
			quartz dolerite - chilled-margin at base of intrusion
			lower baked contact - not exposed on St David's Head but hornfels occur to east of Penlledwen

Fig. 7.10 Generalized section through the St David's Head sill showing the vertical distribution of the various layered gabbroic facies (adapted from Bevins & Roach 1982).

5c St David's Head (south). Continue on the low-level path westwards to St David's Head; protected by the embankments and ditches of an Iron Age fort known as Clawdd-y-Milwyr (warrior's dyke). Note that this locality should be omitted if you dislike steep slopes, if you suffer from vertigo or if the weather is inclement. After crossing the highest inner embankment, leave the path and walk very carefully for 30 m obliquely down the slope to reach the northeastwards-trending contact of the St David's Head intrusion (SM 7234 2789). Here, laminated siltstones have again been baked and metamorphosed into hard flinty hornfels, which contain occasional purple lenses rich in manganiferous garnets (Bevins & Roach 1982). The chilled margin of the intrusion (about 15 m wide) is composed of fine–medium-grain quartz dolerite containing pale-grey phenocrysts of irregular to euhedral plagioclase.

5d St David's Head. Return to the path; cross a grassy depression containing the stone bases of three Iron Age round houses, and continue to the tip of the headland where small platforms expose the coarse grain gabbroic facies of the intrusion. Looking to the northeast, across the deep inlet of Ogof Penmaen, the lower half of the rock face displays many thin (5 cm) dark (mafic) and light (felsic) segregation bands running parallel to the strike of the intrusion (Fig. 7.10). From this point there are excellent views of Ramsey Island to the south, and the small islands of the Bishops and Clerks (with a lighthouse on South Bishop).

5e Porth Llong. Return to the fort and take the path that leads off to the left from between the outer two ramparts. This path skirts around Ogof Crisial where the north side of the inlet is seen to be composed of alternations of quartz leucogabbro and fluxion quartz gabbro. Continue to the top of the ridge above the inlet, and farther northeastwards to join the main path that leads to Porth Llong (SM 7293 2843). From the western side of this inlet, the cliffs to the east display fluxion banding in both the quartz gabbro facies (seaward) and the quartz-magnetite gabbro facies (landward) (Fig. 7.10).

5f Coetan Arthur. On the return journey note the Neolithic burial chamber of Coetan Arthur (SM 7253 2806), with its massive gabbro capstone that is now supported by a single vertical upright. Before the collapse of the other uprights the profile of the capstone would have mimicked the distant profile of Carn Llidi (Fig. 7.11). If the weather is favourable, the walk to the summit of Carn Llidi is exhilarating and offers excellent views of the St David's Peninsula, and Stumble Head and Mynydd Preseli to the northeast.

Fig. 7.11 View of Carn Llidi from Coetan Arthur, a Neolithic burial chamber.

7.3.6 Abereiddy and Porthgain

Take the A487 Fishguard road through St David's; branch left onto the B4583 and continue north on the minor road signposted to Llanrhian. About 1 km southwest of Llanrhian turn left at the crossroads and travel another 1.5 km to the car park above the beach at Abereiddy Bay (Fig. 7.12a). It is possible to get small coaches down to the bay but this is not recommended during busy holiday periods.

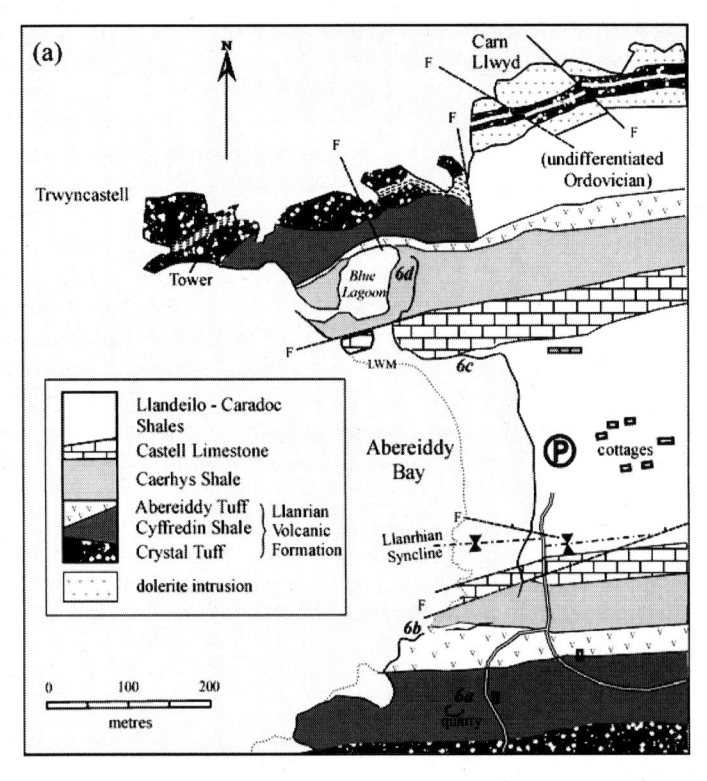

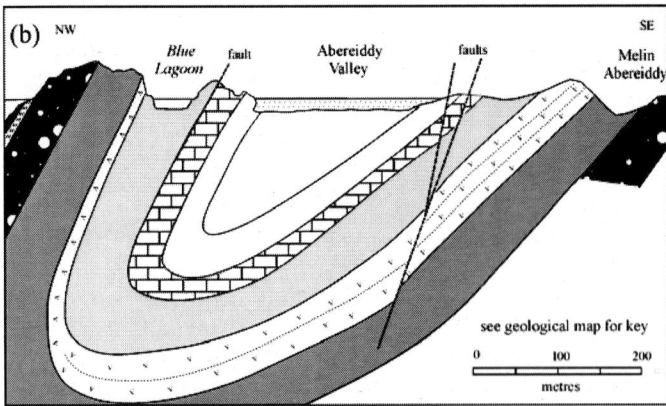

Fig. 7.12 (a) Geology and locality map and (b) cross section of Abereiddy Bay (based on Hughes *et al* 1982).

During the 19th century Abereiddy was an important slate-quarrying centre, and the old quarry on the headland still has the remains of the dressing sheds, the engine house and the stonework for the lift, all of which were once connected by a tramway. After slate production ceased in 1904, a passage was blasted through the quarry face to make a harbour entrance. The flooded quarry, known locally as the "blue lagoon", is a popular diving spot, which has been designated a SSSI for its marine ecosystems. From the quarry the dressed slates were taken in horse-drawn trams to a yard near Ynys Barry Farm and then on to Porthgain for distribution. Other remains of the slate industry include the round gunpowder store, a row of workers houses and the detached and buttressed manager's house, all of which are constructed of slate blocks. The village was largely abandoned by 1938 as a result of a typhoid epidemic and violent storms. More recently, a number of cottages have been renovated as holiday-lets and the bay has become popular with tourists.

The aims and objectives of this itinerary are to examine the lithologies, fossil faunas and structure of the Ordovician strata exposed in a north–south traverse across Abereiddy Bay (Fig. 7.12b). The area is well known for its abundant tuning-fork graptolites (*Didymograptus*) and other fossils preserved in the Ordovician shales (Fig. 7.9). Note that hammering is not allowed on the road leading down to the bay or at the SSSI in the cliffs below the road, and it is suggested that collecting is limited to the scree material at all the other localities. The bay is eroded along the asymmetrical Llanrian Syncline, the axis of which is located at the southern end of the bay (Fig. 7.12b). A further objective of the itinerary is to examine the industrial archaeology of the once important slate and stone-quarrying industries of the Porthgain area.

6a Melin Abereiddy quarry. From the car park walk back up the hill, take the right-hand turning and continue down to Melin Abereiddy cottage where a small quarry is located 50 m to the west. The quarry is cut into shales and tuffs, which dip 50° to the northwest and belong to the Cyffredrin Shale Member of the Llanrian Volcanic Formation. The shales contain a rich graptolite fauna (*Didymograptus, Diplograptus, Cryptograptus* and *Lasiograptus*) and small agnostid trilobites (see Fig. 2.5b), which have been dated as lowermost upper Llanvirn (Hughes *et al.* 1982).

6b Southern corner of bay. Walk down the valley to the beach where the overlying Abereiddy Tuff (about 30 m thick) is exposed in the cliffs and along the slippery foreshore. This tuff, formerly referred to as the *Murchisoni* Ash, contains lapilli and bombs of pumice set in an ash matrix. The erosional base of the unit is succeeded by 27 m of massively bedded tuffs in which the fragments decrease in size upwards. The top 3 m of the unit is composed of well-bedded tuffs, which display basal scours, graded-bedding, occasional ripples and very thin shale laminae. The basal 2 m of the overlying Caerhys Shale also contains thin graded and commonly pyritized pumice tuffs; higher in the sequence (at the SSSI) the shales yield abundant specimens of *Didymograptus murchisoni.*

Farther to the north, just south of a small cave in the cliffs (SM 7970 3112), some 9 m of decalcified micritic limestones and mudstones have been correlated with the Castell Limestone, which is more fully exposed at locality 6c. Along this part of the foreshore the succession is cut by faults some of which are associated with breccias. Here, the slaty cleavage dips at a steeper angle than the bedding, indicating that the northward-dipping strata are the correct way-up. About 75 m above the top of the Castell Limestone the cleavage dips at a lower angle than the bedding, indicating that the beds are inverted and that the axis of the Llanrian Syncline has been crossed.

6c Headland south of the old quarry. The Abereiddy Valley is composed of Llandeilo–Caradoc shales overlain by marshy alluvium. At the north end of the car park go through the gate, ascend the slope and follow the route of the old tramway westwards to the headland. Here the top of the Castell Limestone, exposed in a cutting leading to the old quarry, is composed of a hard impure micritic limestone and more fissile cleaved calcareous shales. The well-developed cleavage in the shales dips at a lower angle than the limestone beds indicating that the strata are still inverted. The overlying Llandeilo–Caradoc Shales, which outcrop seaward of the retaining wall, contain a fauna of graptolites (*Nemagraptus*, and *Diplograptus*) and trilobites (*Trinucleus*).

6d Abereiddy quarry ("blue lagoon"). Continue around the headland, crossing the small bridge across a vertical eastward-trending fault gully, to the old quarry. Slate scree at the southern end of the quarry (Caerhys Shale) contains a fossil fauna of gastropods, orthocone nautiloids, the inarticulate brachiopod *Monobilina*, fragments of trilobites (*Platycalymene* and *Ogygiocarella*) and also small agnostids trilobites (Fig. 7.9). Graptolites are also present near the faulted base of the Castell Limestone, whereas shale scree from the middle part of the quarry contains a graptolite fauna of *Didymograptus murchisoni*, *Diplograptus*, *Cryptograptus* and *Lasiograptus*. The northern, inaccessible, face of the quarry exposes inverted beds belonging to the Abereiddy Tuff.

6e Porthgain. From Abereiddy you can either walk along the coast path to the small harbour of Porthgain (3 km) or you can drive there via the minor road leading northwestwards from the crossroad at Llanrhian. Porthgain is a fascinating small harbour village (Fig. 13a), which developed during the 1830s to 1930s to service the quarrying and its associated industries; both the location of the village, and its industrial archaeology being a direct consequence of the local geology. Today, the vibrant village has an inn, a restaurant, a bistro and art galleries, and survives largely on tourism and fishing.

The Sloop Inn, much enlarged from the original 18th century stone building, serves good food and beer and has an interesting display of photographs and memorabilia of the village in its heyday. The terrace of the inn provides an excellent spot to enjoy lunch and view the layout of the old industrial village (Fig. 7.13). Looking to the left across the car park, there is a row of cottages (*1*) named Y Strydd (The Street) built in the 1850s for the quarrymen and their families. The group of buildings located directly opposite the inn (*2*) were the company offices for the quarries at Porthgain and also those at Abereiddy and Trwyn Llwyd near Trevine. Just to the right is an enclosure (*3*) where a water-powered mill, used to crush the slate and shale for making bricks, was sited. The large building (Ty Mawr) on the opposite side of the road (*4*) was where Porthgain bricks were made; an output of 50 000 bricks per week being achieved during the 1890s. The Shed (now the bistro) adjacent to Ty Mawr was a machine shop. The drying sheds and a kiln were sited outside, and a narrow-gauge railway serviced all of the buildings.

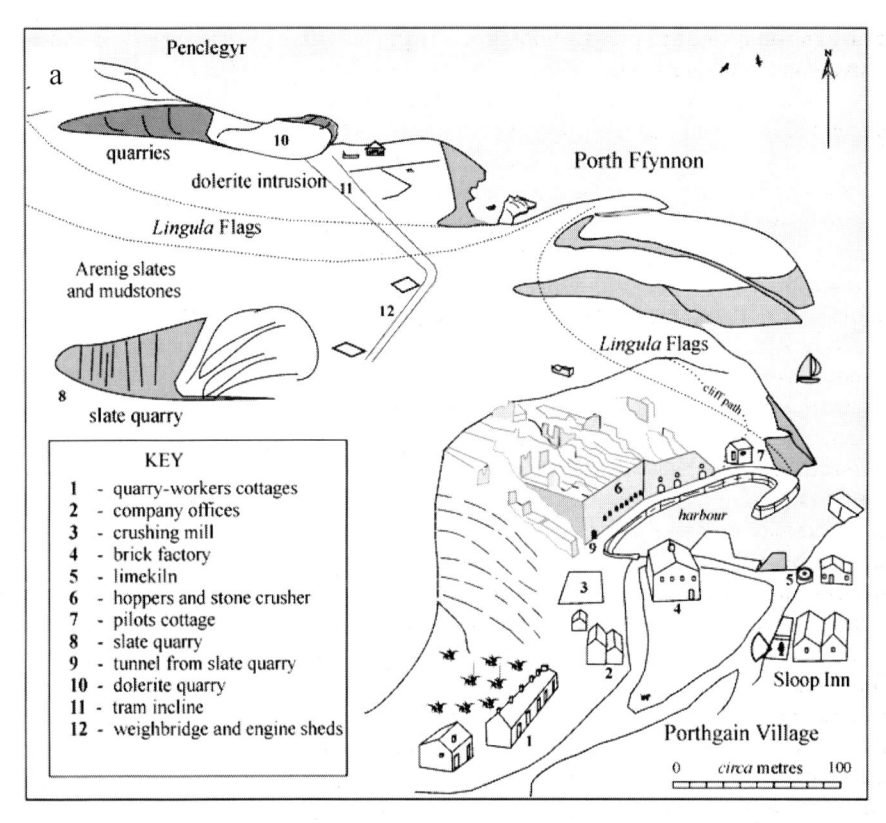

Fig. 7.13 (**a**) Sketch map of the Porthgain area to show the main industrial archaeology sites and localities described in the itinerary (modified from a Pembrokeshire Coast information leaflet and an oblique aerial photograph). (**b**) View of Porthgain Harbour and the remains of the grinding mills and hoppers.

My grandfather worked at the brickworks for a number of years, during which time he cycled daily from St David's. Leave the Sloop Inn, turn right and walk to the harbour slip where a limekiln (**5**) can be seen on the side of the road. In this limekiln, the oldest industrial building in the village, limestone and anthracite coal, shipped in from south Pembrokeshire, were burnt to make agricultural lime to treat the acid soils of the area. On the east side of the harbour the very large buildings, faced with red Porthgain bricks (*6*), were used for the production and storage of roadstone (Fig. 13b). A stream-driven crusher, located at the top of the slope, ground the dolerite rubble into different sizes (0.25 to 2.0 inches in diameter), before being stored in the hoppers. Cargo ships were loaded with roadstone directly from shoots at the bottom of the hoppers. The pilot, who oversaw the arrival, docking and departure of the transporter ships, used the small cottage at the north end of the harbour as an office (*7*). A stepped path, adjacent to the cottage, leads up to a platform at 25–45 m OD where further remains of the quarrying industry are visible. From the top of the path, the two white beacons, seen on adjacent headlands, were used as navigation aids for ships entering the narrow harbour. About 200 m to the southwest is the disused slate quarry (*8*), which is cut into vertically bedded and cleaved silty-mudstones belonging to the Penmaen Dewi Shale Formation (Arenig). The slate was originally brought to the top of the quarry and then transported down an incline to the brick-processing plant at the harbour. As the quarry was deepened, this became impractical, and in the 1880's a tunnel was driven through to the harbour to bring the slate to the crushing mill. The tunnel exit can still be seen at the south end of the hopper wall (*9*). The main quarries, located 500 m to the northwest (*10*), are cut into a dolerite intrusion of Ordovician age. This very hard rock was blasted from the quarry face and then broken manually into smaller blocks, before being loaded onto trams for transport to the stone crusher. The prominent linear depression, leading from the dolerite quarries, marks the position of the tramway (*11*). To the left, the red brick buildings were the quarry workshops, where some of the larger blocks of dolerite were dressed into curbstones. Walking down the incline you come to the remains of the weighbridge and the engine-sheds (*12*). The latter housed two steam locomotives, which provided the power for pulling the trams from the dolerite quarry. No slates were produced after 1910 but the production of roadstone and other products continued until 1931 when all operations suddenly ceased without prior warning to the workers.

7.3.7 Fishguard and Strumble Head

The main aim of this itinerary is to examine and discuss the mode of origin of some of the interesting rocks belonging to the Fishguard Volcanic Group of Ordovician (Llanvirn) age. This group comprises a mixture of mainly subaqueous acid tuffs, rhyolites and acidic–basic pillow lavas, interbedded with black shales and turbidites. These rocks were formed in an elongated caldera, the southern margin of which was defined by the Fishguard–Cardigan Fault Belt. The volcanics reach a maximum thickness of 1800 m on Strumble Head, located to the north of the Goodwick and Pen Caer Faults (Fig. 7.14). These two faults unite in Fishguard Bay, from where they continue across Dinas Island and Newport Bay as the Newport Sands Fault (see Fig. 7.16). Both the volcanic and sedimentary rocks are cut by many high-level doleritic and gabbroic intrusions, which were contemporaneous with the volcanic activity. This 4–5 hour itinerary incorporates a number of stops on a traverse from Fishguard Bay to Strumble Head.

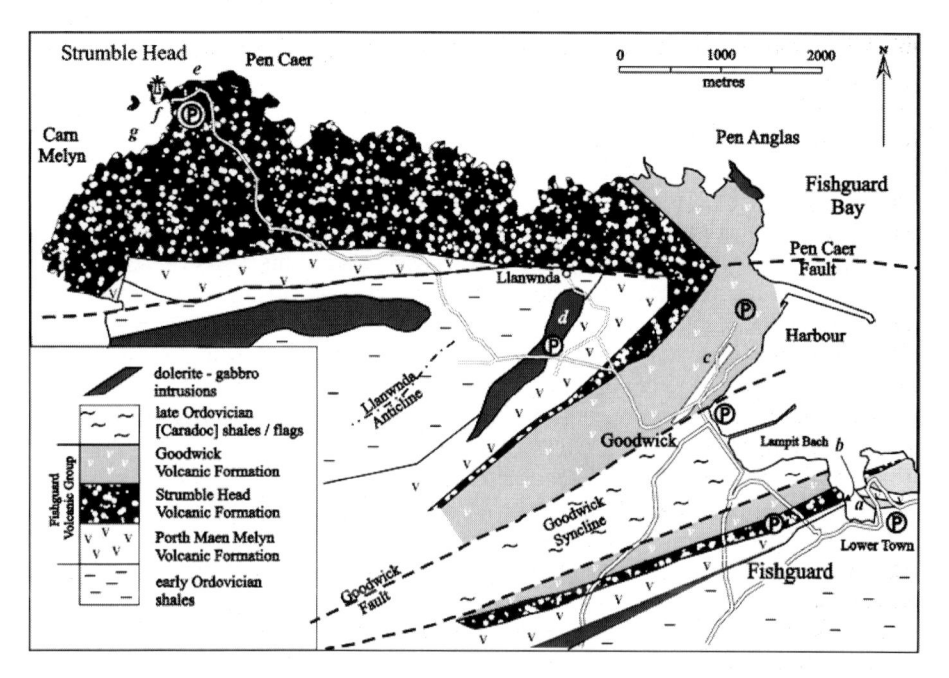

Fig. 7.14 Geology and locality map of the Fishguard and Strumble Head area (modified from Thomas & Thomas 1956).

7a Lower Fishguard. Travel to Fishguard on the A40 and then take the A487 down the hill and over the bridge to the Lower Town (Cwm), where parking is available for cars on the side of the quay. This was the site of the original harbour built at the mouth of the Gwaun Valley – a classic example of a subglacial channel. The quaint row of cottages were used as one of the sets for the film version of *"Under milk Wood"* with Richard Burton and Elizabeth Taylor.

Walk to the end of the quay (just beyond the yacht clubhouse) where the promontory is composed of a hard, greenish-grey rock containing elongate mudstone fragments and altered feldspars set in a matrix, rich in glass shards and occasional pumice (Porth Maen Melyn Volcanic Formation). (Beware this locality is a favourite 'comfort stop' for local dogs). The tuff is interpreted as a submarine ash flow, which ripped-up mudstone fragments during its flow over the sea floor (Bevins & Roach 1982). Some of the textures resemble those of welded tuffs or ignimbrites, which are known to occur in similar strata exposed in the Preseli Hills. Looking across the inlet to the southeast is the small cove of Lampit Bach, where it is possible to see the characteristic shapes of the pillow lavas that are examined at the next locality.

7b Lampit Bach. Return to Fishguard; turn right at the roundabout and continue for 300 m through the town and park in West Street car park. Walk 100 m back along the street; turn left down Marine Walk to the Gorsedd Circle (commemorating the 1936 National Eisteddfod), from where there are good views of Dinas Island and lower Fishguard. Continue onto the footpath; turn right at the bottom of the metal steps and then left after about 100 m to reach a set of brick steps leading down to the boulder-covered and very slippery beach at Lampit Bach (SM 9593 3745).

In the small cove the rocks that dip steeply to the northwest, consist of volcaniclastics overlain by pillow lavas with interbedded basic ashes and tuffs (Strumble Head Volcanic Formation). Many of the pillows have irregular or flat bases, rounded tops, and glassy chilled margins with concentric zones of vesicles.

7c Goodwick. From Fishguard the A40 road descends the steep hill to the Goodwick Valley, eroded into late Ordovician (early Caradoc) shales preserved in a syncline. Good parking facilities are available at the northern end of the beach adjacent to the embarkation zone for the ferry terminal. The site of Fishguard Harbour was developed in the early 1900s by blasting 2 million tonnes of the Goodwick Volcanic Formation from the cliffs. The rock waste was used to construct the north breakwater, the quarry-floor being used for the boat and railway terminals. From the car park walk over the railway bridge to the Rose and Crown; cross the road and ascend 300 m up New Hill. Please do not hammer the outcrops and beware of traffic descending the hill. The rhyolite crags above the road are overgrown but the outcrops on the left-hand side of the road improve after New Hill Villas (SM 7472 3856). The outcrops are composed of hard, fine-grain rhyolitic lavas, which display dark and lighter greeny-grey flow bands with minor folds and contortions (Goodwick Volcanic Formation). Small phenocrysts of feldspar and white recrystallized spherulites can be seen with a hand lens. At the time of their extrusion these rhyolitic lavas were very viscous and they would have built-up small thick domes over the vent. In other parts of the area (e.g. Y Penrhyn 2 km to the northwest) the domes occur in association with rhyolitic 'crumble breccias', formed as a result of *in situ* chilling and fracturing (autobrecciation) leading to the accumulation of scree on the flanks of the domes (Bevins & Roach 1982). These volcanic rocks occur in association with volcaniclastics and graptolitic mudstones.

7d Henner Cross. From Goodwick take the minor road up the steep hill to Stop-and-Call and on to Henner Cross, where parking is available in the layby on the right-hand side of the road (about 100 m to the west of the crossroad). Walk back towards the crossroad and take the track that leads north past the old school, to reach the northeastwards-trending crags of Carnwnda (SM 9335 3918). These crags are composed of a medium to coarse-grain, basic igneous rock intruded into early Ordovician (Arenig) shales. The sheet-like intrusion, referred to as the Llanwnda Gabbro, is composed primarily of varying amounts of altered plagioclase feldspar and clinopyroxene, with some hornblende and black to brownish-red accessory mineral ilmenite (iron-titanium oxide). In addition the gabbro includes minor amounts of pumpellyite, prehnite and epidote, formed as a result of low grade Caledonian metamorphism (Bevins & Roach 1982).

7e Pencaer. Continue westwards on the minor road taking the right-hand fork through Pontiago, and travel a further 4.5 km to the car park at Strumble Head; coaches must be parked at Tresinwen Farm. The Strumble Shuttle (Fishguard to St David's coastal bus) also operates a regular service to the headland. From the car park walk to the old lookout station, now used for bird watching, and make a careful descent to the wave-worm cliff platform to the east (SM 8955 4138). The basaltic pillow lava pile at this locality is very thick, and composed of well-formed pillows with long diameters of 0.25–4.0 m (average 1 m) (Fig. 7.15). Their upper surfaces are rounded and convex upwards, while their lower surfaces conform to the shape of

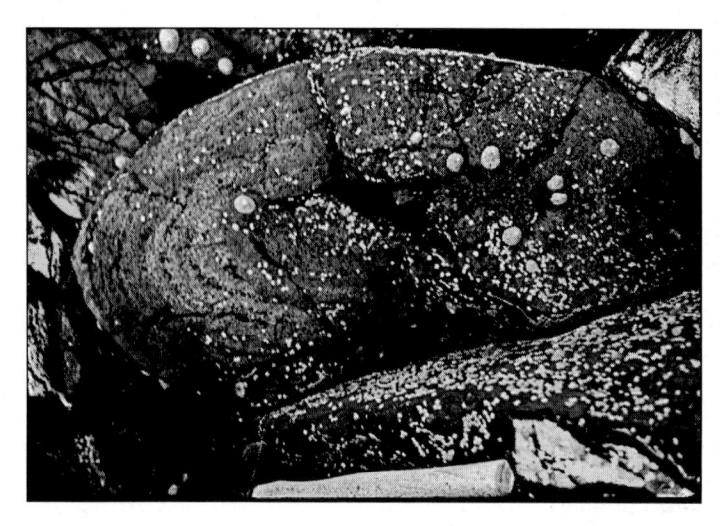

Fig. 7.15 Pillow lava showing concentric lines of vesicles at Stumble Head.

the pillows below and characteristically form downward-projecting cusps. A thin glassy lava skin, representing a chilled margin, surrounds each pillow and internally they display concentric zones of vesicles, which often increase in size and abundance outwards (Fig. 7.15). Some of the vesicles are filled with calcite or dark green chlorite, but often these amygdales have been removed as a result of recent weathering. The spaces between adjacent pillows are sometimes filled with cherty sediments. Hyaloclastites composed of aggregates of volcanic glass formed as a result of the rapid quenching of subaqueous basalt flows and volcaniclastic turbidites also occur with the succession.

These pillow lavas were formed as a result of extrusions of basaltic lava from fault-defined submarine fissures. On contact with the sea water the basaltic lava was rapidly chilled resulting in the formation of a glassy outer skin, which insulated the interior of the pillow from further rapid heat loss. This skin prevented gas from escaping, which was preserved as concentric zones of vesicles and amygdales. As further molten magma flowed through the centre of the pod, pressure built up until the skin ruptured and a new pillow was extruded. The formation of pillow lavas have been filmed along the coast of Hawaii where subaerial basalts flow into the sea, and in deeper-water locations where basalts are extruded at constructive plate margins.

7f Strumble Head lighthouse. Return to the car park and walk to the end of the road opposite the lighthouse. From this point a columnar jointed dolerite sheet can be seen below the lighthouse and further examples occur in the cliffs to the southwest. These sheets, which were intruded into the pillow lavas just after their eruption, are referred to as high-level intrusions.

7g Carn Melin. Follow the coast path southwards for 1 km to the inlet just east of Carn Melyn (SM 8867 4056). The cliffs to the west expose a 15 m-thick unit of breccio-conglomerate, composed of complete and broken pillow lavas held in a partly silicified matrix. These breccio-conglomerates were formed during theprocess of flow budding when pillows become detached from the parent body and accumulated on the flanks of the pillow flow.

7.3.8 Newport Bay and Ceibwr Bay

This itinerary is intended as a brief introduction to the highly deformed Ordovician (Caradoc) turbidites exposed at Newport Bay and Ceibwr Bay. These strata were deformed during the Caledonian Orogeny, which resulted from the progressive closure of the Iapetus Ocean. The extended phase of tectonic deformation generated tightly packed northeast-trending folds and the Fishguard–Cardigan Fault Belt (Fig. 7.16).

8a Newport Sands. Travel to Newport on the A487; then take the minor road through Pen-y-Bont where you can make a short stop to visit the Iron Age burial chamber of Careg Coetan. Continue across the iron bridge over the River Nevern (Afon Nyfer); then a further 4 km (following the signs for the golf club and beach) to the car park adjacent to the coastguard station. In the foreground, across the Afon Nyfer estuary is Parrog, which was the original port serving Newport, prior to the silting-up of the estuary. Looking due west you can see Dinas Island, the profile of which slopes gently inland to the beautiful wooded valley of Cwm-yr-Eglwys (valley of the church); interpreted as an east–west-trending subglacial channel. The name of the valley refers to the Celtic sailors' chapel of St Brynach, which was largely destroyed by gales during 1859. Dinas Island, the type locality for the Caradoc (Ordovician) of the area, is composed of mudstones and distal turbidites of the Cwm-yr-Eglwys Mudstone Formation, which are succeeded to the north by proximal turbidites belonging to the Dinas Island Formation (Fig. 7.16).

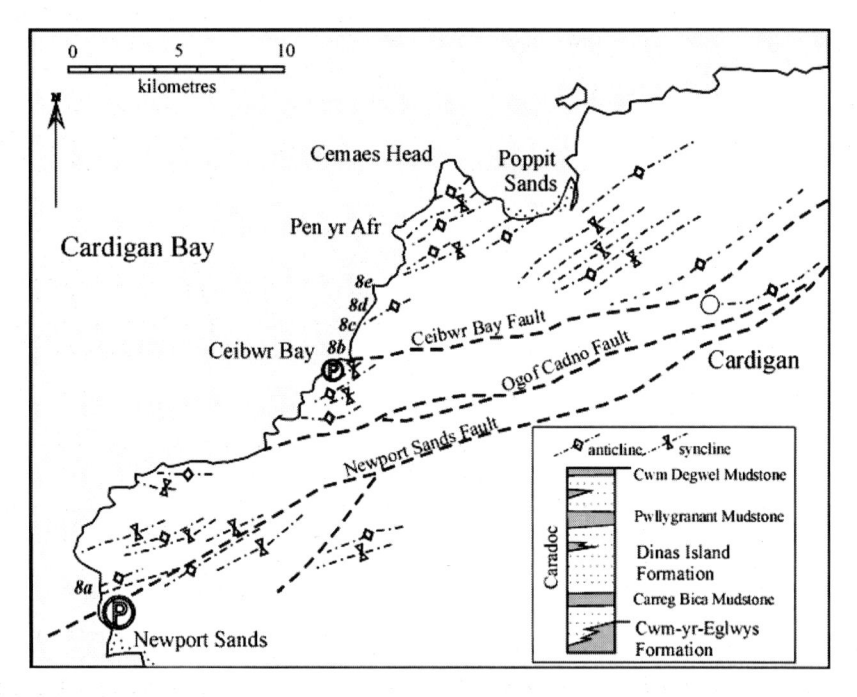

Fig. 7.16 Trends of folds and major faults in the Ordovician strata of the Cardigan area (modified from BGS Sheet193).

The Newport Sands Fault intersects the coast just north of the car park from where it extends inland towards the golf clubhouse. From the car-park walk due north to the rocky shore and cliffs composed of highly folded Ordovician shales and interbedded distal turbidites belonging to the Cwm-yr-Eglwys Mudstone Formation. A minor splay of the fault intersects the cliffs near some caves and brings in strata that are close to the transitional base of the Dinas Island Formation. Farther north (SN 054 408) is a spectacular northeast-trending chevron anticline with fan cleavage, minor limb thrusts and sandstone beds that are thickened in the axial zone of the structure (see Frontispiece B). A thinning- and fining-upwards turbidite cycle (6 m thick) defines the core of the fold. In this unit the coarser-grain amalgamated turbidites contain rip-up clasts, but most of the characteristic Bouma structures have been destroyed as a result of dewatering and other forms of soft-sediment deformation. In the next small cove the strata are deformed into a more open and symmetrical anticline–syncline pair, which plunge to the northeast. Here, the thicker turbidites contain dewatering dish and pillar structures, while many of the thinner turbidites display convolute laminations and dyke-injection structures. The occurrence of these deformation structures suggests that the sediments were deposited very rapidly from fluidized flows (Appendix 7).

8b Ceibwr Bay. From Newport Sands follow the minor road to Moylgrove; take a sharp left for Ceibwr Bay and park in one of the National Trust lay-bys. From the south side of the inlet there is a magnificent view of the folded and faulted Ordovician strata (Dinas Island Formation) extending northeastwards to the headland of Pen yr Afr; many of the limbs of these structures display minor parasitic folds. The narrow gully, seen at the northern end of the inlet, marks the position of the east–west-trending Ceibwr Bay Fault, which has a downthrow of about 600 m to the north. To the south of the fault the cliffs are composed predominantly of argillaceous facies (Carreg Bica Mudstone Member), developed near the base of the Dinas island Formation. Here the cliffs display impressive coastal erosion features, including natural arches, stacks and the collapsed cave system of Pwll y Wrach (the witches cauldron) (SN 102 450).

8c Pen-castell. Follow the track to the beach; cross the clapper bridge, near to the limekiln, and ascend the cliff path to Pen-castell. From here walk a further 250 m along the path to a point where a northeast-trending anticline is exposed in the cliffs (SN 1108 4618). Note the parasitic folds in the core of the anticline and its faulted western limb.

8d Pen-y-Graig. Continue 450 m farther along the coast path to a small hollow eroded in Quaternary glacial deposits (SN 1125 4651). Looking north the cliffs in the foreground expose a prominent channelized sandbody (3–5 m thick), which has an erosive base that is covered with large flute casts. The top of the channel is capped by a thinning- and fining-upwards turbidite cycle (Fig. 7.18). This type of sandbody geometry is typical of sinuous turbidite channels, which distribute mass flows to sand-rich submarine fans

8e Foel Hendre. Walk 1 km northeastwards on the coast path to the small headland, located to the south of Pwllygranant (SN 1155 4740), which is a good position to view the intricate folding in the cliffs at Pen yr Afr. Here the Dinas Island

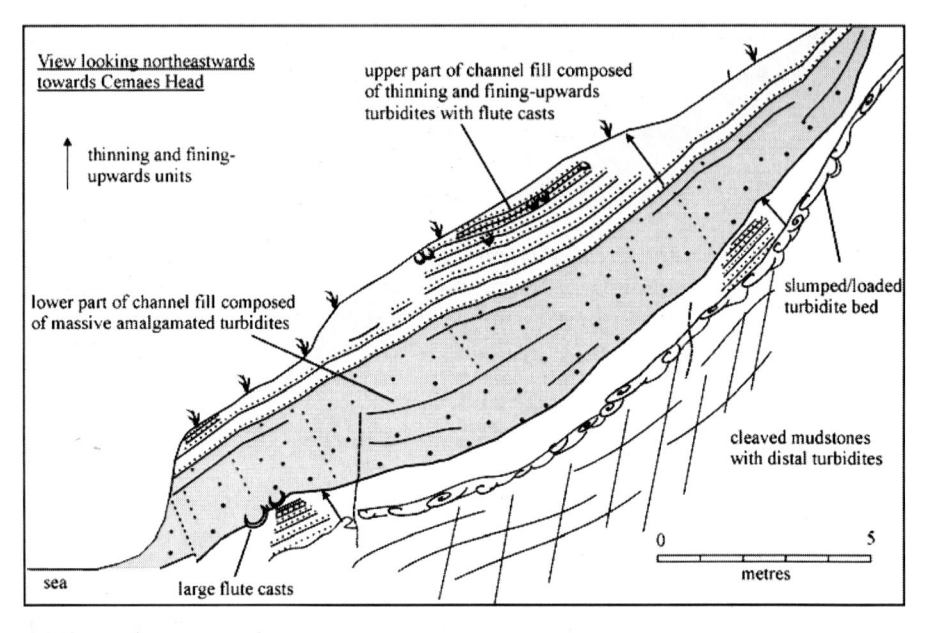

Fig. 7.18 Field sketch of a thinning- and fining- upwards turbidite channel fill (Dinas Island Formation) exposed in the cliffs to the north of Ceibwr Bay.

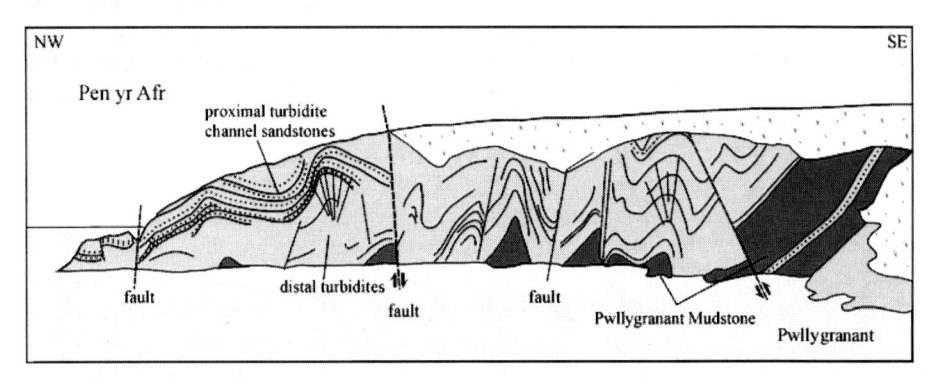

Fig. 7.19 Sketch of the disharmonic folding seen in the Dinas Island Formation exposed at Pen yr Arf.

Formation exhibits complex disharmonic folds, illustrated by the change in their geometry from, (a) open, upright and near symmetrical structures in the northwest to (b) tighter, asymmetrical structures with vertical to overturned limbs in the southeast (Fig. 7.19). The sequence is cut by at least three faults running parallel to the northeast-trending fold axes. Seawards of the central fault, the cliffs are composed of light grey proximal turbidites, some of which get thinner and subdivide laterally, suggesting that they represent turbidite channel fills. Below the sand-rich facies the succession is composed of distal turbidites, followed by mudstones belonging to the Pwllygranant Mudstone.

Recent work on the sedimentology of the Ordovician strata of the Cardigan and Dinas Island area (Davies *et al.* 2003) has shown that the turbidites were deposited in a northeastwards-trending graben, the southern margin of which was defined by the Cardigan Bay–Fishguard Fault Belt. The sand-rich facies of the Dinas Island

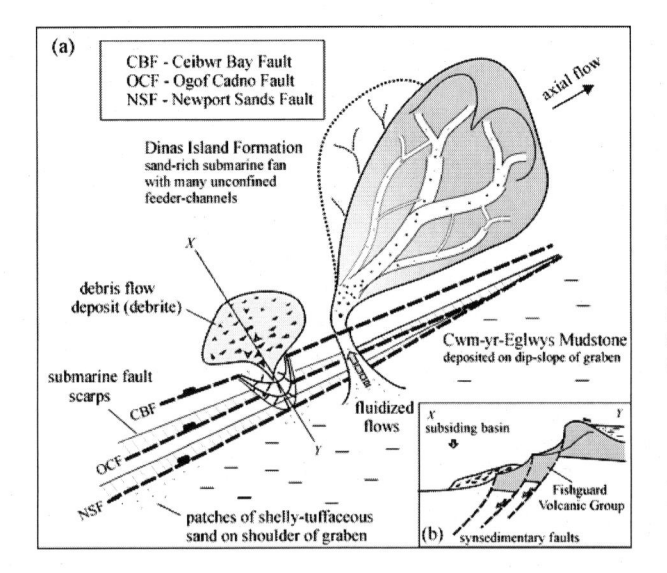

Fig. 7.20 (a) Depositional model for a small sand-rich submarine fan developed in the Dinas Island Formation. (b) Cross section to show the synsedimentary faults (based on Waters et al. 2003).

Formation are interpreted as debris and fluidized flow deposits (Appendix 7), which were sourced from the uplifted shoulders of the graben (Fig. 7.20). Deposition occurred in sinuous–braided fan-channels that were filled very rapidly with sand and debris, causing constant channel switching. In some cases the channels migrated laterally resulting in the deposition of fining- and thinning-upwards units. These processes resulted in the construction of relatively small submarine fans, composed predominantly of sandy channel fills with only minor fan lobe and levee deposits (Appendix 8).

7.3.9 New Quay and Aberaeron

Travel to New Quay via the A487 and A486 and if you arrive by car or minibus you can park at the harbour, adjacent to the Tourist Information Centre. New Quay was one of Dylan Thomas's favourite towns and he made frequent visits to the "cliff-perched, toppling town" and lived here during 1944–45. This was one of his most productive writing periods and many local people and places inspired the characters and locations in *Under Milk Wood*. Leaflet guides (Dylan Thomas' New Quay and Ceredigion) can be obtained from the tourist office; the Black Lion Hotel, at the end of Glanmor Terrace, has an interesting collection of photographs of the poet.

The aims and objectives of this itinerary are to examine and interpret the excellent examples of turbidite facies developed in the Aberystwyth Grits Group (Llandovery), which are exposed in the cliffs at New Quay and Aberaeron. Both of the localities require medium to low tides, and 3 hours field time should be allocated for the New Quay itinerary and 2 hours for the Aberaeron itinerary. The car journey between the two localities takes about 25 minutes.

9a Carreg Walltog foreshore. From the harbour walk up Rock Street to the seafood-processing factory at New Quay Head, noting the adjacent abandoned quarry where about 30 m of turbidites, similar in facies to those seen on the foreshore, are exposed. Continue through the factory yard (with permission) to the edge of the cliffs where a steep and rocky path leads down to the beach, which is composed entirely of shells from the processing plant. The waste outlet from the factory is located on the crest of a southwest-trending anticline; consequently this locality is very popular with seagulls. The bay itself is composed of highly folded and faulted turbidites belonging to the Mynydd Bach Formation (Aberystwyth Grits Group). Continue past the sea stack (Carreg Walltog) to a series of small coves where impressive successions of over six coarsening- and thickening-upwards turbidite cycles (2–3 m thick) are exposed (Fig. 7.21). It is possible to subdivide some of the cycles into minor units, which also display this distinctive vertical trend (Fig. 7.22).

Fig. 7.21 Well developed coarsening- and thickening-upwards turbidite packets in the Aberystwyth Grits at New Quay Head (clipboard scale 35 cm).

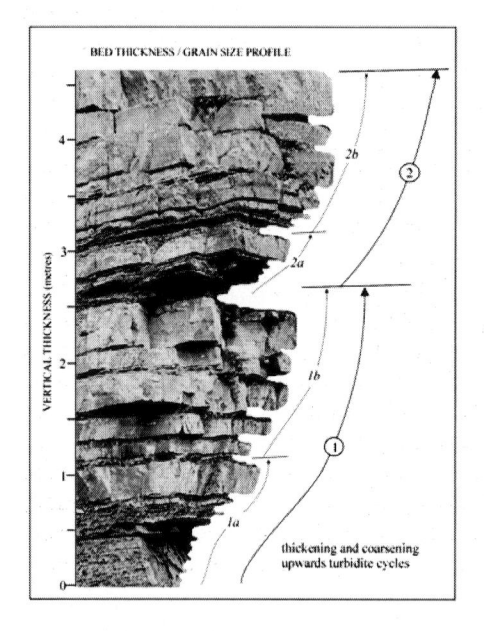

Fig. 7.22 Profile photograph illustrating the compound coarsening- and thickening-upwards packets shown in Figure 7.21.

Individual beds have basal erosional surfaces with flute casts and occasional rip-up clasts, while the thickest turbidites of each cycle are amalgamated and often display dyke-injection structures. Palaeocurrents are difficult to measure from the cross sections of the flute casts, but current ripples indicate transport to the northeast. A 3 m-thick unit of bioturbated (oxic) hemipelagic mudstone containing thin distal turbidites (Tcde) occurs in the middle part of the succession. Farther to the southwest a mudstone that infills a depression in a thick turbidite bed has been interpreted as a slump scar (Bates 1982, see also Dobson 1995).

9b Foreshore north of New Quay pier. Return to the car park, walk down the ramp to the beach and continue about 250 m northwestwards to the low cliffs (SN 3885 6025). Here the foreshore and cliffs that extend to New Quay Head expose a further, slightly younger succession in the Mynydd Bach Formation of the Aberystwyth Grits Group. This succession consists of packages of sand-rich turbidites and occasional debrites that are organized into coarsening- and thickening-upwards cycles, fining- and thinning-upwards cycles and more randomly distributed beds (Fig. 7.23a, b). The thicker amalgamated beds display scoured bases, wedge bedding and partial Bouma grading (mainly Tab or Tbc) and some of the units appear to have been homogenized as a result of dewatering. Flute casts can be seen in cross section on the bases of some of the beds, and on fallen blocks. To the southeast a mud-rich unit (about 3 m thick) is developed above a thinning- and fining-upwards cycle (Fig. 7.23b).

Just north of the pier the strata are deformed into a tight periclinal (whaleback) anticline. Here the vertical facies architecture of the Mynydd Bach Formation is similar to that described previously. One conspicuous unit has a 15 cm graded sandstone base (Tab), which is abruptly overlain by a 20 cm mud-clast debrite. This bed is similar to the cogenetic units identified in modern and ancient distal fans including the Aberystwyth Grits (Gilbert Kelling pers. comm.; see also Talling *et al.* 2004).

The relatively thin coarsening- and thickening-upwards cycles and fining- and thinning-upwards cycles developed at New Quay, are interpreted as the deposits of prograding and laterally migrating submarine sandy-lobe systems. The randomly distributed sandy turbidites were either deposited as a result of both processes operating within a clustered lobe system, or they could have been deposited as more elongate mounds and drifts. Bed amalgamation commonly observed within these facies resulted from the deposition of two discrete turbidites from density currents closely spaced in time and space. In this case the second density flow eroded and reworked the non-cohesive top of the underlying bed. A number of models have been proposed to account for the deposition of cogenetic debrite–turbidite units (see summary in Talling *et al.* 2004). One model invokes the down-slope dilution of the head of a debris flow to form a sandy turbidity current, which outruns the parent mass flow and deposits the turbidite before the debrite. Conversely the debris flow could have been generated as a result of the erosion of irregularities in the sea floor during the passage of a high-density turbidity current. This process would again result in the turbidite being deposited prior to the debrite. Within this fairly complex depositional setting the mud-rich units, developed in the Mynydd Bach Formation, are interpreted as mixtures of lobe fringe, levee and hemipelagic deposits. This depositional model is discussed further after the facies at Aberaeron are examined.

Fig. 7.23 (a) Sand-rich turbidites displaying a coarsening- and thickening- upwards motif passing up into random amalgamated beds, capped by a thinning- and fining- upwards cycle. (b) Mud-rich unit developed above a thinning- and fining- upwards cycle capped by random beds. Both photographs from the Aberystwyth Grits at New Quay.

9c Aberaeron. Depart from New Quay on the B4342; join the A487 and travel 8 km northeastwards to the vibrant harbour and seaside town of Aberaeron. Park at the South Beach car park where tide-tables and weather forecasts are posted outside the yacht club and harbourmasters office. From the car park walk 1 km southwestwards along the path above the pebble beach to the cliffs at SN 4512 6254. At the start of the cliff section the Aberystwyth Grits dip 30° eastwards, and are composed of turbidites (up to 60 cm thick) interbedded with darker coloured mudstones. On polished sections the beds display classic examples of complete Bouma graded units (intervals Tabcde, Fig. 7.24) and attenuated units (intervals Tabc). Many of the thicker beds contain internal erosional surfaces that are typical of amalgamated turbidites. Low amplitude flute casts and other bottom structures (Fig. 7.25) indicate northeasterly palaeoflow directions.

Farther along the section the strata are faulted and deformed into a NE–SW-trending anticline with a vertical to slightly inverted western limb. The wave polished turbidites, at beach level, display amalgamation, mudstone rip-up clasts, load casts and flame structures. About 20 m from the core of the anticline a relatively thick (about 15 m) multistorey thinning- and fining-upwards turbidite cycle is exposed, the basal bed of which is channelled and contains pockets of debrite composed of mudstone rip-up clasts (see Davies *et al.*, 2003, Plate 3 for a very similar channel facies from the Dinas Island Formation). This surface is succeeded by 5 m of amalgamated turbidites displaying flutes, load casts and flame structures. Higher in the cycle, the amalgamated beds are replaced by thinner bedded (5–30 cm) turbidite packets. The complete 15 m-cycle is interpreted as a submarine fan-channel fill. Within the succession some of the argillaceous-bedded units display basal flame structures, well-developed current ripples (types A, B and S) and bioturbated tops, features that are characteristic of submarine fan levee deposits (Fig. 7.26). Walk to the point where the turbidite succession is deformed

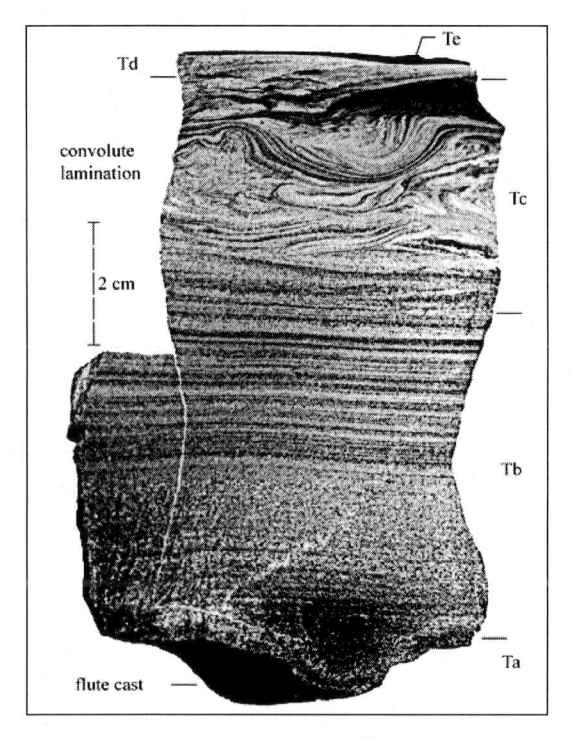

Fig. 7.24 Polished sample of a Bouma turbidite unit, beginning with a basal erosional surface and succeeded by the graded (Ta), lower parallel (Tb), rippled and deformed (Tc), upper parallel (Td) and pelitic (Te), intervals, Aberystwyth Grits at Aberaeron.

Fig. 7.25 Photograph of bottom structures on a graded turbidite (not from the Aberystwyth Grits), including a flute cast, prod casts (a, b) and bounce cast (c); the palaeocurrent direction is left to right.

Fig. 7.26 Turbidite levee facies with basal flame structures (F) succeeded by very fine grain silty sandstones with types A, B and S ripples, capped by a highly bioturbated unit.

into a series of tight southwest-plunging asymmetrical anticlines and synclines. The trace-fossil *Palaeodictyon* can be seen on the bases of some of the mudstone interbeds.

On the walk back to the harbour examine the large blocks of rock (rip-rap), protecting South Beach from erosion. The blocks are composed of relatively thick (often over 1 m), granule-rich, graded and amalgamated turbidite beds and thin debrites, which have basal erosional surfaces displaying large irregular flute casts, groove casts, rip-up clasts, and internal convolute lamination. These turbidites are similar to the very thick and granule-rich facies that are well exposed in the cliff section located 1 km northeast of Aberarth (Fig. 7.27). At this locality erosionally based amalgamated, thinning- and fining- upwards packets (up to 5m thick) are interpreted as proximal/axial turbidite facies that were deposited from high concentration sand-charged flows and highly fluid slurry flows. They appear to represent a channel fill similar to that illustrated in Figure 7.18, although they have also been interpreted as the deposits of prograding sandy lobes (Dobson 1995).

In the past the Aberystwyth Grits have been interpreted as a mixed sand/mud submarine fan with northeastwards flowing distributary channels that supplied fan lobe systems. More recently the Aberystwyth Grits Group has been interpreted as a submarine fan composed almost entirely of NNE-prograding sandy lobes deposited from high concentration turbidity currents and dilute debris flows (Mynydd Bach Formation). These pass laterally and distally into finer grain NE-prograding sandy lobe fringe facies (Trefechan Formation) and muddy lobe fringe facies (Borth Mudstone Formation) (Davies *et al*. 1997; Fig. 7.28). Farther north at Aberystwyth the succession consists of a thick pile of thinly bedded, graded turbidites, interbedded with a relatively high proportion of mudstone facies (Trefechan Formation). Here the individual turbidite beds are laterally extensive and are usually composed of the upper intervals of the Bouma sequence (Tcde). Palaeocurrents measured from flute casts indicate transport to the NNE. Farther north these distal lobe-fringe deposits pass into the mudstone-dominated Borth Mudstones facies.

Fig. 7.27 Examples of turbidite facies belonging to the Mynydd Back Formation exposed in the cliff 1 km northeast of Aberarth. (**a**) Example of one of the many slumped and deformed beds exposed towards the base of the section. (**b**) A prominent amalgamated turbidite packet (~5m thick), seen in the upper part of the photo, which displays basal and internal erosional planes with large flute casts and rip-up clasts and is interpreted as a fan channel-fill.

There can be little doubt that the lobe-fringe turbidites developed between Aberystwyth and Borth are more distal than the coarsening- and thickening-upward and thinning- and fining upward turbidites examined at Aberaeron and New Quay (Mynydd Bach Formation).

This large submarine fan depositional system was supplied from a southerly source now represented by the sub-Coralliferous (intra-Telychian) unconformity, and later by the incised and shelf-edge fluvio-deltaic systems of the Coralliferous Group (Wenlock). The source area was activated as a result of syn-sedimentary extensional faulting on the Wenall and Benton faults (Hillier 2002), which resulted in the uplift and exposure of the contemporaneous shelf of southern Pembrokeshire

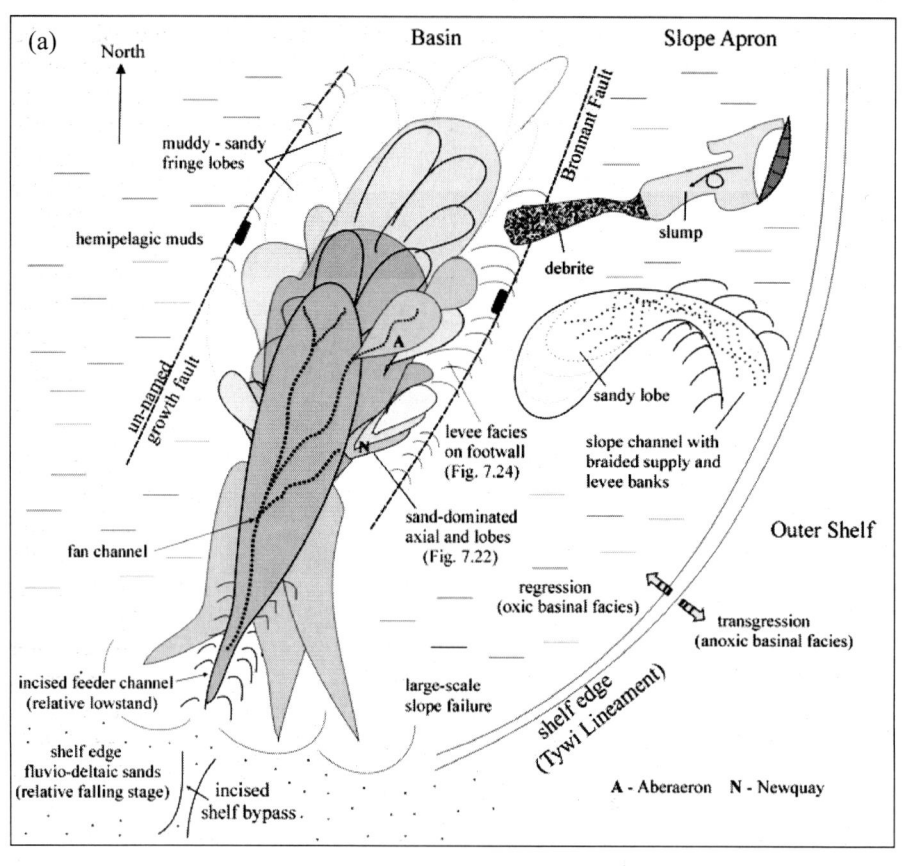

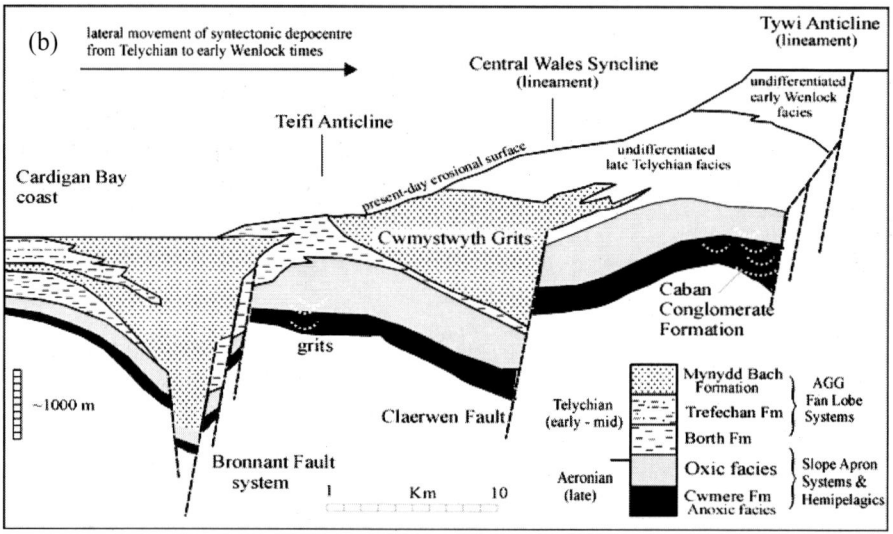

Fig. 7.28 (a) Depositional model for the Aberystwyth Grits. (b) Cross section illustrating the facies architecture of the Aberystwyth Grits Group (AGG) and other Llandovery sequences (both adapted and modified from Davies *et al.* 1997).

throughout most of the Upper Llandovery (Telychian) time interval. The development of this sand-rich source also coincided with active growth faulting in the basin, which provided accommodation space for the abundant sand supplies and also confined the sandy lobe systems to their footwalls. A number of, progressively more easterly located, intra-basinal growth faults and lineaments were activated during the Llandovery–Wenlock time interval (e.g. the Bronnant Fault and later faults defining the Central Wales and Twyi lineaments; see Figs 2.8, 7.28b). During periods of relative tectonic quiescence slope-apron systems became important components of the basin fill. These facies, derived from the eastern margin of the basin, are composed of wedge-shaped packets of hemipelagic mudstones, turbidites, debrites, and conspicuous vertically–offset-stacked, leveed channel-fill conglomerates (e.g. the Caban Conglomerate Formation, see Kelling & Woollands 1969; Fig. 7.28). These slope apron systems are generally composed of a lower anoxic facies assemblage and an upper oxic facies, related to alternating episodes of eustatic transgression and regression respectively. This trend is well seen in hemipelagic mudstones where the anoxic facies are dark grey–black and contain abundant graptolites, and the oxic facies are lighter coloured and bioturbated. The quite complex vertical and lateral facies changes that occurred within the basin during Llandovery to Wenlock times have been discussed in detail Davies *et al.* (1997) and much of this data has been summarized by Cherns *et al.* (2006).

This new fan-lobe model appears to be appropriate for the coarsening- and thickening-upwards turbidites recorded at New Quay, although the occurrence of sandy debrites, common bed amalgamation and basal scouring, indicative of high sediment flux and high flow energy, appears to be more representative of a channel-lobe transition zone (Gilbert Kelling, unpublished field guide). This setting would also account for what appear to be good examples of fining- and thinning-upwards fan channels seen at Aberaeron and Aberarth. Some of these coarse grain, channel fills occur within ripple-laminated and bioturbated distal turbidite facies, interpreted as levee deposits. The scarcity of fan channels could be the result of a high sediment input from mass wasting and slope-failure processes, which would have moderated the influence of any incised (point source) supply system. Another possibility is that a bypass zone existed at the break of slope, downdip of the main incised feeder channel, resulting in the detachment of the sandy lobes (e.g. Type 1 system of Mutti 1985). In this scenario occasional channels could be developed during periods of relative (intra-basinal) sea-level fall (Fig. 7.28a).

Appendix 1

New stratigraphical terms for the Upper Carboniferous of South Wales.

During the preparation of this book, new stratigraphical terms for the Carboniferous successions of southern Britain were proposed and published by the British Geological Survey (Waters *et al.* 2007). Those terms applicable to South Wales, which have been adopted in this field guide, are summarized in the stratigraphic column below. In this new scheme the Carboniferous Limestone has been divided into the Avon Group (previously the Lower Limestone Shales) and the Pembroke Limestone Group. The Namurian strata, traditionally referred to as the Millstone Grit, has been renamed the Marros Group, after the excellent coastal section in western Carmarthenshire, which extends from Ragwen Point across Marros Sands to Telpyn Point. This group has been subdivided into the Twrch Sandstone Formation (previously the Basal Grit), the Bishopston Mudstone Formation (previously the Middle Shales or Shale Group) and the Telpyn Point Sandstone (previously the Upper Sandstone). Only minor modifications have been made to the succeeding Lower Coal Measures Formation.

	New Groups	New Formations	Previous Names	Zones
WESTPHALIAN A	SOUTH WALES COAL MEASURES GROUP	*Vanderbeckei* (Amman) M B — South Wales Lower Coal Measures Formation (Farewell Rock)	Lower Coal Measures	
NAMURIAN	MARROS GROUP	*Subcrenatum* Marine band — Telpyn Point Sandstone Formation	Upper Sandstone Group / Farewell Rock	
		Cancellatum Marine Band — Bishopston Mudstone Formation	Middle Shale Group / Shale Group	
		Superbilinguis Marine Band — Twrch Sandstone Formation	(*Superbilingue* M B) / Basal Grit Group / Basal Grits	
DINANTIAN	PEMBROKE LIMESTONE GROUP	Oystermouth Formation	Upper Limestone Shales	D
		Oxwich Head Limestone Formation		S_2
		Hunts Bay Oolite Formation	Seminula Oolite / Stackpole Limestone	$C_2 S_1$
		High Tor Limestone Formation		
		Caswell Bay Mudstone Fm	*Modiola* Phase	
		Gully Oolite Formation	Caswell Bay Oolite / Caninia Oolite	C_1
		Black Rock Limestone Subgroup	Tairs Point Limestone / Brofiscan Oolite / Shipway Limestone	Z
	AVON GROUP	Avon Group (undivided)	Lower Limestone Shales / Cefn Bryn Shales	K

Appendix 2

Key for all graphic logs.

SEDIMENTARY STRUCTURES		FOSSILS / TRACE FOSSILS	
	Horizontal stratification	MB	Marine Band
	Low angle stratification	M	Marine fossils
	Trough cross stratification	A	*Anthracoceras*
	Planar cross stratification	L	*Lingula*
	Swaley cross stratification (SCS)		Marine shell debris
	Hummocky cross stratification (HCS)		Non-marine bivalves
	Hummocky cross lamination (HCS)		Lycopod stumps/ *Stigmaria*
	Current ripple lamination		Rootlets (Horsetails) with thin coal
	Symmetrical ripple lamination		Plants
	Streaky to flaser lamination		*Lepidodendron / Calamites* casts
	Flute and gutter casts		Bioturbation (*Skolithos* types)
	Calcretes / Nodules		*Thalassinoides*
	Soft-sediment deformation		*Zoophycos*
	Slumps	**LITHOLOGIES**	
	Load casts		Limestone / Oolitic Limestone
	Dewatering pipes		Dolomitic Limestone
	Stylolites		Lime Mudstone / Marl
	Desiccation cracks		Shale / Mudstone
PALAEOCURRENT DATA			Siltstone
	Channel axis orientation		Sandstone
	Current lineation		Conglomerate
	Log casts		Breccia
	Gutter or flute cast trend		Coal with seatearth
	Cross stratification and ripples	**SEQUENCE STRATIGRAPHY**	
	grouped readings (rose diagram) n = number vm = vector mean direction	HST	Highstand Systems Tract
		LST	Lowstand Systems Tract
		TST	Transgressive Systems Tractl
		SB	Sequence Boundary
	individual readings	ISB	Interfluve Sequence Boundary
		RSME	Regressive Surface of Marine Erosion
		TSE	Transgressive Surface of Erosion
		FS	Flooding Surface
		MFS	Maximum Flooding Surface

Appendix 3

Determination of grain size and textural features in the field.

In the field the relative grain size and textural features (sorting and roundness) of sand and sandstones can be estimated by reference to a comparison chart. These are normally small plastic cards containing printed dots (to scale) or actual sand grains, which can be examined and compared with a sample of the rock using a hand lens. An example of a grain size and textural comparator designed by the author is illustrated below. Clasts larger than sand grade can be measured with a ruler and classified using the scale summarized in the Table.

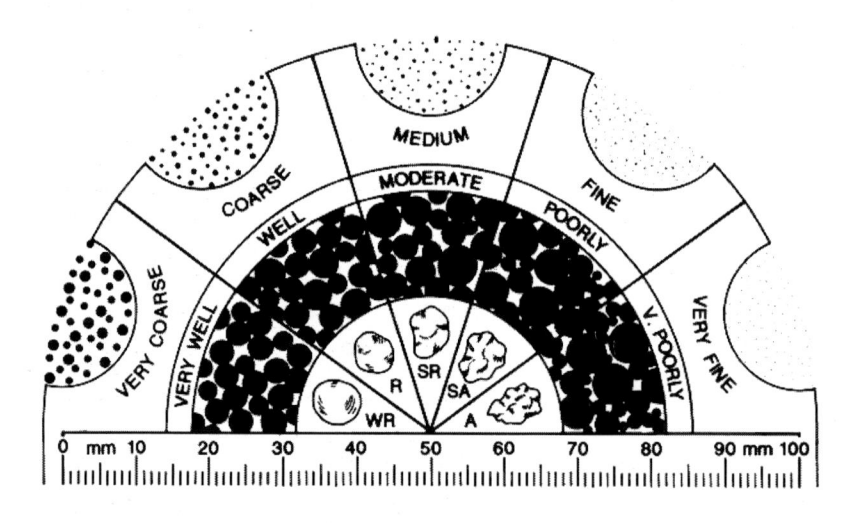

Clast Type	Size (mm)	Rock Names / Terms	
		Clastics	Carbonates
boulder	256	*rudaceous*	*calcirudite*
cobble		conglomerate (rounded clasts)	conglomerate
	16	breccia (angular clasts)	breccia
pebble	4	breccio-conglomerate (rounded & angular clasts)	breccio-conglomerate
granule	2	diamictite (tillite) gritstone	debrite
sand (very coarse)	1	*arenaceous*	*calcarenite*
sand (coarse)	0.5	sandstone arenite quartzite greywacke	bioclastic limestone
sand (medium)	0.25		oolitic limestone
			grainstone (grain support)
sand (fine)	0.125		packstone (>10% grains)
sand (very fine)	0.0625		wackestone (<10% grains)
silt	0.0039	*argillaceous* siltstone claystone mudstone shale	*calcilutite* lime mudstone micrite
clay		lutite marl- hemipelagite - pelagite	

Appendix 4

Summary of Silesian clastic facies (modified from George 2001, 2002)

FACIES ASSOCIATION	FACIES CODE	FACIES DESCRIPTOR	FACIES INTERPRETATION
Fluvial Channel (FC)	FC 1	Erosional surfaces with coarse-grain lags	Channel base
	FC 2	Massive medium-grain sandstones	Mass-flow channel fill
	FC 3	Cross stratified medium-grain sandstones	Braided channel fill
	FC 4	Ripple-laminated fine-grain sandstones	Sinuous channel fill
	FC 5	Rooted sandstone surfaces	Channel abandonment
	FC 6	Ganisters/seatearths/coals	Interfluve / Floodplain
Fluvial Floodplain (FP)	FP 1	Plant-bearing shales	Lake (distal)
	FP 2	Silty-mudstones	Lake (proximal)
	FP 3	Ripple-laminated v. fine-grain sandstones	Crevasse splay
Fluvio-deltaic (FD)	FD 1	Channelized pebbly quartz arenites	Braided channel fill
	FD 2	Ganisters with *in situ* root systems	Palaeosol (well drained)
	FD 3	Rooted thin-bedded arenites	Palaeosol (stagnant)
	FD 4	Carbonaceous–coaly shales	Floodplain swamp
Estuary Channel (EC)	EC 1	Channelized massive to erosionally bedded arenites with abundant lithic clasts	Fluvio-estuarine channel
	EC 2	Channelized medium-grain arenites, erosionally bedded, bimodal cross strata	Bayhead delta distributary channel
	EC 3	Rooted heterolithic seatearths	Bayhead delta levee
	EC 4	Inclined heterolithic facies (IHS)	Distal estuarine channel
Estuary Bay & Back-barrier (EB)	EB 1	Plant-bearing, shales with palaeosols	Estuarine central bay
	EB 2	Silty mudstones with *Carbonicola*	Freshwater lagoon
	EB 3	Silty – sandy mudstones	Back-barrier bay
	EB 4	Parallel and HCS laminated siltstones and fine-grain sandstones	Back-barrier washover fan/lobe
	EB 5	Channelized coarse–fine-grain sandstones (F-U), HCS, log clasts, polymodal vectors	Back-barrier channel
Shoreface & Foreshore (SF)	SF 1	Heterolithic mudstones–very fine-grain sandstones, parallel-laminated with HCS	Shelf–shoreface transition
	SF 2	Fine-grain sandstones with HCS and thin (cm) shale drapes	Lower shoreface
	SF 3	Medium-grain quartz arenites with low angle cross laminations and SCS	Lower–middle shoreface
	SF 4	Medium-grain quartz arenites with cross laminations	Upper shoreface
	SF 5	Medium grain quartz arenites with parallel and low angle laminations	Foreshore/beach
	SF 6	Ganisters with *in situ* root systems	Palaeosol (backshore)
Marine Shelf (MS)	MS 1	Bioturbated, granule–pebble lags	Transgressive lag
	MS 2	Dark grey shales with brackish-marine fossils, *Lingula*, pyritic	Nearshore shelf/bay
	MS 3	Black marine shales with thick-shelled goniatites and Anthracoceras	Outer shelf
	MS 4	Barren grey mudstones/shales with *Planolities*	Outer shelf to prodelta transition zone
	MS 5	Striped silty-sandy mudstones	Prodelta
	MS 6	HCS/SCS medium–fine sandstones, erosive/gutter cast bases, F-U trends	Transgressive shoreface

Appendix 5

Autocyclic fluvio-deltaic depositional processes and deposits.

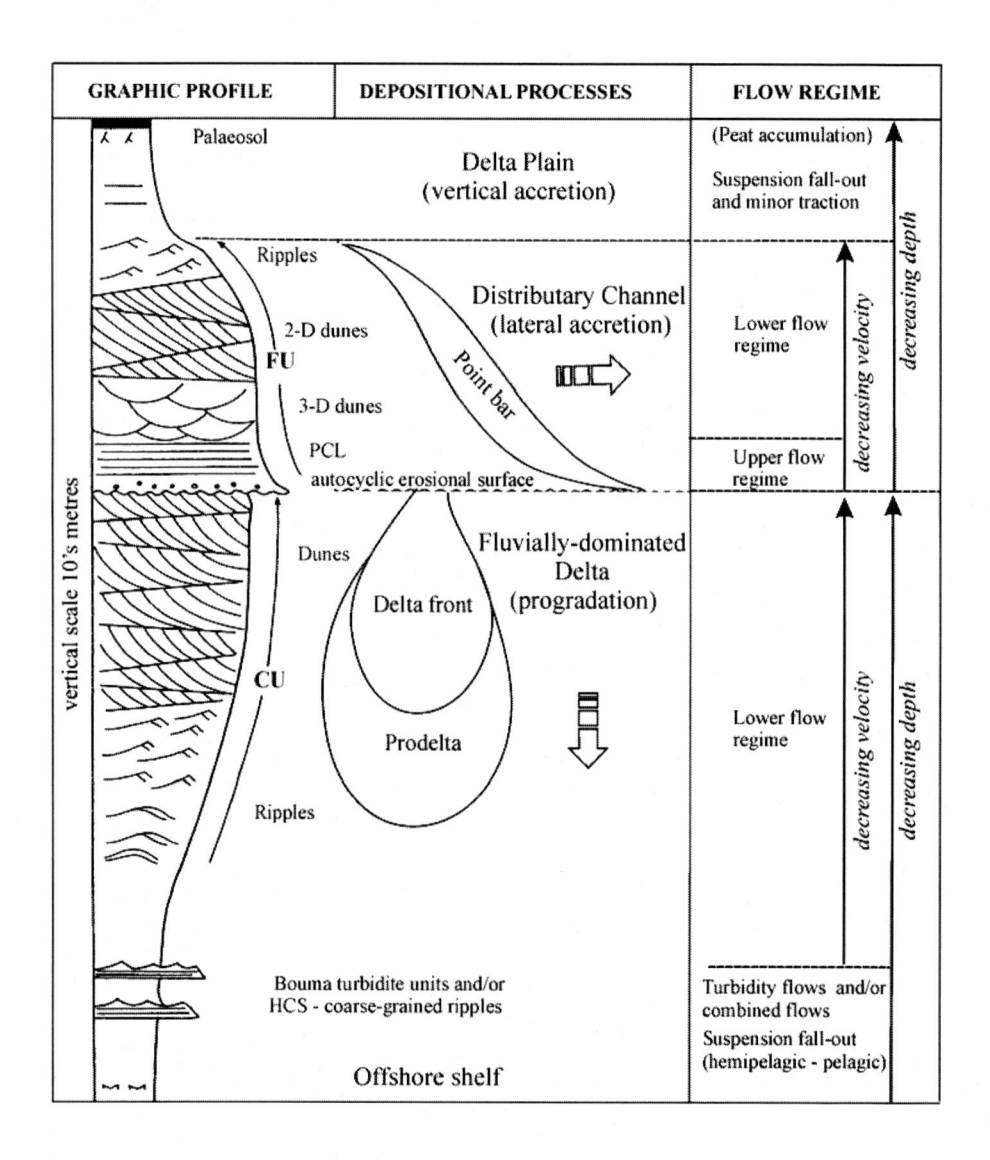

Appendix 6

Provisional sedimentology and sequence stratigraphy of the Namurian strata encountered in the Gelli Isaf borehole (gamma-ray log and palaeontology supplied by the BGS).

Appendix 7

Summary of sediment density-flow processes and deposits (based on Middleton & Hampton 1976).

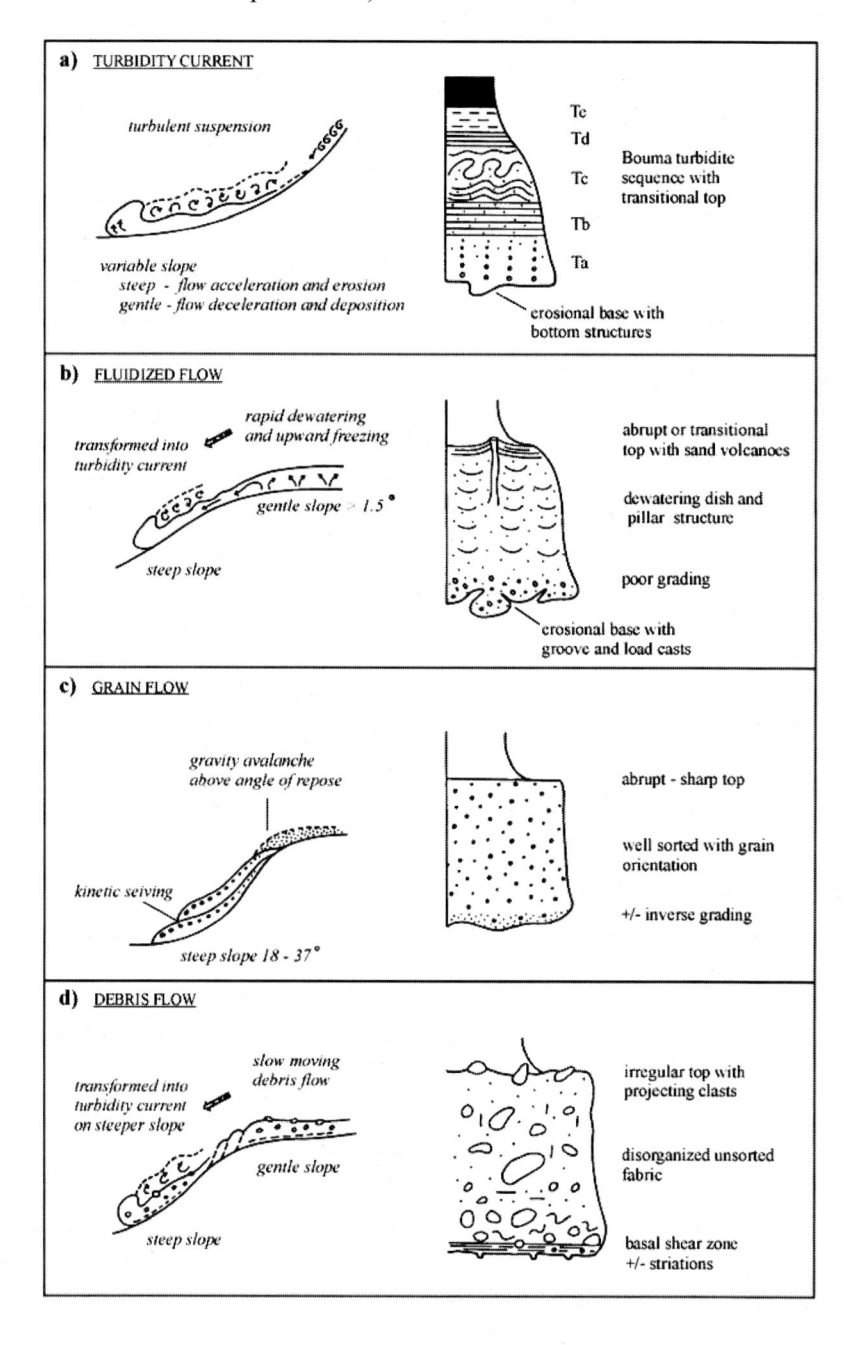

Appendix 8

Generalized classification of submarine fans (based on many sources).

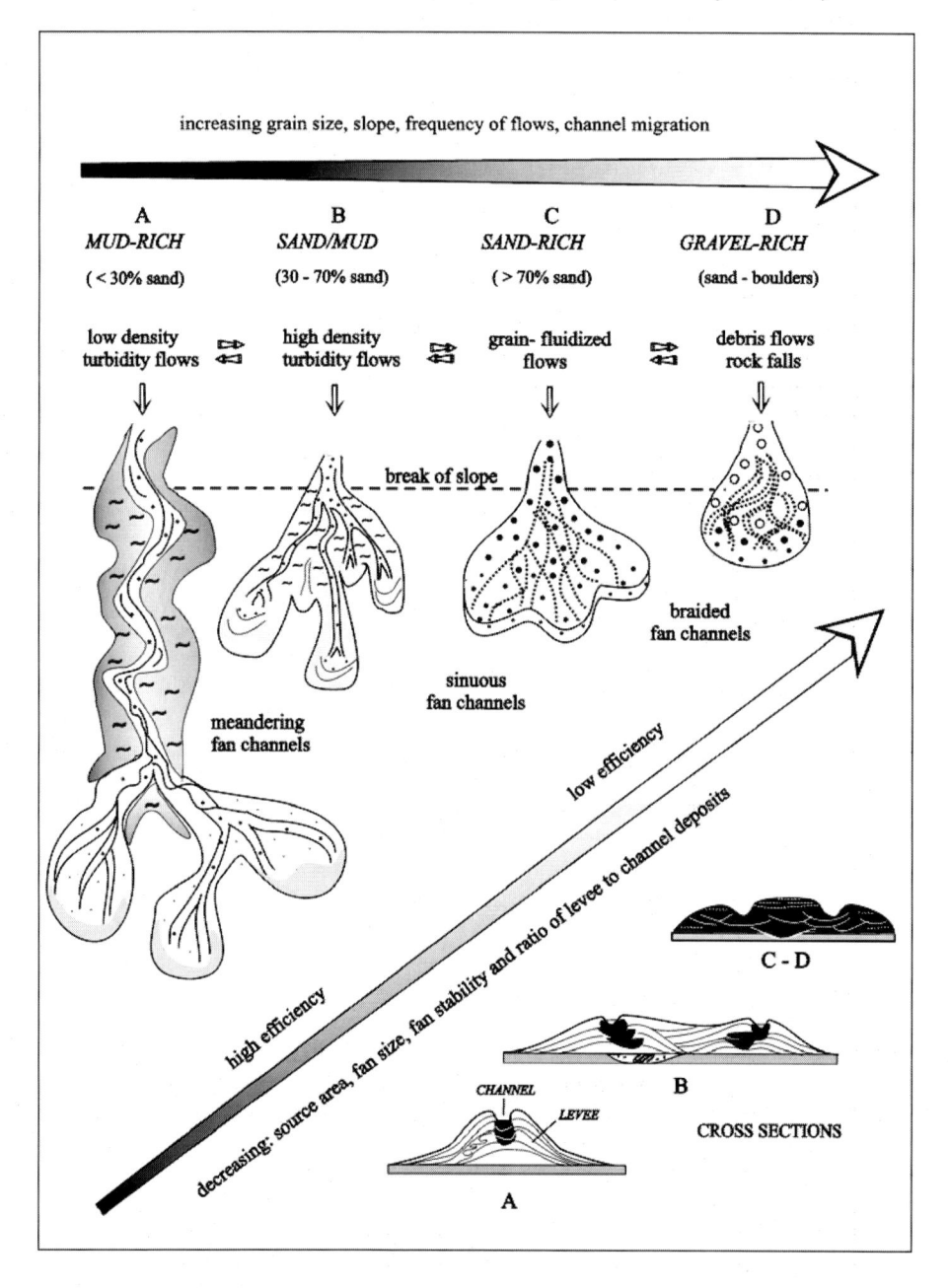

Glossary

accommodation Term used in sequence stratigraphy to define the space available for sediment to accumulate. It is controlled by base level datum, which is taken as sea level in marine and marginal marine environments, and the graded river profile or lake-level in continental environments.

allostratigraphy The study and subdivision of sedimentary strata by reference to discontinuities and sequence stratigraphy concepts.

amygdale A vesicle in a lava filled with a secondary mineral such as calcite or quartz.

angle-of-repose Depositional angle at which unconsolidated sediments can accumulate in air or water and remain stable.

anticlinorium A regionally developed antiform containing smaller scale folds.

antidune High velocity bedform created when a standing wave breaks upstream and deposits up current-inclined cross laminae; they are not common because they have a low preservation potential.

arenaceous Grain size term applied to sediments composed of sand; the term arenite is often used as a synonym for a sandstone.

argillaceous Grain size term applied to sediments composed of clay and silt; argillite can be used as a general term for mudstones and siltstones.

autocyclic Sedimentary processes that occur within the depositional basin (e.g. lateral migration of rivers, delta progradation), and are not controlled by external factors such as eustatic changes in sea-level or tectonic events.

avulsion The process which results in the rapid abandonment of a channel to a new course, usually during bank-full to flood stage.

bioturbation The reworking of a sedimentary facies as a result of the burrowing and feeding activities of organisms, usually invertebrates.

Bouma graded unit Sequence of sedimentary structures/facies deposited from a decelerating turbidity current. Such turbidites (T) display overall grading and up to five intervals: Ta–massive to graded, Tb–lower parallel laminated, Tc–ripple laminated to convolute laminated, Td–upper parallel laminated, Te–pelagic interval.

calc-alkaline Term applied to igneous rocks in which the dominant feldspar is calcium rich; such volcanic and intrusive igneous rocks are typically developed in subduction zones and island arc systems.

calcarenite A carbonate sediment composed of sand size (0.06–2.00 mm) particles.

calcirudite A carbonate sediment composed of coarse grain particles over 2 mm in diameter.

calcrete A concretionary carbonate horizon formed in the soil profile in semi-arid environments; also referred to as caliche.

carbonate ramp A gently seaward-inclined shallow marine depositional environment, where shallow water carbonate facies pass gradually offshore into deeper-water facies.

clints and grykes Weathering features of limestone pavements comprising irregular upward-directed protections (clints) separated by hollows (grykes).

concentric folds Anticlines and synclines in which individual beds maintain a more or less uniform thickness perpendicular to the folded surface; also referred to as parallel folds.

condensed bed   A facies that is much thinner than its lateral equivalent due to an extremely slow sedimentation rate; good examples are the marine bands of the Upper Carboniferous, which usually contain high densities of goniatites and other pelagic faunas.

conjugate joints   A cross-cutting set of joints or fault planes that intersect at angles of between 60–120°.

coprolite   Fossilized faeces and faecal pellets, often phosphatic in composition.

cryoturbation   Term used to describe the movement of permafrost soils as a result of freeze-thaw and frost-heaving etc.

cryptalgal laminae   Fine organic laminations, characteristic of lake and lagoon sediments, believed to be of algal origin.

crypto-crystalline   An extremely fine aggregate of crystals, which often cannot be identified under the petrological microscope.

crystal fractionation   See magmatic differentiation.

dextral   A term referring to tear, wrench or strike-slip faults where one block is displaced to the right of the block from which the observation is being made; where the relative displacement is to the left the fault has a **sinistral** displacement.

diagenesis, diagenetic   Low temperature and pressure chemical changes, such as compaction, cementation and recrystallization, which take place in a sediment after deposition; these changes result in the gradual conversion of unconsolidated sediment into a sedimentary rock.

disconformity   An unconformity where the beds above and below the surface have the same dip.

disturbance   A complex fault zone with a long history of reactivation and synsedimentary movement; the term is most frequently used in coal mining regions.

drag folds   See fault drag.

drusy   Texture resulting from the infilling of a cavity by well-formed crystals; drusy crystal fills are commonly found in amygdales of igneous rocks, in nodules and solution cavities of sedimentary rocks, and in minerals veins.

event bed   Generally a lithologically distinctive bed deposited from a single physical event such as a storm, a rapid sea-level rise or fall, a volcanic eruption etc.

facies dislocation   A term used in sequence stratigraphy to indicate a surface (e.g. regressive surfaces of erosion and sequence boundaries) where facies of a continental or shallow water origin rest on facies of a significantly deeper-water origin; in such cases Walther's law is not followed but it is dislocated.

falling stage systems tract   Term used in sequence stratigraphy for the sedimentary facies deposited during falling sea level.

fault drag   Small-scale flexuring or folding of bedding (**drag folds**), caused during relative movements along a fault planes.

fenestrae   Elongate to irregular cavities present in carbonate sediments of intertidal origin, formed as a result of gas generation from decomposing organic material such as algae.

fireclay   Clay-rich fossil soil (**palaeosol**) found in association with coal seams; often used to make refractory bricks.

firmground   A specific horizon which has been partially lithified near or just below the water–sediment interface; these surfaces are often colonized by a distinctive trace-fossil assemblage (*Glossifungites*). See also **hardground**.

flaser bedding   Sandy ripple cross laminations with well-defined drapes of silt and mud, often concentrated in the ripple troughs; this bedform is common in tidal environments where flow velocities fluctuate between the ebb and flood tides.

fluidized flow Gravity-induced flow, initiated when water saturated cohesionless sediment is subjected to a sudden shock that causes the particles to loose contact and to become suspended in pore water; such grain dispersions have negligible friction and can therefore flow down very gentle slopes. Their deposits frequently display dewatering dish and pillar structures.

fluxion banding Segregation of minerals in a rock due to flow (flow banding).

footwall See hanging-wall.

forced regressive Usually applied to facies deposited during an eustatic fall in sea level e.g. forced regressive shoreface sandbody. The term is used to distinguish between other regressive facies deposited as a result of progradation during periods of more stable sea-level conditions.

foreland basin Depositional basin developed on the continental crust along a tectonic plate margin; the part of the basin nearest the stable craton is referred to as a **peripheral foreland basin**.

foreland bulge Area of a foreland basin on its cratonic side, uplifted as a result of flexure during tectonic shortening.

gamma-ray profile Gamma-ray readings of sedimentary facies presented as a profile, usually on a vertical log; the measurements can be taken in the field with a hand-held spectrometer and measured in API (American Petroleum Institute) units.

ganister A very hard, silica-rich **palaeosol**; the high silica content is caused by the removal (leaching) of unstable minerals from the soil profile by humic acids derived from decaying vegetation. The rock is often used to make refractory bricks.

geopetal A small-scale feature preserved in a sediment that records the **way-up** of the bed at the time of deposition. Commonly they are cavities and fossil voids, the lower part of which is filled with clay (bottom) and the upper part filled by later cement (top).

graben Downthrown block between two approximately parallel, inward facing normal faults; the upstanding shoulders are referred to as horsts. In a **half-graben** only one side of the structure is bounded by a normal fault.

grainstone A limestone composed entirely of framework-supported grains (bioclasts, ooids etc) with no micrite matrix; these clean limestones are characteristic of high-energy carbonate environments such as barrier bars, tidal channels and shoals.

gutter cast Steep-sided, elongate erosional scour on the base of a sharp-based sandstone, particularly common on the bases of forced-regressive sandbodies.

hanging-wall The part of a fault block that lies above an inclined fault plane, as opposed to the **footwall**, which lies below the fault plane.

hardground A specific horizon that has undergone early lithification near or just below the sediment-water interface; such surfaces often contain burrows belonging to both the *Glossifungites* and *Trypanites* trace fossil assemblages.

head deposit An unsorted mass of angular debris formed as a result of the down-slope movement (solifluxion) of superficial deposits in periglacial regions.

highstand systems tract Term used in sequence stratigraphy for the sedimentary facies deposited during a highstand of sea level.

hummocky cross stratification, hummocky set A sedimentary structure composed of bundles of concave-upward laminae; the name is often contracted to **HCS**. Most occurrences of his bedform and **swaley cross stratification (SCS)** have been recorded from relatively shallow storm-dominated shelf environments.

ichnofacies A sedimentary lithology (facies) containing a distinctive trace-fossil assemblage.

ignimbrite A volcanic deposit formed from fragments of viscous lava, ash and gas formed during explosive eruption; where the temperature and thickness of the pyroclastic flow is high, welding may occur e.g. a welded-tuff.

imbrication A structure seen in conglomerates where elongate clasts, aligned parallel to the flow, are stacked on one another and inclined (generally < 20°) up-current.

incised valley fill Sequence stratigraphy term used to describe a valley deepened during sea level fall and subsequently filled by sediment.

interfluve sequence boundary Sequence stratigraphy term applied to the surface formed along the margins of an incised valley; such surfaces represent periods of non-deposition and soil formation when the valleys are being eroded and filled.

intraclast Eroded fragment of carbonate or clastic sediment redeposited in a similar lithology.

intraformational conglomerate A conglomerate formed as a result of the reworking of contemporary sedimentary facies. Conglomerates formed from clasts derived from outside the depositional system are referred to as extra-formational or extra-basinal.

lag deposit A coarse-grain or dense residue left behind after the finer particles have been removed by a flowing current.

Last Glacial Maximum A name used worldwide for the Late Devensian glaciation (Dimlington Stadial).

lateral accretion surface Inclined depositional surface formed during lateral sedimentation, particularly during the lateral migration of a meandering river.

listric normal fault A curved extensional fault which flattens out at depth. They are also referred to as syn-sedimentary growth faults and are often associated with rollover anticlines.

lowstand systems tract Term used in sequence stratigraphy for the sedimentary sequences deposited during a lowstand of sea level.

magmatic differentiation A process leading to the formation of a variety of igneous rock types from a single parental magma.

mass flows Gravity-induced high-density flows including slumps, grain flows, fluidized flows and turbidity currents.

massif A structural or topographic terrain composed of older, usually more resistant or more crystalline rocks.

maximum flooding surface Key stratigraphical surface marking the position where the maximum rate of relative sea-level rise was achieved; the surface delineates the **transgressive** from the **highstand systems tracts**.

micrite Very fine grain (microcrystalline) calcite or lime mud, formed as a result of organic or inorganic processes; the process of micritization can also occur after deposition due to biochemical processes.

Milankovitch cyclicity High frequency climatic cycles brought about by variations in solar radiation, which result from changes in the rotation and orbit of the Earth (*viz.* eccentricity, obliquity and precession).

multi-lateral The term applied to channel-fill sandbodies where individual erosional storeys display a prominent lateral stacking pattern.

multi-storey The term applied to channel-fill sandbodies composed of two or more vertically-stacked erosional storeys; this type of geometry is typical of braided rivers, whereas meandering rivers are usually **single-storey**.

offlap A stratal relationship where facies, deposited during regressive phases, are inclined and young down palaeo-slope.

onlap A stratal relationship where successive facies, deposited during transgressive phases, extend further inland and pinch-out against an inclined shoreline.
ooid (oolith) Spherical sand-size carbonate particle, which forms by accretion around a nucleus; ooids are most readily formed in shallow (<2 m) tropical marine environments where the sea-water is saturated with calcium carbonate.
outsize boulders Abnormally large clasts developed in a breccia or conglomerate.
overstep An unconformable stratal relationship where the younger rocks rest on progressively older tilted strata.
oxic Term used to describe depositional environments that contain sufficient free oxygen (aerobic) to sustain benthonic communities; environments depleted in oxygen are described as **anoxic** (anaerobic).
packstone A limestone composed of framework-supported grains (bioclasts, ooids etc) held within a lime mud (**micrite**) matrix.
palaeokarst A fossilized limestone surface, which displays subaerial weathering (karstic) features such as **clints and grykes**, potholes and a veneer of soil.
palaeosol A fossil soil deposit usually characterized by *in situ* plant remains.
pedogenesis The natural processes that produce soils.
periclinal fold A fold where the limbs dip uniformly inwards (basin) or outwards (dome), or a more elongate fold with its axis plunging in two diametrically opposed directions.
phenocryst A relatively large and well-shaped (euhedral) crystal in a fine grain igneous rock.
photic zone The zone measured from the surface of water to the maximum depth of light penetration.
pinch-out Lateral thinning and disappearance of a facies against an inclined surface; often associated with coastal (transgressive) **onlap**.
Plinian An explosive volcanic eruption that produces thick airfall ashes and pyroclastic flows, similar to those recorded by Pliny the Elder at Pompeii in AD 79.
plunge Inclination of a fold axis or axial surface measured from a horizontal datum.
polymict conglomerate A conglomerate composed of a mixed suite of pebbles; conversely an oligomict conglomerate has only one type of pebble.
porphyritic A common texture of igneous rocks where large well-shaped crystals (phenocrysts) occur in a finer-grain groundmass.
pressure-solution cleavage A cleavage defined by mineral segregations and darker seams of insoluble material.
radiolarian chert Amorphous quartz sediment composed of a high proportion of the siliceous tests of marine zooplankton called Radiolaria.
ravinement An erosional surface caused by a high-energy rise in sea level, also known as a transgressive surface of erosion.
regressive cycle Facies sequence formed during a relative fall in sea level, caused by depositional processes such as delta or beach-shoreface progradation; compare **forced regressive.**
rhizocretion Deposition of minerals (iron oxide, calcium carbonate) around a root during pedogenesis or later.
rhizolith Fossil remains of a root or root system.
ria Drowned river valley commonly formed during a post-glacial rise in sea level.

sabkha   Coastal plain bordering a lagoon or restricted shelf sea in arid environments such as the Trucial Coast of Arabia; evaporation from the surface of the supra-tidal flats result in the precipitation of evaporites such as nodular (chicken-wire) anhydrite, and halite.

sequence boundary   A key stratigraphical surface marking a major fall in sea level, recorded by a regionally developed unconformity / erosional surface and a facies dislocation.

sheet geometry   Term used to describe the geometry of sandbodies that have flat bases and tops, and retain a more or less constant thickness at outcrop.

slickensides   Grooves, striations and abrasions formed on fault planes as a result of frictional movement; the fault plane is often mineralised with calcite or quartz.

solifluxion   Term applied to the down-slope movement of unconsolidated wet sediment, particularly soil and scree; common in cold periglacial environments.

sphaero-siderite   Iron carbonate concretion common in coal measure sequences in which the siderite occurs as **ooids** and larger circular aggregates.

stadial   A period of extremely cold climate, usually associated with glaciation; a period of warmer climate, associated with a glacial melt-out, is referred to as an **interstadial**.

standing wave   A high flow velocity stationary wave, developed on the surface of water, that generates an in-phase sinusoidal bedform on the underlying sediment surface. At higher flow velocities the waves become unstable and break up-stream to form **antidunes**.

strike direction   The compass direction of a horizontal line on an inclined bedding plane; the dip direction is the angle of inclination, measured perpendicular to the strike.

stromatolite   Laminated mounds of organic material and sediment built up by bacteria or algae.

stylolite (stylolitic)   A very irregular (sutured) seam or bed contact formed as a result of pressure solution during burial; the structure is defined by a concentration of insoluble clay residues.

swaley cross stratification (swaley sets)   A sedimentary structure composed of bundles of concave-downward laminae; the term is often contracted to **SCS**. See also **hummocky cross stratification**.

synaeresis cracks   The subaqueous shrinkage of clays often resulting in the formation of small cracks. The process occurs due to the loss of pore water in certain clay minerals as a result of either changes in salinity or the compaction of flocculated clays

synclinorium   A regionally developed synform deformed into smaller scale folds.

syn-sedimentary   Processes occurring contemporaneously with sedimentation, particularly tectonic processes such as faulting (e.g. growth faulting).

systems tract   Three-dimensional sedimentary package deposited during a specific time interval on the sea-level curve e.g. highstand, falling-stage lowstand and transgressive systems tracts.

terrane   A small crustal plate or fault-bounded fragment of a larger plate, which has distinctive geological characteristics; it may be displaced from its original site and added to another plate during global-tectonic processes.

tension gash   A lensoid–sigmoidal fracture, which forms as a result of tensional forces developed during deformation. They are often filled with fibrous quartz, occur in en échelon patterns and are sometimes associated with pressure solution cleavage.

tholeiite (tholeiitic)   A type of basalt, poor in alkali elements and relatively rich in silica, produced from the shallow mantle, and typical of mid-oceanic ridges and hot spots.

traction carpet   A high concentration of sand grains being transported at relatively high velocities in a layer close to the depositional surface.

transgressive systems tract   Term used in sequence stratigraphy for the sedimentary sequence deposited during rising sea level; such sequences display **onlap** (retrogradational) relationships.

travertine   Calcium carbonate deposited by precipitation from hot springs in volcanic regions and from percolating water in caves (stalagmite, stalactite)

tufa   Calcium carbonate precipitate with a spongy and porous form.

types A , B and sinusoidal climbing ripples   Type A ripples have angles of climb <15° (sub-critical), and only the lee-side laminae are preserved; type B ripples have angles of climb 15-75° (supercritical) and unequal preservation of the lee-side (thicker) and stoss-side (thinner) laminae; sinusoidal climbing ripples have angles of climb 75-90° and equal preservation of both the lee-side and stoss-side laminae.

vug   A cavity in a rock or vein, which often contains a lining of crystalline minerals; they are commonly formed in limestones as a result of dissolution and dolomitization.

wackestone   A limestone composed of more than 10 per cent, matrix-supported grains; when the proportion of grains is less than 10 per cent the limestone is classified as a carbonate mudstone or lime mudstone.

Wales–Brabant High   A Carboniferous landmass, which extended from Leinster across the Irish Sea to Central Wales, into Essex and East Anglia and across the North Sea into Belgium; also known as the Anglo-Welsh High or Wales–London–Brabant High.

Waltherian   Sedimentary sequences that conform to Walther's law of correlation of facies, which implies that vertical sequences of facies were deposited in environments related to each other in space and time. Note that this is not the case where sea level changes or syn-sedimentary tectonics have resulted in breaks in the depositional record (see **facies dislocation** and **sequence boundary**).

way-up criteria   Any features, particularly sedimentary structures and geopetals, that allow the original top and bottom (facing direction) of a bed or sequence to be determined.

Bibliography and further reading

1. South Wales field itineraries

Bassett, D A & Bassett M G (eds). 1971. *Geological excursionsi in South Wales and the Forest of Dean.* Geologists' Association, South Wales Group, Cardiff.

Cited chapters from the above field guide:

Baker, J W. The Pre-Cambrian rocks of Pembrokeshire, 170–179.
Bassett, M G. Silurian rocks of the south Pembrokeshire coast, 206–221.
Bloxam, T W. Haverfordwest, Strumble Head and Abereiddy Bay, 199–205.
Bowen, D Q. The Quaternary succession of south Gower, 135–142.
Cope, J C W. Mesozoic rocks of the southern part of the Vale of Glamorgan, 114–124.
Kelling, G. Upper Carboniferous sedimentation in the central part of the South Wales Coalfield, 85–95.
Kelling, G & George G T. Upper Carboniferous sedimentation in the Pembrokeshire coalfield, 240–259.
Owen, T R. The headwater region of the River Neath, 74–84.
Owen, T R. The Gower peninsula, 125–134.
Stead, J T G & Williams B P J. The Cambrian rocks of north Pembrokeshire, 180–198.
Thomas, T M. The geology and geomorphology of the upper Swansea Valley area with particular reference to karstic landforms, 96–105.

Bassett, M G (ed.). 1982. *Geological excursions in Dyfed, south-west Wales.* National Museum of Wales, Cardiff.

Cited chapters from the above field guide:

Allen, J R L, Thomas, R G & Williams B P. The Old Red Sandstone north of Milford Haven, 123–150.
Baker, J W. The Precambrian rocks of south-west Dyfed, 15–25.
Bassett, M G. Silurian rocks of the Marloes and Pembroke peninsulas, 103–122.
Bates, D E B. The Aberystwyth Grits, 81–90.
Bevins, R E & Roach, R A. Ordovician igneous activity in south-west Dyfed, 65–80.
Bowen, D Q. Pleistocene deposits and fluvioglacial landforms of north Preseli, 289–295.
Cope, J C W. The geology of the Llanstephan peninsula, 259–269.
George, G T. Sedimentology of the Upper Sandstone Group (Namurian G1) in south-west Dyfed: a case study, 203–214.
George, G T & Kelling, G. Stratigraphy and sedimentation of Upper Carboniferous sequences in the coalfield of south-west Dyfed, 175–202.
Hancock, P L, Dunne, W M & Tringham M E. Variscan structures in south-west Dyfed, 215–248.
Hughes, C P, Jenkins, C J & Rickards R B. Abereiddy Bay and the adjacent coast, 51–64.
Williams, B P J, Allen, J R L & Marshall J D. Old Red Sandstone facies of the Pembrokeshire peninsula, south of the Ritec Fault, 151–174.
William, B P J & Stead, J T G. The Cambrian rocks of the Newgale–St David's area, 27–49.

2. Other South Wales guides

Allen, J R L, Elliot, T & Williams B P J. 1981. Old Red Sandstone and Carboniferous fluvial sediments in South Wales. In *Field guide to modern and ancient fluvial systems in Britain and Spain,* Elliot, T (ed). Proceedings of the 3rd International Symposium on fluvial sedimentology: Keele, 1–39.

Barrett, J H. 1974. *The Pembrokeshire Coast Path.* HMSO.

Bowen, D Q & Henry A (eds). 1984. *Wales: Gower, Preseli, Fforest Fawr.* Quaternary Research Association Field Guide, Cambridge.

Davies, J H, Holroyd, J, Lumley, R G & Owen-Roberts, D. 1978. *Geology of Powys in outcrop.* Powys County Council, Llandrindod.

Dobson, M R (ed.). 1995. *The Aberystwyth district.* Guide No. 54, Geologists' Association.

Howe, S R. 1998. *Geology around Bridgend.* Geologists' Association, South Wales Group.

Howe, S, Owen, G & Sharp T. 2004. *Walking the rocks – six walks discovering scenery and geology along the Glamorgan coast.* Geologists' Association, South Wales Group.

John, B. 2001. *Pembrokeshire Coast Path.* Aurum Press.

Kokelaar, B P, Howells, M F, Bevins, R E & Roach R A. 1984. Volcanic and associated sedimentary and tectonic processes in the Ordovician marginal basin of Wales: a field guide. In *Marginal basin geology: volcanic and associated sedimentary and tectonic processes in modern and ancient marginal basins,* B P Kokelaar & M F Howell (eds), 291-322. Special Publication **16**, Geological Society London.

Owen, T R. 1973. *Geology explained in South Wales.* David & Charles, Newton Abbott.

Owen, T R, Rhodes, F H T, Jones, D G & Kelling, G. 1965. Summer (1964) field meeting in South Wales. *Proceedings of the Geologists' Association,* **76**, 463–496.

Parker, M, &. Whitefield, P. 2003. *The rough guide to Wales.* Rough Guides.

Perkins, J W, Gayer, R. A, Baker, J W & Williams, G D. 1979. *Glamorgan Heritage Coast: A guide to its geology.* Glamorgan Heritage Coast.

Ridge, M. 1999. Conservation. In *A guide to Gower,* D Strawbridge & P. J. Thomas (eds). The Gower Society, Swansea.

Walker, M J C & McCarroll D (eds). 2001. *The Quaternary of West Wales: Field Guide.* Quaternary Research Association, London.

Warrington, G & Ivimey-Cook, H C. 1995. The Late Triassic and Early Jurassic of coastal sections in west Somerset and South and Mid-Glamorgan. In *Field Geology of the British Jurassic,* P D Taylor (ed.), 9-30. Geological Society of London.

Williams, A, Davies, P & Caldwell N. 1997. *Coastal Processes and Landforms: the Glamorgan Heritage Coast.* Glamorgan Heritage Coast.

Williams, B P J (ed). 1978. The Old Red Sandstone of the Welsh Borderlands and South Wales. International Symposium on the Devonian System (P.A.D.S. 78), September 1978. The Palaeontology Association

Woodcock, N H & Bassett M G. 1993. *Geological excursions in Powys Central Wales.* University of Wales Press, National Museum of Wales.

3. Geological Conservation Review Series

Aldridge, R J, Siveter, D J, Siveter, D J, Lane, P D, Palmer, D & Woodcock, N H. 2000. *British Silurian stratigraphy.* Geological Conservation Review Series 19, Joint Nature Conservation Committee.

Barclay, W J, Browne, M A E, McMillan, A A, Pickett, E A, Stone, P & Wilby P R (eds). 2005. *The Old Red Sandstone of Great Britain,* Geological Conservation Review Series 31, Joint Nature Conservation Committee.
Sites from the above review
Barclay, W J. Albion Sands and Gateholm Island, Pembrokeshire, 281–284: Barclay, W J & Williams, B P J. Freshwater West, Pembrokeshire, 291–30: Barclay, W J. Freshwater East – Skrinkle Haven, Pembrokeshire, 301–308.

Benton, M J, Cook, E. & Turner, P. (eds). 1998. *Permian and Triassic red beds and the Penarth Beds of Great Britain.* Geological Conservation Review Series 24, Joint Nature Conservation Committee.
Sites from the above review
Benton, M J, Cook, E. & Turner, P. Sutton Flats, mid Glamorgan, 191,–194: Barry Island, south Glamorgan, 195–198: Hayes Point to Bendrick Point, south Glamorgan 198–202: Sully Island, south Glamorgan, 202–205: Lavernock Point to Penarth, south Glamorgan, 226–232.

Carney, J.N, Horak, T C Pharoh, T C et al. (eds). 2000. *Precambrian Rocks of England and Wales.* Geological Conservation Review Series 20, Joint Nature Conservation Committee.
Sites from above review
Bevins, R E. Llangynog, 122–126: Cope, J C W. Coed Cochion, 196–198: Jones, K. A. Hanter Hill, 118–122: Woodcock, N. H. Dolyhir and Strinds quarries, 114–118.

Rushton, A W A, Owen, A W, Owens, R M & Prigmore J K. 1999. *British Cambrian to Ordovician Stratigraphy.* Geological Conservation Review Series, Joint Nature Conservation Committee.
Chapter 4 from above review
Prigmore, J K & Rushton, A W A. 1999. Cambrian of South Wales: St David's area, 53–67.

Simms, M J, Chidlaw, N, Morton, N & Page K N (eds). 2004. *British Lower Jurassic Stratigraphy.* Geological Conservation Review Series 30, Joint Nature Conservation Committee.
Sites from the above review
Simms, M J. Lavernock to St Mary's Well Bay, Glamorgan, 113–119: Simms, M J. Pant y Slade to Witches Point, Glamorgan, 119–128.

4. General field guides

British Museum (Natural History) 1969. *British Palaeozoic fossils.* Trustees of the British Museum (Natural History), London.
Britsh Museum (Natural History) 1967. *British Mesozoic fossils.* Trustees of the British Museum (Natural History), London.
Collinson, J D, Mountney, N & Thompson, D B. 2006. *Sedimentary structures.* Terra.
Goldring, R. 1999. *Field Palaeontology.* Longman.
McClay, K R. 1987. *The mapping of geological structures.* John Wiley & Sons, Chichester.
Stow, D A V. 2005. *Sedimentary rocks in the field: a colour guide.* Manson Publishing.

5. Geological Survey publications

Barclay, W J, Davies, J R, Humpage, A J, Waters, R A, Wilby, P R, Williams M & Wilson D. 2005. Geology of the Brecon district–a brief explanation of the geological map. *Sheet Explanation of the British Geological Survey.* 1:50 000 Sheet 213 Brecon (England and Wales).
Barclay, W J, Taylor, K, Thomas L P. 1988. Geology of the South Wales Coalfield, Part V, the country around Merthyr Tydfil. *Memoirs of the British Geological Survey,* Sheet 231 (England and Wales), HMSO, London.
Barclay, W J & Wilby, P R. 2003. Geology of the Talgarth district – a brief explanation of the geological map. *Sheet Explanation of the British Geological Survey.* 1:50 000 Sheet 214 Talgarth (England and Wales).

Cantrill, T C, Dixon, E E L, Thomas, H H & Jones, O T. 1916. *The geology of the South Wales Coalfield. Part XII: The country around Milford.* Memoir of the Geological Survey, UK.

Davies, J R, Schofield, D I, Sheppard, T H, Waters, R A, Williams, M & Wilson, D. 2006. Geology of the Lampeter district – a brief explanation of the geological map. *Sheet Explanation of the British Geological Survey.* 1:50 000 Sheet 195 Lampeter (England and Wales).

Davies, J R, Sheppard, T H, Waters, R A & Wilson, D. 2006. Geology of the Llangranog district – a brief explanation of the geological map. *Sheet Explanation of the British Geological Survey.* 1:50 000 Sheet 194 Llangranog (England and Wales).

Davies, J R, Waters, R A, Wilby, P R, Williams, M & Wilson D. 2003. Geology of the Cardigan and Dinas Island district – a brief explanation of the geological map. *Sheet Explanation of the British Geological Survey.* 1:50,000 Sheet 193 Cardigan and Dinas Island (England and Wales).

Dixon, E E L. 1921. *The geology of the South Wales Coalfield. Part XIII: The geology of the country around Pembroke and Tenby.* Memoir of the Geological Survey, UK.

George, T N. 1970. *British Regional Geology: South Wales.* HMSO, London.

Howells, M F. 2007. *British Regional Geology: Wales.* British Geological Survey, Keyworth, Nottingham.

Jackson, D I, Jackson A A, Evans D, Wingfield, R T R, Barnes, R P & Arthur, M J. 1995. *The geology of the Irish Sea.* United Kingdom Offshore Regional Reports, British Geological Survey, HMSO, London.

Schofield, D I, Davies, J R, Waters, R A, Wilby, P R, Williams, M & Wilson D. 2004. Geology of the Builth Wells district – a brief explanation of the geological map. *Sheet Explanation of the British Geological Survey* 1:50 000 Sheet 196 Builth Wells (England and Wales).

Tappin, D R, Chadwick, R A, Jackson, A A, Wingfield, R T R & Smith N J P. 1994. *The geology of Cardigan Bay and the Bristol Channel.* United Kingdom Offshore Regional Report, British Geological Survey, HMSO, London.

Waters, R A & Lawrence D J D. 1987. Geology of the South Wales Coalfield, Part III, the country around Cardiff. *Memoirs of the British Geological Survey*, Sheet 263 (England and Wales), HMSO, London.

Waters, C N, Waters, R A J, Barclay, W & Davies J. 2007. Stratigraphical framework for Carboniferous successions of southern Great Britain (onshore). *British Geological Survey Research Report.* British Geological Survey, Keyworth, Nottingham.

Wilson, D, Davies, J R, Fletcher, C J N & Smith M. 1990. Geology of the South Wales Coalfield, Part VI, the country around Bridgend. *Memoirs of the British Geological Survey*, Sheet 261 & 262 (England and Wales), HMSO, London.

6. Textbooks

Brenchley P J & Rawson P F (eds). 2006. *The Geology of England and Wales* (2nd edition). The Geological Society, London.

Coe, A L (ed.). 2003. *The sedimentary record of sea-level change.* The Open University, Cambridge University Press.

Doyle, P & Bennett, M R. 1998. *Unlocking the stratigraphic record: advances in modern stratigraphy.* John Wiley & Sons.

Duff, P McL D & Smith, A J (eds). 1992. *Geology of England and Wales.* Geological Society, London.

Emery, D & Myers, K (eds). 1966. *Sequence stratigraphy.* Blackwell Science, Oxford.

Nichols, G. 1999. *Sedimentology and stratigraphy.* Blackwell Science, Oxford.

Reading, H G (ed.). 1986. *Sedimentary environments and facies.* Blackwell Scientific Publications, Oxford.

Reading, H G (ed.). 1996. *Sedimentary environments: processes, facies and stratigraphy.* Blackwell Science, Oxford.

Tucker, M E & Wright, V P. 1990. *Carbonate sedimentology.* Blackwell Scientific Publications, Oxford.

Woodcock, N. & R. Strachan (eds). 2000. *Geological history of Britain and Ireland.* Blackwell Science.

7. Research articles

Ager, D V. 1974. The Jurassic Period in Wales. In *The Upper Palaeozoic and post-Palaeozoic rocks of Wales*, T R Owen (ed.), 323 – 339. University of Wales Press, Cardiff.

Bassett, M G, Bluck, B J, Cave, R, Holland, C H & Lawson J D. 1992. Silurian. In *Atlas of palaeogeography and lithofacies,* J. C. W. Cope, J. K. Ingham, P. F. Rawson (eds), 37–56. Memoir **13**, Geological Society, London.

Bluck, B J. 1965. The sedimentary history of some Triassic conglomerates in the Vale of Glamorgan, South Wales. *Sedimentology*, **4**, 225–245.

Bowen, D Q. 1999. Wales. In *A revised correlation of Quaternary deposits in the British Isles.* D Q Bowen (ed.), 79–90. Geological Society Special Report, **23**.

Brasier, M D, Ingham, J K & Rushton, A W A. 1992. Cambrian. In *Atlas of palaeogeography and lithofacies,* J. C. W. Cope, J. K. Ingham, P. F. Rawson (eds), 13–18. Memoir **13**, Geological Society, London.

Brenchley, P J, Rushton, A W A, Howells, M & Cave, R. 2006. Cambrian and Ordovician: the early Palaeozoic tectonostratigraphic evolution of the Welsh Basin, Midland and Monian Terranes of eastern Avalonia. In *The Geology of England and Wales*, P J Brenchley & P F Rawson (eds), 25–74. The Geological Society, London.

Burgess, P M & Gayer, R A. 2000. Late Carboniferous tectonic subsidence in South Wales: implications for Variscan basin evolution and tectonic history in SW Britain. *Journal of the Geological Society, London,* **157**, 93–104.

Cherns, L, Cocks, L R M, Davies, J R, Hillier, R D Water, R A & Williams, M. 2006. Silurian: the influence of extensional tectonics and sea-level changes on sedimentation in the Welsh Basin and on the Midland Platform. In *The Geology of England and Wales*, P J Brenchley & P F Rawson (eds), 75–102. The Geological Society, London.

Cope, J C W. 2006. Jurassic: the returning seas. In *The Geology of England and Wales*, P J Brenchley & P F Rawson (eds), 325–364. The Geological Society, London.

Cope, J C W & Bevins, R E. 1993. The stratigraphy and setting of the Precambrian rocks of the Llangynog Inlier, Dyfed, South Wales. *Geological Magazine,* **130**, 101–111.

Fletcher, C J N. 1988. Tidal erosion, solution cavities and exhalative mineralisation associated with the Jurassic unconformity at Ogmore, South Glamorgan. *Proceedings of the Geologists' Association,* **99**, 1–7.

George, G T. 1982. *Sedimentary features of a Pleistocene kame delta sequence, Mullock Bridge, Dyfed.* The Open University.

George, G T. 2000. Characterisation and high resolution sequence stratigraphy of storm-dominated braid delta and shoreface sequences from the Basal Grit Group (Namurian) of the South Wales Variscan peripheral foreland basin. *Marine and Petroleum Geology,* **17**, 445–475.

George, G T. 2001. Late Yeadonian (Upper Sandstone Group) incised valley supply and depositional systems in the South Wales peripheral foreland basin: implications for the evolution of the Culm Basin and for the Silesian hydrocarbon plays of onshore and offshore UK. *Marine and Petroleum Geology,* **18**, 671–705.

George, T N. 1940. The structure of Gower. *Geological Society London, Quarterly Journal* **96**, 131–198.

Gradstein, F M Oog, J G Smith, A G et al. 2004. *A geological timescale.* Cambridge University Press.

Hallam, A. 1960. A sedimentary and faunal study of the Blue Lias of Dorset and Glamorgan. *Philosophical Transactions of the Royal Society of London*, series **B, 243**, 1–44.

Hampson, G J. 1998. Evidence for relative sea-level falls during deposition of the Upper Carboniferous Millstone Grit, South Wales. *Geological Journal,* **33**, 243–266.

Hillier, R D. 2000. Silurian marginal marine sedimentation and the anatomy of the marine–Old Red Sandstone transition in Pembrokeshire, SW Wales. In *New perspectives on the Old Red Sandstone,* P F Friend & B P J Williams (eds), 343–354. Special Publication **180**, Geological Society, London.

Hillier, R D. 2002. Depositional environment and sequence architecture of the Silurian Coralliferous Group, southern Pembrokeshire, UK. *Geological Journal*, **37**, 247–268.

Hillier, R D & Williams, B P J. 2006. The alluvial Old Red Sandstone: fluvial basins. In *The Geology of England and Wales*, P J Brenchley & P F Rawson (eds), 154–171. The Geological Society, London.

Hillier, R D & Williams, B P J. 2007. The Ridgeway Conglomerate Formation of SW Wales and its implications. The end of the Lower Old Red Sandstone? *Geological Journal*, **42**, 55–83.

Holdsworth, R E, Woodcock, N H & Strachan, R A. 2000. Geological framework of Britain and Ireland. In *Geological history of Britain and Ireland*, N Woodcock & R Strachan (eds), 19–37. Blackwell Science.

Hounslow, M W & Ruffell, A H. 2006. Triassic: seasonal rivers, dusty deserts and saline lakes. In *The Geology of England and Wales*, P J Brenchley & P F Rawson (eds), 295–324. The Geological Society, London.

James, D M D. 1997. Llanvirn–Llandovery activity on the Llangranog Lineament in southwest Ceredigion, Wales. *Mercian Geologist*, **14**, 68–78.

James, D M D & James, J. 1969. The influence of deep structures on some areas of Ashgillian–Llandoverian sedimentation in Wales. *Geological Magazine*, **106**, 562–582.

Jenkins, T B H. 1962. The sequence and correlation of the Coal Measures of Pembrokeshire. *Geological Society of London, Quarterly Journal* **118**, 65–101.

Johnson, M E & McKerrow, W S. 1995. The Sutton Stone: an Early Jurassic rocky shore deposit in South Wales. *Palaeontology*, **38**, 529–541.

Jones, D G & Owen, T R. 1957. The rock succession and geological structure of the Pyrddin, Sychryd and Upper Cynon valleys, South Wales. *Proceedings of the Geologists' Association,* **67**, 232–250.

Kelling, G. 1974. Upper Carboniferous sedimentation in South Wales. In *The Upper Palaeozoic and Post-Palaeozoic Rocks of Wales*, T. R. Owen (ed.), 185–22. University of Wales Press, Cardiff.

Kelling, G. 1988. Silesian sedimentation and tectonics in the South Wales Basin: a brief review. In *Sedimentation in a synorogenic basin complex: The Upper Carboniferous of Northwest Europe,* B Besley & G Kelling (eds), 31–42. Blackie, Glasgow.

Kelling, G & Woollands, M A. 1969. The stratigraphy and sedimentation of the Llandovery rocks of the Rhayader district. In *The Pre-Cambrian and Lower Palaeozoic rocks of Wales*, A Wood (ed.), 255–282. University of Wales, Cardiff.

Kokelaar, B P, Howells, M F, Bevins, R E, Roach, R A & Dunkley, P N. 1984. The Ordovician marginal basin of Wales. In *Marginal basin geology: volcanic and associated sedimentary and tectonic processes in modern and ancient marginal basins,* B P Kokelaar & M F Howells (eds), 245–269. Special Publication **16**, Geological Society London.

Leveridge, B E & Hartley, A J. 2006. The Variscan Orogeny: the development and deformation of Devonian/Carboniferous basins in SW England and South Wales. In *The Geology of England and Wales*, P J Brenchley & P F Rawson (eds), 225–255. The Geological Society, London.

Love, S E & Williams, B P J. 2000. Sedimentology, cyclicity and floodplain architecture in the Lower Old Red Sandstone of SW Wales. In *New perspectives on the Old Red Sandstone,* P. F. Friend & B. P. J. Williams (eds), 371–388. Special Publication, **180** Geological Society, London.

Marshall, J D. 2000. Fault-bounded basin fill: fluvial response to tectonic controls in the Skrinkle Sandstone of SW Pembrokeshire, Wales. In *New perspectives on the Old Red Sandstone,* P. F. Friend & B. P. J. Williams (eds), 401–416. Special Publication **180**, Geological Society, London.

McIlroy, D & Horak, J M. 2006. Neoproterozoic: the late Precambrian terranes that formed Eastern Avalonia. In *The Geology of England and Wales,* P J Brenchley & P F Rawson (eds), 9–24. The Geological Society, London.

Middleton, G V & Hampton, M A. 1976. Subaqueous sediment transport and deposition by sediment gravity flows. In *Marine sediment transport and environmental management,* D J Stanley & D J P Swift (eds), 197–218. Wiley, New York.

Morrissey, D J &. Braddy S J. 2004. Terrestrial trace fossils from the Lower Old Red Sandstone southwest Wales. *Geological Journal,* **39**, 337–358.

Mutti, E. 1985. Turbidite systems and their relations to depositional sequences. In *Provenance of arenites,* G G Zuffa (ed.), 65–93. Reidel, Amsterdam.

Nemec, W. 1992. Depositional controls on plant growth and peat accumulation in a braidplain delta environment: Helvetiafjellet Formation (Barremian–Aptian), Svalbard In *Controls on the distribution and quality of Cretaceous Coals,* P J McCabe & J T Parish (eds), 209–226. Special Paper 267, Geological Society America.

Plint, A G & Nummedal, D. 2000. The falling stage systems tract: recognition and importance in sequence stratigraphic analysis. In *Sedimentary responses to forced regressions,* D Hunt & R L Gawthorpe (eds), 1–17. Special Publication **172**, Geological Society, London.

Ramsay, A T S. 1987. Depositional environments in the Dinantian Limestones of Gower, South Wales. In *European Dinantian environments,* J Miller, A E Adams & V P Wright (eds), 265–308. John Wiley, Chichester.

Rawson, P F. 2006. Cretaceous: sea levels peak as the North Atlantic opens. In *The Geology of England and Wales,* P J Brenchley & P F Rawson (eds), 154–171. The Geological Society, London.

Sheppard, T H. 2006. Sequence architecture of ancient rocky shorelines and their response to sea-level change: an Early Jurassic example from South Wales, UK. *Journal of the Geological Society, London,* **163** (4), 595–606.

Sheppard, T H. 2007. Life's a beach: lessons from the Earth's rarest sedimentary rocks. *Geology Today,* **23** (3), 108–113.

Sheppard, T H, Houghton, R D & Swan, A R H. 2006. Bedding and pseudobedding in the Early Jurassic of Glamorgan: deposition and diagenesis of the Blue Lias in South Wales. *Proceedings of the Geologists' Association,* **117**, 249–264.

Simms, M, Little, C T S & Rosen, B R. 2002. Corals not serpulids: mineralised colonial fossils in the Lower Jurassic marginal facies of South Wales. *Proceedings of the Geologists' Association,* **113**, 31–36.

Talling, P J, Amy, L A, Wynn, R B, Peakall, J, Robinson, M. 2004. Beds comprising debrite sandwiched within cogenetic turbidite: origin and widespread occurrence in distal depositional environments. *Sedimentology,* **51**, 163–194.

Thomas, G E & Thomas, T M. 1956. The volcanic rocks of the area between Fishguard and Strumble Head, Pembrokeshire. *Geological Society of London, Quarterly Journal,* **112**, 291–314.

Trueman, A E. 1920. The Liassic rocks of the Cardiff district. *Proceedings of the Geologists' Association,* 31, 93–107.

Trueman, A E. 1922. The Liassic rocks of Glamorgan. *Proceedings of the Geologists' Association,* **33**, 245–284.

Trueman, A E. 1930. The Lower Lias (*bucklandi* Zone) of Nash Point, Glamorgan. *Proceedings of the Geologists 'Association,* **41,** 148 – 159.

Tucker, M E. 1977. The marginal Triassic deposits of South Wales: continental facies and palaeogeography. *Geological Journal* **12**, 169-188.

Walker R G & Plint, A G. 1992. Wave- and storm-dominated shallow marine systems. In *Facies Models: response to sea level change,* R G Walker & N P James (eds), 219–238. Geological Association of Canada.

Williams, B P J & Hillier, R D. 2004. Variable alluvial sandstone architecture within the Lower Old Red Sandstone, southwest Wales. *Geological Journal*, **39**, 257–275.

Williams, B P J, Hillier, R D, Marriott, S B (eds). 2004. The Lower Old Red Sandstone of the Welsh Basin. *Geological Journal,* **39.**

Williams, P F. 1968. The sedimentology of Westphalian (Ammanian) Measures in the Little Haven–Amroth Coalfield, Pembrokeshire. *Journal of Sedimentary Petrology,* **38,** 332–362.

Wobber, F J. 1965. Sedimentology of the Lias (Lower Jurassic) of South Wales. *Journal of Sedimentology,* **35**, 683–703.

Woodcock, N H. 1990. Sequence stratigraphy of the Welsh Basin. *Journal of the Geological Society, London,* **147,** 537–547.

Woodcock, N H. 2000. The Quaternary history of an ice age. In *Geological history of Britain and Ireland*, N Woodcock & R Strachan (eds), 392–411. Blackwell Science.

Woodcock, N H & Stoper. N J. 2006. The Acadian Orogeny: the mid-Devonian phase of deformation that formed slate belts in England and Wales. In *The Geology of England and Wales*, P J Brenchley & P F Rawson (eds), 131–146. The Geological Society, London.

Wu, X-T. 1982. Storm-generated depositional types and associated trace fossils in Lower Carboniferous shallow-marine carbonates of Three Cliffs Bay and Ogmore-by-Sea, South Wales. *Palaeogeography, Palaeoclimatology, Palaeoecology*, **39**, 187–202.

8. General interest and novels

Atkinson, R. J. C. 1959. *Stonehenge and Avebury*. HMSO, London.
Borrow, G. H. 2002. *Wild Wales: its people, language and scenery.* Bridge Books.
Cordell, A. 1959. *Rape of the fair country.* Blorenge, UK 320 pp.
------------ 1960. *Hosts of Rebecca.* Blorenge, UK, 256 pp.
------------ 1969. *Song of the Earth.* Blorenge, UK, 332 pp.
------------ 1972. *The fire people.* Paperback. 384 pp
Mullard, J. 2006. *Gower.* Collins
Pressdee, C . 2005. *Food Wales.* Graffeg.
Stickings, T G. 1970. *The story of Saundersfoot.* H G Walters Ltd., Tenby.
Thomas, D. 1954. *Collected Stories* and *Under Milk Wood; a play for voices.* Everyman.
Vaughan-Thomas, W. 1976. *Portrait of Gower.* Robert Hales, London.
Wallace , A R. 1905. *My life; a record of events and opinions.* Chapman & Hall, London.

9. Dictionaries

Allaby, A & Allaby, M. 2003. *A Dictionary of Earth Sciences.* Oxford University Press.
Evans, M A & Thomas, W O. 1997. *The new Welsh dictionary.* Christopher Davies.

10 Maps

British Geological Survey (BGS) maps:

1:625 000 scale
Bedrock Map of Wales and Adjacent Area (in Howells 2007)

1:250 000 scale
Geological Map of Wales. 1994.

1:50 000 scale
Bridgend (Sheet 261 / 262). 1990.
Cardiff (Sheet 263). 1988.
Carmarthen (Sheet 229). 1967.
Llangranog (Sheet 194). 2006
Merthyr Tydfil (Sheet 231).1979.
Pembroke and Linney Head (244/245). 1983.
St David's (Sheet 209). 1992.
Worms Head (Sheet 246). 2003.

Builth Wells (Sheet 196). 2005.
Cardigan & Dinas Head (193/210). 2003.
Haverfordwest (Sheet 228). 1976.
Llanilar (Sheet 178). 1994.
Milford (Sheet 226 / 227). 1978.
Rhyader (Sheet 179). 1994.
Swansea (Sheet 247). 1977.

Ordnance Survey (OS) maps:

1:25 000 scale (Explorer series)
Cardiff and Bridgend (Sheet 151).
Cardigan and Newquay (Sheet 198)
North Pembrokeshire (Sheet OL35).

Carmarthen Bay (Sheet 177).
Gower (Gwyr) (Sheet 164)
South Pembrokeshire (Sheet OL36).

Errata
Page 17 Fig. 2.5a – <u>Monian</u> Terranes: Page 18 line 18 – Davies *et al*. <u>2002</u>: Page 76 line 29 – <u>Alpine</u> Orogeny: Page 82 – note that only locality *3a* can be examined at high tide: Page 140 Fig. 6.21 – looking due <u>east towards</u> locality (ii): Page 209 line 15 – to the <u>southwest</u>: *Appendix 5* flow regime column for **C-U** cycle – <u>increasing</u> velocity: *Bibliography* (add reference)
Davies, J R, Fletcher, C J N, Waters, R A, Wilson, D, Woodhall, D G & Zalasiewicz, J A. 1997. *Geology of the country around Llanilar and Rhyader*. Memoir of the British Geological Survey, Sheets 178 & 179 (England and Wales), HMSO, London.